An Introduction to Scientific Computing with MATLAB® and Python Tutorials

An Introduction to Scientific Computing with MATLAB® and Python Tutorials

Sheng Xu

CRC Press

Taylor & Francis Group
Boca Raton London New York

CRC Press is an imprint of the
Taylor & Francis Group, an **informa** business

A CHAPMAN & HALL BOOK

First edition published 2022
by CRC Press
6000 Broken Sound Parkway NW, Suite 300, Boca Raton, FL 33487-2742

and by CRC Press
4 Park Square, Milton Park, Abingdon, Oxon, OX14 4RN

CRC Press is an imprint of Taylor & Francis Group, LLC

Library of Congress Cataloging-in-Publication Data
A catalog record has been requested for this book.

ISBN: 978-1-032-06315-7 (hbk)
ISBN: 978-1-032-06318-8 (pbk)
ISBN: 978-1-003-20169-4 (ebk)

DOI: 10.1201/9781003201694

Typeset in Nimbus Roman
by KnowledgeWorks Global Ltd.

To my family and friends.

Contents

Preface   **xiii**

Author   **xv**

1 An Overview of Scientific Computing   **1**
    1.1 What Is Scientific Computing? 1
    1.2 Errors in Scientific Computing 3
        1.2.1 Absolute and Relative Errors 3
        1.2.2 Upper Bounds 4
        1.2.3 Sources of Errors 4
    1.3 Algorithm Properties 5
    1.4 Exercises 6

2 Taylor's Theorem   **9**
    2.1 Polynomials 9
        2.1.1 Polynomial Evaluation 10
    2.2 Taylor's Theorem 11
        2.2.1 Taylor Polynomials 12
        2.2.2 Taylor Series 13
        2.2.3 Taylor's Theorem 14
    2.3 Alternating Series Theorem 19
    2.4 Exercises 20
    2.5 Programming Problems 22

3 Roundoff Errors and Error Propagation   **25**
    3.1 Numbers 25
        3.1.1 Integers 26
    3.2 Floating-Point Numbers 26
        3.2.1 Scientific Notation and Rounding 27
        3.2.2 DP Floating-Point Representation 29
    3.3 Error Propagation 32
        3.3.1 Catastrophic Cancellation 32
        3.3.2 Algorithm Stability 33
    3.4 Exercises 35
    3.5 Programming Problems 37

4   Direct Methods for Linear Systems   **43**
       4.1   Matrices and Vectors   44
       4.2   Triangular Systems   45
       4.3   GE and A=LU   46
              4.3.1   Elementary Matrices   51
              4.3.2   A=LU   52
              4.3.3   Solving $Ax = \mathbf{b}$ by A=LU   55
       4.4   GEPP and PA=LU   57
              4.4.1   GEPP   59
              4.4.2   PA=LU   61
              4.4.3   Solving $Ax = \mathbf{b}$ by PA=LU   65
       4.5   Tridiagonal Systems   67
       4.6   Conditioning of Linear Systems   69
              4.6.1   Vector and Matrix Norms   72
              4.6.2   Condition Numbers   76
              4.6.3   Error and Residual Vectors   77
       4.7   Software   78
       4.8   Exercises   79
       4.9   Programming Problems   82

5   Root Finding for Nonlinear Equations   **85**
       5.1   Roots and Fixed Points   86
       5.2   The Bisection Method   90
       5.3   Newton's Method   94
              5.3.1   Convergence Analysis of Newton's Method   96
              5.3.2   Practical Issues of Newton's Method   99
       5.4   Secant Method   99
       5.5   Fixed-Point Iteration   105
       5.6   Newton's Method for Systems of Nonlinear Equations   109
              5.6.1   Taylor's Theorem for Multivariate Functions   110
              5.6.2   Newton's Method for Nonlinear Systems   114
       5.7   Unconstrained Optimization   117
       5.8   Software   121
       5.9   Exercises   122
       5.10   Programming Problems   127

6   Interpolation   **131**
       6.1   Terminology of Interpolation   132
       6.2   Polynomial Space   133
              6.2.1   Chebyshev Basis   135
              6.2.2   Legendre Basis   138
       6.3   Monomial Interpolation   141
       6.4   Lagrange Interpolation   143
       6.5   Newton's Interpolation   145
       6.6   Interpolation Error   149

		6.6.1	Error in Polynomial Interpolation	150
		6.6.2	Behavior of Interpolation Error	153
			6.6.2.1 Equally-Spaced Nodes	153
			6.6.2.2 Chebyshev Nodes	156
	6.7	Spline Interpolation		159
		6.7.1	Piecewise Linear Interpolation	159
		6.7.2	Cubic Spline	161
		6.7.3	Cubic Spline Interpolation	163
	6.8	Discrete Fourier Transform (DFT)		167
	6.9	Exercises		173
	6.10	Programming Problems		177

7   Numerical Integration   183
	7.1	Definite Integrals		184
	7.2	Numerical Integration		187
		7.2.1	Change of Intervals	189
	7.3	The Midpoint Rule		191
		7.3.1	Degree of Precision (DOP)	192
		7.3.2	Error of the Midpoint Rule	195
	7.4	The Trapezoidal Rule		200
	7.5	Simpson's Rule		205
	7.6	Newton-Cotes Rules		209
	7.7	Gaussian Quadrature Rules		209
	7.8	Other Numerical Integration Techniques		216
		7.8.1	Integration with Singularities	216
		7.8.2	Adaptive Integration	217
	7.9	Exercises		218
	7.10	Programming Problems		221

8   Numerical Differentiation   225
	8.1	Differentiation Using Taylor's Theorem		225
		8.1.1	The Method of Undetermined Coefficients	227
	8.2	Differentiation Using Interpolation		229
		8.2.1	Differentiation Using DFT	231
	8.3	Richardson Extrapolation		232
	8.4	Exercises		233
	8.5	Programming Problems		235

9   Initial Value Problems and Boundary Value Problems   237
	9.1	Initial Value Problems (IVPs)		238
		9.1.1	Euler's Method	239
			9.1.1.1 Local Truncation Error and Global Error	242
			9.1.1.2 Consistency, Convergence and Stability	244
			9.1.1.3 Explicit and Implicit Methods	247
		9.1.2	Taylor Series Methods	248

9.1.3   Runge-Kutta (RK) Methods   249
9.2   Boundary Value Problems (BVPs)   252
    9.2.1   Finite Difference Methods   253
        9.2.1.1   Local Truncation Error and Global Error   254
        9.2.1.2   Consistency, Stability and Convergence   255
9.3   Exercises   256
9.4   Programming Problems   259

10 Basic Iterative Methods for Linear Systems   261
10.1   Jacobi and Gauss-Seidel Methods   263
    10.1.1   Jacobi Method   264
    10.1.2   Gauss-Seidel (G-S) Method   267
10.2   Convergence Analysis   270
10.3   Other Iterative Methods   272
10.4   Exercises   273
10.5   Programming Problems   274

11 Discrete Least Squares Problems   277
11.1   The Discrete LS Problems   278
11.2   The Normal Equation by Calculus   281
11.3   The Normal Equation by Linear Algebra   284
11.4   LS Problems by A=QR   286
11.5   Artificial Neural Network   287
11.6   Exercises   289
11.7   Programming Problems   292

12 Monte Carlo Methods and Parallel Computing   295
12.1   Monte Carlo Methods   297
12.2   Parallel Computing   298
12.3   Exercises   300
12.4   Programming Problems   302

Appendices   305

A   An Introduction of MATLAB for Scientific Computing   307
A.1   What is MATLAB?   307
    A.1.1   Starting MATLAB   307
    A.1.2   MATLAB as an Advanced Calculator   308
    A.1.3   Order of Operations   308
    A.1.4   MATLAB Built-in Functions and Getting Help   309
    A.1.5   Keeping a Record for the Command Window   310
    A.1.6   Making M-Scripts   310
A.2   Variables, Vectors and Matrices   311
    A.2.1   Variables   311
    A.2.2   Suppressing Output   312
    A.2.3   Vectors and Matrices   312

	A.2.4	Special Matrices	313
	A.2.5	The Colon Notation and `linspace`	314
	A.2.6	Accessing Entries in a Vector or Matrix	315
A.3	Matrix Arithmetic		317
	A.3.1	Scalar Multiplication	317
	A.3.2	Matrix Addition	317
	A.3.3	Matrix Multiplication	318
	A.3.4	Transpose	319
	A.3.5	Entry-wise Convenience Operations	320
A.4	Outputting/Plotting Results		321
	A.4.1	`disp`	321
	A.4.2	`fprintf`	322
	A.4.3	`plot`	322
A.5	Loops and Decisions		325
	A.5.1	`for` Loops	325
	A.5.2	Logicals and Decisions	327
	A.5.3	`while` Loops	329
A.6	Functions		329
	A.6.1	M-Functions	330
	A.6.2	Anonymous Functions	332
	A.6.3	Passing Functions to Functions	333
A.7	Creating Live Scripts in the Live Editor		334
A.8	Concluding Remarks		335
A.9	Programming Problems		336

B An Introduction of Python for Scientific Computing   **339**
B.1	What is Python?		339
	B.1.1	Starting Python	339
	B.1.2	Python as an Advanced Calculator	340
	B.1.3	Python Programs	342
B.2	Variables, Lists and Dictionaries		342
	B.2.1	Variables	342
	B.2.2	Lists	343
	B.2.3	Dictionaries	345
B.3	Looping and Making Decisions		347
	B.3.1	`for` Loops	347
	B.3.2	`if` Statements	349
	B.3.3	`while` Loops	351
	B.3.4	`break` and `continue` in Loops	352
B.4	Functions		354
	B.4.1	Passing Arguments	355
	B.4.2	Passing Lists	357
B.5	Classes		358
	B.5.1	Attributes	359
	B.5.2	Methods	361

	B.5.3	Inheritance	362
	B.5.4	Objects as Attributes	364
B.6	Modules		366
B.7	numpy, scipy, matplotlib		370
	B.7.1	numpy	370
	B.7.2	scipy	374
	B.7.3	matplotlib	374
B.8	Jupyter Notebook		375
B.9	Concluding Remarks		376
B.10	Programming Problems		376

Index   **379**

Preface

Features of the Book

We write this book to have the following features.

- We cover fundamental numerical methods in the book. The book can be used as a textbook for a **first** course in scientific computing. The prerequisites are calculus and linear algebra. A college student in science, math or engineering can take the course in sophomore year.
- We write this book **for students**. We present the material in a self-contained accessible manner for a student to easily follow and explore. We use motivating examples and application problems to catch a student's interest. We add remarks at various places to satisfy a student's curiosity.
- We provide short **tutorials on MATLAB® and Python**. We give pseudocodes of algorithms instead of source codes. With the tutorials and pseudocodes, a student can enjoy writing programs in MATLAB or Python to implement, test and apply algorithms.
- We balance the underlying idea, algorithm implementation and performance analysis for a fundamental numerical method so that a student can gain **comprehensive** understanding of the method.
- We review necessary material from calculus and linear algebra at appropriate places. We also make the **connection** between the fundamental numerical methods with advanced topics such as machine learning and parallel computing.
- We design **paper-and-pen exercises and programming problems** so that students can apply, test and analyze numerical methods toward comprehensive understanding and practical applications.

Sample Syllabi

Below is a sample syllabus for a two-semester course sequence.

Semester 1
1. Appendix A or B: MATLAB or Python Tutorial

2. Chapter 1: An Overview of Scientific Computing
3. Chapter 2: Taylor's Theorem
4. Chapter 3: Roundoff Errors and Error Propagation
5. Sections 5.1–5.5: Root Finding for Nonlinear Equations
6. Sections 6.1–6.6: Interpolation
7. Sections 7.1–7.6: Numerical Integration
8. Chapter 8: Numerical Differentiation

Semester 2
1. Chapter 4: Direct Methods for Linear Systems
2. Sessions 5.6–5.7: Nonlinear Systems and Optimization
3. Sessions 6.6–6.7: Spline Interpolation and DFT
4. Sessions 7.7–7.8: Gaussian Quadrature Rules
5. Chapter 9: IVPs and BVPs
6. Chapter 10: Iterative Methods for Linear Systems
7. Chapter 11: LS Problems
8. Chapter 12: Monte Carlo Methods and Parallel Computing

Acknowledgments

We acknowledge our colleagues Daniel Reynolds and Johannes Tausch for detailed critiques of this first edition. We would appreciate any comments, suggestions and corrections that readers may wish to send us using the email address *sxu@smu.edu.*

Author

Sheng Xu is associate professor of mathematics at Southern Methodist University (SMU). He holds a Ph.D. in mechanical engineering from Cornell University. He conducts research on development and application of computational methods for problems in fluid mechanics, including aerodynamics of insect flight, two-fluid flows, supersonic turbulence, and turbulence control. His published work has appeared in *Journal of Computational Physics, SIAM Journal on Scientific Computing, Physics of Fluids,* and *Journal of Fluid Mechanics.*

1

An Overview of Scientific Computing

In this chapter, we address what scientific computing is about and why it is important. We give a brief introduction to algorithms and errors. We emphasize the importance of error upper bounds.

1.1 What Is Scientific Computing?

Let's consider two examples first.

Example 1.1 (Apple in free fall) An apple falls from a tree under gravity. How long does it take to reach the ground (or Newton's head)?
[Solution:] Assume that the apple has a zero initial velocity and falls without drag. Denote the height of the apple as H and Earth's gravity as g. By Newton's second law, the time T that the apple takes to reach the ground satisfies

$$\frac{1}{2}gT^2 = H$$

This equation can be solved exactly (analytically) to give

$$T = \sqrt{\frac{2H}{g}}$$

■

Example 1.2 (Particle chasing) Two particles A and B move from the same location along a straight path in the same direction. The particle A moves with the constant speed 2, and the particle B starts with a zero speed and accelerates with a time-dependent speed $e^t - 1$, where t is the time. How long does the particle B take to catch the particle A?
[Solution:] When the particle B catches up with the particle A at $t = T > 0$, they travel the same distance. The distance traveled by the particle A is $2T$, and the distance traveled by the particle B is $\int_0^T (e^t - 1)dt$. The time T therefore satisfies

$$2T = \int_0^T (e^t - 1)dt, \quad T > 0$$

DOI: 10.1201/9781003201694-1

which gives

$$3T + 1 = e^T, \quad T > 0$$

This equation cannot be solved analytically using elementary functions, but we clearly know it has a positive real solution from the physics intuition or the plots of $f(T) = 3T + 1$ and $g(T) = e^T$ in Fig. 1.1. ■

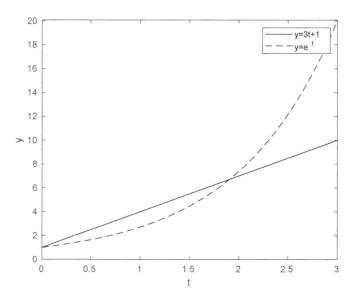

FIGURE 1.1
Graphs of $f(T) = 3T + 1$ and $g(T) = e^T$ for $T \in [0, 3]$.

In the first example, we can find an **analytical solution** of the equation. An analytical solution is a closed-form expression for an unknown variable in terms of the known variables. However, the analytical solutions of many mathematical problems may not be available or may be difficult to obtain, such as in the second example. We therefore resort to scientific computing to find numerical solutions of those mathematical problems. A **numerical solution** is an approximate solution of a problem, and it appears as numerical values instead of closed-form expressions.

Scientific computing involves the development and study of step-by-step procedures to find numerical solutions of mathematical problems. The step-by-step procedures are called **(numerical) algorithms**. A **numerical/computational method** refers to both the idea underlying an algorithm and the fulfillment of the idea as an algorithm. We sometimes do not distinguish a method with an algorithm.

For the second example above, we may devise an algorithm to approximate the solution as follows. We notice that the solution is between 1 and 3 because $f(1) = 3 \times 1 + 1 > g(1) = e^1$ while $f(3) = 3 \times 3 + 1 < g(3) = e^3$ (the solid curve is above the dashed curve at 1 while below at 3 in Fig. 1.1). We can then bisect the interval $[1, 3]$ by the middle point 2 and look at the two subintervals $[1, 2]$ and $[2, 3]$. We can use the same strategy to narrow down the solution to one subinterval, in this case $[1, 2]$. We can repeatedly apply this strategy until we find a subinterval that contains the solution and is small enough such that its middle point is a good approximation of the solution. Actually, this procedure is called bisection **iteration**, which will be learned in more details later in Section 5.2 of Chapter 5.

Scientific computing, theoretical study and experiments are three pillars to support scientific research and technological applications. The diagram in Fig. 1.2 shows where scientific computing comes into play.

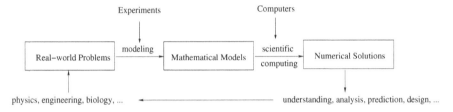

FIGURE 1.2
Scientific computing in solving real-world problems.

With the advancement of modern computers, scientific computing plays a heavier and heavier role in almost all areas. This introductory text will give you a taste of scientific computing and prepares you for your broader and deeper exploration and application of scientific computing.

1.2 Errors in Scientific Computing

The results from scientific computing are typically **approximate**. One important goal in scientific computing is to ensure that the approximate results are close enough to the true/exact results.

1.2.1 Absolute and Relative Errors

Definition 1.1 *Let T denote the true/exact value of a scalar/number, and A an approximation of the scalar. We define the error E, the absolute error $|E|$ and the relative error R of the approximation A as follows, respectively*
- *The error:* $E = T - A$
- *The absolute error:* $|E| = |T - A|$
- *The relative error:* $R = \dfrac{|T - A|}{|T|} = \dfrac{|E|}{|T|}, \quad$ *if $T \neq 0$*

```
┌─────────────────────────────────────────────────────────────┐
│                                                             │
└─────────────────────────────────────────────────────────────┘
```

The relative error is often used to compare approximations of numbers of widely different scales.

Example 1.3 (Absolute and relative errors) The absolute errors in the measurements of the thickness of a book 5 cm thick and a paper 0.2 mm thick are both 0.1 mm. Are the two measurements of the same quality? Which one is better?
[Solution:] The absolute errors are the same, but the relative errors are quite different. The relative error in the measured book thickness is $0.1/50 = 0.002 = 0.2\%$, which is much smaller than the relative error in the measured paper thickness, $0.1/0.2 = 0.5 = 50\%$. The two measurements are not of the same quality, and the measurement of the book thickness has better quality. ∎

1.2.2   Upper Bounds

In real practice, we know the approximation but do not know the true value (otherwise we may not need the approximation), and we therefore do not know the exact value of an error (absolute or relative), but we can find an **upper bound** for the absolute or relative error, which is a value that the error cannot exceed. The error may reach an upper bound if the upper bound is **sharp**. For example, if we know the absolute error in an approximation ranges in $[0, 1]$ and can be 1, then 1 is a sharp upper bound, and it is true that the absolute error is also less than the upper bound 100, but the upper bound 100 may not be quite useful. So we want to find an upper bound as sharp as possible (i.e. the minimum upper bound).

Example 1.4 (Upper bounds) Let $A = 3.14$ be an approximation of π. Find a meaningful upper bound of the absolute error in the approximation.
[Solution:] The true value of π is

$$\pi = 3.14159\cdots$$

The approximation A is
$$A = 3.14$$
The absolute error $|E|$ in the approximation is

$$|E| = |\pi - A| = |3.14159\cdots - 3.14| = 0.00159\cdots < 0.0016$$

We can choose 0.0016 to be an upper bound, which is "sharper" ("better") than an upper bound 0.002. An upper bound 0.1 is not wrong, but not meaningful. ∎

1.2.3   Sources of Errors

Where are errors from? According to their sources, we can categorize errors into the following three types.

- **Modeling errors:** the errors due to the simplifications and hypotheses in the modeling process that convert a real-world problem into a mathematical model. For example, the error due to the neglect of air resistance in the apple's free fall problem above. Modeling errors are not in the scope of scientific computing.
- **Mathematical approximation errors:** the errors due to the approximation of an actual quantity by an approximate formula.

Example 1.5 We know that $f'(a)$, the derivative of $f(x)$ at $x = a$, is the slope of the tangent line to the curve $y = f(x)$ at $x = a$; and $\dfrac{f(a+h) - f(a)}{h}$, where h is finite, is the slope of the secant line through the points $(a, f(a))$ and $(a+h, f(a+h))$ on the curve $y = f(x)$. If we approximate the tangent slope $T = f'(a)$ by the secant slope $A = \dfrac{f(a+h) - f(a)}{h}$, we have the mathematical approximation error:

$$|T - A| = \left| f'(a) - \frac{f(a+h) - f(a)}{h} \right|$$

Later, we will analyze how this error depends on the value of h using Taylor's theorem and introduce the big-O notation to describe this dependence. ■

- **Roundoff errors:** the errors due to finite-precision representation of real numbers.

Example 1.6 Let $\hat{x} = 1.2345679$ be a 8-digit representation of $x = 1.23456789$. The absolute error $|x - \hat{x}| = 0.00000001 = 10^{-8}$ is caused by rounding x to $\hat{x}$ and is the roundoff error in the representation. Later, we will describe how computers represent real numbers and learn that roundoff errors are inevitable in computer representations. ■

1.3 Algorithm Properties

A numerical algorithm is a step-by-step procedure to find the numerical solution of a mathematical problem. We care about the following three properties of a numerical algorithm.

Accuracy Accuracy concerns about the magnitude of the error in a numerical solution. The error needs to be controlled in an acceptable range.

Efficiency Efficiency includes time efficiency and storage efficiency. Time efficiency depends on how many arithmetic operations are needed in an algorithm. Storage efficiency depends on how many memory space is needed to execute an algorithm.

Stability An algorithm is stable if small errors (inevitable in practice) introduced in the algorithm stay small. If the small errors get amplified to be out of control, the algorithm is unstable.

Later we will use specific algorithms to illustrate and discuss these properties.

1.4 Exercises

Exercise 1.1 A_1 is an approximation of $T_1 = 1$ with the absolute error 1, and A_2 is an approximation of $T_2 = 100$ with the absolute error 1. Which approximation (A_1 or A_2) has better accuracy? Why?

Exercise 1.2 (1) Suppose the true value $T = 2$, and the relative error of its approximation A is less than 5%. What is the range for A? (2) Suppose an approximation $A = 99$, and the relative error in A is 10%. What are the possible values of the true value T?

Exercise 1.3 If $A = 10$ is an approximation of a true value T which is in the range $8 \le T \le 11$, find an upper bound for the absolute error and an upper bound for the relative error of the approximation.

Exercise 1.4 Euler's number e can be defined as $e = \lim_{n \to \infty} \left(1 + \dfrac{1}{n}\right)^n$. It is a famous transcendental number with infinitely many digits $e = 2.718282828459....$ Suppose one approximates e by $A = \left(1 + \dfrac{1}{2}\right)^2$ with $n = 2$. (1) Find a meaningful upper bound for the absolute error of the approximation. (2) Find a meaningful upper bound for the relative error of the approximation.

Exercise 1.5 If the function $y = \sin(x)$ is approximated by the function $y = x$ for $x \in [-\pi/2, \pi/2]$, which type does the error of the approximation belong to? How does the absolute error change as $|x|$ increases toward $\pi/2$?

Exercise 1.6 Let A_h be an approximation of the true value T. Suppose the approximation A_h depends on a small parameter h, and the absolute error $|T - A_h|$ depends on h as $|T - A_h| = Ch^p$, where $C > 0$ and $p > 0$ are constants. (1) To make the error small, do we want a small or large value of p? (2) If we double the value of h, how many times larger does the absolute error become for $p = 1$, $p = 2$ and $p = 3$, respectively?

Exercise 1.7 Count the total number of multiplications in the nested loops of the following pseudocodes.
(1)
```
for m = 1 to 20 do
    for n = 1 to 20 do
        bₘ = bₘ − G_{m,n}bₙ
    end for
end for
```
(2)
```
for m = 1 to 20 do
    for n = m to 20 do
        bₘ = bₘ − G_{m,n}bₙ
```

end for
  end for

Exercise 1.8 If you are asked to evaluate $p_4(x) = 1 + 2x + 3x^2 + 4x^3 + 5x^4$ at a given value of x, how many multiplications do you use? (Note that evaluating x^n as $x \cdot x \cdots x$ needs $n - 1$ multiplications.) Can you use less?

2

Taylor's Theorem

In this chapter, we presents Taylor's theorem, which is used later for development and analysis of some numerical methods. We show how to approximate functions by Taylor polynomials and how to analyze the errors in such approximations. The big-O notation is introduced to describe efficiency and accuracy.

2.1 Polynomials

A polynomial $p_n(x)$ in the variable x of degree n, where n is a non-negative integer, can be written in the general form as

$$p_n(x) = c_0 + c_1 x + c_2 x^2 + \cdots + c_{n-1} x^{n-1} + c_n x^n \qquad (2.1)$$

where $c_0, c_1, c_2, ..., c_{n-1}$ and c_n are constant coefficients (if $c_n \neq 0$, n is the true/exact degree). The summation in $p_n(x)$ can be written in the sigma notation as

$$c_0 + c_1 x + c_2 x^2 + \cdots + c_{n-1} x^{n-1} + c_n x^n \equiv \sum_{k=0}^{n} c_k x^k \qquad (2.2)$$

Note that $p_n(x)$ is built from constant coefficients and the variable x using only addition, multiplication and exponentiation of x to non-negative integer powers (repeated multiplication). So polynomials can be easily evaluated, differentiated and integrated, and are good candidates to approximate functions or data.

The polynomial $p_n(x)$ can be regarded as a linear combination of the monomial **basis functions** $1, x, x^2, ..., x^{n-1}$ and x^n using the constant weights $c_0, c_1, ..., c_{n-1}$ and c_n, respectively. Later we will construct a polynomial of degree at most n as a **linear combination** of other basis functions.

A polynomial centered at the number a has the form

$$p_n(x) = c_0 + c_1(x-a) + c_2(x-a)^2 + \cdots + c_{n-1}(x-a)^{n-1} + c_n(x-a)^n \qquad (2.3)$$

Below are a few examples of polynomials
- $z(x) = 0$: a zero polynomial (of undefined degree)
- $p_0(x) = -5$: a constant polynomial (of degree 0)

DOI: 10.1201/9781003201694-2

- $p_1(x) = 2 + 3x$: a linear polynomial (of degree 1)
- $p_2(x) = \pi + 1.3x - 2.7x^2$: a quadratic polynomial (of degree 2)
- $p_3(x) = -6 + 2(x - 7) - 5(x - 7)^2 + (x - 7)^3$: a cubic polynomial (of degree 3) centered at 7
- $p_4(x) = c_0 + c_1x + c_2x^2 + c_3x^3 + c_4x^4$: a polynomial of degree ≤ 4 ($= 4$ if $c_4 \neq 0$)

Later, we will define the orthogonality of polynomials and introduce special polynomials such as Chebyshev polynomials and Legendre polynomials.

A very important theorem regarding polynomials is the fundamental theorem of algebra.

Theorem 2.1 (Fundamental theorem of algebra) *Let* $p_n(x) = c_nx^n + c_{n-1}x^{n-1} + \cdots + c_2x^2 + c_1x + c_0$, $c_n \neq 0$, *be a polynomial of degree* n. *Then* $p_n(x)$ *can be factorized as*

$$p_n(x) = c_n(x - r_1)(x - r_2)\cdots(x - r_n) \tag{2.4}$$

i.e. the polynomial equation $p_n(x) = 0$ *has* n *roots* $r_1, r_2, \ldots, r_n$.

Remark The roots can be repeated and can be complex numbers. The theorem can be proved using complex analysis.

2.1.1 Polynomial Evaluation

Let's count the numbers of additions and multiplications in evaluating

$$p_4(x) = 1 + 2x + 3x^2 + 4x^3 + 5x^4$$

at a given value of x using different methods.

Naive method: x^k, $k = 2, 3, 4$, in $p_4(x)$ is calculated as $x \cdot x \cdots \cdot x$, the repeated multiplication of x for $k - 1$ times. So the term c_kx^k needs k multiplications. The total number of multiplications in evaluating $p_4(x)$ is

$$0 + 1 + 2 + 3 + 4 \equiv \sum_{k=0}^{4} k = 10.$$

The total number of additions in evaluating $p_4(x)$ is 4.

Horner's method: We can reduce the number of multiplications by calculating x^k as $x^{k-1} \cdot x$ if x^{k-1} is known. To use this fact, we can write $p_4(x)$ in the form of **nested multiplication**

$$p_4(x) = 1 + x \cdot (2 + x \cdot (3 + x \cdot (4 + 5 \cdot x)))$$

where only 4 multiplications are required. The number of additions is still 4. The evaluation procedure using nested multiplication is known as Horner's method.

The above analysis can be easily extended to a polynomial of degree n, $p_n(x) = \sum_{k=0}^{n} c_k x^k$, for which the naive method needs

$$\sum_{k=0}^{n} k = 0 + 1 + 2 + 3 + \cdots + n = \frac{n(n+1)}{2} \tag{2.5}$$

multiplications and n additions, while Horner's method requires n multiplications and n additions. So Horner's method is more efficient (in terms of computational time).

We may use **the big O notation** to describe the **efficiency** of a method. Denote the size of a problem as n and the computational time of a method for the problem as w. We say that the computational time is of order $f(n)$, i.e.

$$w = O(f(n)) \tag{2.6}$$

which means that the growth of the computational time w is bounded from above for large enough n as

$$w \leq C f(n) \tag{2.7}$$

where C is a positive constant. For example, the computational time of the naive method for polynomial evaluation is $O(n^2)$, while Horner's method is $O(n)$, where n is the degree of a polynomial.

The polynomial $p_n(x)$ centered at a in the form of nested multiplication is

$$p_n(x) = c_0 + (x-a) \cdot (c_1 + (x-a) \cdot (c_2 + \cdots + (x-a) \cdot (c_{n-1} + (x-a) \cdot c_n) \cdots)) \tag{2.8}$$

which can be evaluated by Horner's method by starting with the innermost parentheses and working outward. The pseudocode of Horner's method is

---

Algorithm 1 Horner's method to evaluate the polynomial $p_n(x) = c_0 + c_1(x-a) + c_2(x-a)^2 + \cdots + c_{n-1}(x-a)^{n-1} + c_n(x-a)^n$

$p \leftarrow c_n$
$z \leftarrow x - a$
for k from $n-1$ down to 0 **do**
    $p \leftarrow c_k + z \cdot p$
end for

2.2 Taylor's Theorem

Taylor's theorem is very important for us in scientific computing to construct and analyze numerical approximations. Below we distinguish Taylor polynomials and Taylor series, and then review Taylor's theorem.

2.2.1 Taylor Polynomials

Let $f(x)$ be a function with continuous first n derivatives (i.e. continuous f', f'', ...,
$f^{(n)}$) in an interval I containing the number a (we may say $f \in C_I^n$, where C_I^n denotes
the **function space** consisting of all functions with continuous first n derivatives in
the interval I). The Taylor polynomial of degree n for the function $f(x)$ about a is

$$
\begin{aligned}
T_n(x) &= f(a) + f'(a)(x-a) + \frac{f''(a)}{2!}(x-a)^2 + \cdots + \frac{f^{(n)}(x)}{n!}(x-a)^n \\
&\equiv \sum_{k=0}^{n} \frac{f^{(k)}(a)}{k!}(x-a)^k
\end{aligned}
\tag{2.9}
$$

where $k! = 1 \cdot 2 \cdots \cdot k$ is the factorial of k and $0! := 1$.

> **Remark** $T_1(x) = f(a) + f'(a)(x-a)$ is the **linearization** of $f(x)$ at a (the
> tangent line equation).

The Taylor polynomial $T_n(x)$ is constructed using the derivatives of $f(x)$ at a such
that $T_n(x)$ and $f(x)$ satisfy the following matching conditions at a

$$T_n(a) = f(a) \tag{2.10}$$
$$T_n'(a) = f'(a) \tag{2.11}$$
$$T_n''(a) = f''(a) \tag{2.12}$$
$$\cdots$$
$$T^{(n)}(a) = f^{(n)}(a) \tag{2.13}$$

Example 2.1 (Taylor polynomials) Find the Taylor polynomial of degree 2 for
$f(x) = 1 + 2x + 3x^2$ about $a = 1$.
[Solution:] In this problem, $f(x) = 1 + 2x + 3x^2$ and $a = 1$. The Taylor polynomial is

$$T_2(x) = f(1) + f'(1)(x-1) + \frac{f''(1)}{2!}(x-1)^2$$

We have

$$
\begin{aligned}
f(x) &= -1 + 2x + 3x^2, & f(1) &= 4 \\
f'(x) &= 2 + 6x, & f'(1) &= 8 \\
f''(x) &= 6, & f''(1) &= 6
\end{aligned}
$$

So the Taylor polynomial is

$$T_2(x) = 4 + 8(x-1) + 3(x-1)^2$$

∎

2.2.2 Taylor Series

If $f(x)$ is infinitely differentiable over an interval containing the number a, we can write down Taylor series/expansion of $f(x)$ about a as

$$f(x) \sim f(a) + f'(a)(x-a) + \frac{f''(a)}{2!}(x-a)^2 + \cdots \equiv \sum_{k=0}^{\infty} \frac{f^{(k)}(a)}{k!}(x-a)^k \quad (2.14)$$

Note that
(1) the Taylor series (on the right) can be regarded as $\lim_{n \to \infty} T_n(x)$, and a Taylor polynomial is a truncated Taylor series;
(2) the symbol $\sim$ is used to indicate that the Taylor series may not equal $f(x)$; If the Taylor series converges to $f(x)$ for any x in an interval, we can replace $\sim$ by $=$ with the specification of the convergence interval;
(3) the Taylor series is also called the Maclaurin series if $a = 0$.

Below are some familiar convergent Maclaurin series with their convergence intervals.

$$e^x = 1 + x + \frac{x^2}{2!} + \frac{x^3}{3!} + \cdots \equiv \sum_{k=0}^{\infty} \frac{x^k}{k!}, \quad |x| < \infty \quad (2.15)$$

$$\sin x = x - \frac{x^3}{3!} + \frac{x^5}{5!} - \frac{x^7}{7!} + \cdots \equiv \sum_{k=0}^{\infty} \frac{(-1)^k x^{2k+1}}{(2k+1)!}, \quad |x| < \infty \quad (2.16)$$

$$\cos x = 1 - \frac{x^2}{2!} + \frac{x^4}{4!} - \frac{x^6}{6!} + \cdots \equiv \sum_{k=0}^{\infty} \frac{(-1)^k x^{2k}}{(2k)!}, \quad |x| < \infty \quad (2.17)$$

$$\frac{1}{1-x} = 1 + x + x^2 + x^3 + \cdots \equiv \sum_{k=0}^{\infty} x^k, \quad |x| < 1 \quad (2.18)$$

$$\tan^{-1} x = x - \frac{x^3}{3} + \frac{x^5}{5} - \frac{x^7}{7} + \cdots \equiv \sum_{k=0}^{\infty} \frac{(-1)^k x^{2k+1}}{2k+1}, \quad |x| \leq 1 \quad (2.19)$$

$$\ln(1+x) = x - \frac{x^2}{2} + \frac{x^3}{3} - \frac{x^4}{4} + \cdots \equiv \sum_{k=1}^{\infty} \frac{(-1)^{k-1} x^k}{k}, \quad -1 < x \leq 1, \quad (2.20)$$

$$\begin{aligned}
(1+x)^p &= 1 + px + \frac{p(p-1)}{2!} x^2 + \frac{p(p-1)(p-2)}{3!} x^3 + \cdots \\
&\equiv \sum_{k=0}^{\infty} \binom{p}{k} x^k, \quad |x| < 1 \quad (2.21)
\end{aligned}$$

In the last binomial series, the power p is real, and the binomial notation is defined by

$$\binom{p}{k} := \frac{p(p-1)\cdots(p-k+1)}{k!} \tag{2.22}$$

We may substitute x by new variables in the above series or integrate or differentiate the above series to obtain new series.

Example 2.2 Substitute x by $-t^2$ in the Maclaurin series for e^x, we get

$$e^{-t^2} = 1 - t^2 + \frac{t^4}{2!} - \frac{t^6}{3!} + \cdots \equiv \sum_{k=0}^{\infty} \frac{(-1)^k t^{2k}}{k!}, \quad |t| < \infty \tag{2.23}$$

■

Example 2.3 Substitute $1+x$ by t in the Maclaurin series for $\ln(1+x)$, we get the Taylor series of $\ln t$ about $a = 1$ as

$$
\begin{aligned}
\ln t &= (t-1) - \frac{(t-1)^2}{2} + \frac{(t-1)^3}{3} - \frac{(t-1)^4}{4} + \cdots \\
&\equiv \sum_{k=1}^{\infty} \frac{(-1)^{k-1}(t-1)^k}{k}, \quad |t-1| < 1\,(0 < t < 2) \tag{2.24}
\end{aligned}
$$

■

2.2.3 Taylor's Theorem

Polynomials can be easily evaluated with only additions, subtractions and multiplications. In addition, polynomials can be easily differentiated and integrated. Many other functions, for example $\sin x$, e^x, $\ln x$ and $\sqrt{x}$, cannot be evaluated exactly using only these arithmetic operations. It is thus desirable to approximate a function by a polynomial. Here we approximate a function by its Taylor polynomials and use Taylor's theorem to analyze the errors. Later, we will also approximate a function by other kinds of polynomials.

Fig. 2.1 shows the graphs of $y = \cos x$ and the Taylor polynomials $T_0(x)$, $T_2(x)$ and $T_4(x)$ of the degrees 0, 2 and 4, respectively, for $f(x) = \cos x$ about $a = 0$ on $[-\pi, \pi]$, where

$$
\begin{aligned}
T_0(x) &= 1 \\
T_2(x) &= 1 - \frac{x^2}{2!} \\
T_4(x) &= 1 - \frac{x^2}{2!} + \frac{x^4}{4!}
\end{aligned}
$$

which are obtained by truncating the Maclaurin series of $y = \cos x$. In this case, the Taylor polynomials approximate $f(x)$ well if x is close to a; and a higher-order Taylor polynomial approximates $f(x)$ better for a fixed value of x on $[-\pi, \pi]$.

The accuracy of Taylor polynomial approximation can be analyzed using Taylor's theorem.

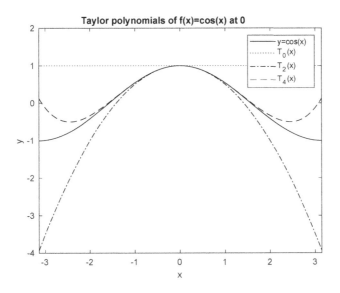

FIGURE 2.1

Approximations of $y = \cos x$, $x \in [-\pi, \pi]$ by its Taylor polynomials of different degrees.

Theorem 2.2 (Taylor's Theorem) *Assume $f(x) \in C_{[\alpha,\beta]}^{n+1}$, and let a be a number in (α, β). Then*

$$f(x) = T_n(x) + R_n(x), \quad x \in [\alpha, \beta] \tag{2.25}$$

where

$$T_n(x) = \sum_{k=0}^{n} \frac{f^{(k)}(a)}{k!}(x-a)^k \tag{2.26}$$

is the n-th order Taylor polynomial of f at a, and $R_n(x)$ is called Taylor's remainder (or the error term in approximating $f(x)$ by $T_n(x)$). Taylor's remainder $R_n(x)$ in Lagrange mean value form is

$$R_n(x) = \frac{f^{(n+1)}(c)}{(n+1)!}(x-a)^{n+1} \tag{2.27}$$

where c is a point (generally unknown to us) between a and x.

Remark Taylor's theorem can be proved by repeatedly applying Rolle's theorem to the function

$$g(t) = f(t) - T_n(t) - (f(x) - T_n(x))\frac{(t-a)^{n+1}}{(x-a)^{n+1}}$$

which is a function of t that satisfies $g(a) = g'(a) = \cdots = g^{(n)}(a) = 0$ and $g(x) = 0$.

Remark Taylor's remainder $R_n(x)$ also has an integral form as

$$R_n(x) = \int_a^x \frac{f^{n+1}(t)}{n!}(t-x)^n dt \tag{2.28}$$

which can be revealed using integration by parts as follows

$$f(x) = f(a) + \int_a^x f'(t)dt = f(a) + \int_a^x f'(t)(t-x)'dt$$

$$= f(a) + f'(a)(x-a) - \int_a^x f''(t)(t-x)dt$$

$$= f(a) + f'(a)(x-a) - \int_a^x f''(t)\left(\frac{(t-x)^2}{2}\right)' dt$$

$$= f(a) + f'(a)(x-a) + f''(a)\frac{(x-a)^2}{2} + \int_a^x f'''(t)\frac{(t-x)^2}{2}dt$$

$$= f(a) + f'(a)(x-a) + f''(a)\frac{(x-a)^2}{2} + \int_a^x f'''(t)\left(\frac{(t-x)^3}{3!}\right)' dt$$

$$= f(a) + f'(a)(x-a) + f''(a)\frac{(x-a)^2}{2} + f''(a)\frac{(x-a)^3}{3!} + \int_a^x f^{(4)}(t)\frac{(t-x)^3}{3!}dt$$

$$= \cdots = T_n(x) + \int_a^x \frac{f^{n+1}(t)}{n!}(t-x)^n dt$$

Remark There is also Taylor's theorem for multivariate functions in high dimensions (see Section 5.6.1 of Chapter 5).

Below are a few notes regarding the theorem.

- Considering $n = 0$ in Taylor's theorem, we get

$$f(x) = f(a) + f'(c)(x-a) \quad \text{or} \quad \frac{f(x) - f(a)}{x-a} = f'(c) \tag{2.29}$$

which is **the mean value theorem (MVT)**. The MVT has a clear geometry interpretation: existence of a tangent line parallel to the secant line through the two ending points.

- Let $h = x - a$, then $x = a + h$ and we can write Taylor's theorem in the form in terms of h as

$$f(a+h) = \sum_{k=0}^{n} \frac{f^{(k)}(a)}{k!} h^k + \frac{f^{(n+1)}(c)}{(n+1)!} h^{n+1} \tag{2.30}$$

where c is an unknown point between a and $a + h$.
Similarly, if we let $h = a - x$, then $x = a - h$ (and $x - a = -h$) and we can write Taylor's theorem in the form in terms of h as

$$f(a-h) = \sum_{k=0}^{n} \frac{f^{(k)}(a)}{k!} (-h)^k + \frac{f^{(n+1)}(c)}{(n+1)!} (-h)^{n+1} \tag{2.31}$$

- Taylor series $\lim_{n \to \infty} T_n(x)$ converges to $f(x)$ if $\lim_{n \to \infty} R_n(x) = 0$.

Below are two examples in which Taylor's theorem is applied to analyze errors in mathematical approximations.

Example 2.4 (Approximation by a Taylor polynomial) $f(x) = e^x$ is approximated by $T_n(x)$, the n-th order Taylor polynomial of f about 0, for $x \in [-2, 2]$. (1) Find an upper bound of the absolute error in terms of only the degree n. (2) Then determine the degree n such that the absolute error is at most 10^{-4}.
[Solution:] According to Taylor's theorem, the absolute error $|f(x) - T_n(x)|$ is given by Taylor's remainder (the error term) as

$$|f(x) - T_n(x)| = \left| \frac{f^{(n+1)}(c)}{(n+1)!} (x-0)^{n+1} \right| = \frac{e^c}{(n+1)!} |x|^{n+1}$$

where the unknown number c is between 0 and x.
(1) For $x \in [-2, 2]$ and $a = 0$, we have

- $c \in (-2, 2)$, i.e. c falls in the same interval $[-2, 2]$ as x, as illustrated in Fig. 2.2, and thus $e^c < e^2$.
- $|x| \le 2$ and thus $|x|^{n+1} \le 2^{n+1}$

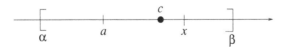

FIGURE 2.2
Relative positions of a, x and c if $a \in [\alpha, \beta]$ and $x \in [\alpha, \beta]$.

So we obtain the upper bound of the absolute error in terms of only n as

$$|f(x) - T_n(x)| = \frac{e^c}{(n+1)!} |x|^{n+1} < \frac{e^2}{(n+1)!} 2^{n+1}$$

(2) If we set the upper bound be at most 10^{-4}:

$$\frac{e^2}{(n+1)!}2^{n+1} \leq 10^{-4}$$

then we can guarantee the absolute error is at most 10^{-4} for $x \in [-2,2]$ as

$$|f(x) - T_n(x)| \leq \frac{e^2}{(n+1)!}2^{n+1} \leq 10^{-4}$$

Since $2^{n+1}/(n+1)!$ is decreasing with n for $n = 1,2,3,\ldots$ (see Exercise 2.14), the inequality can be satisfied if n is large enough, and in this case we require $n \geq 11$. ∎

Example 2.5 (Finite difference approximation) We can approximate $T = f'(a)$ (the slope of a tangent line) by the so-called forward finite difference (more on such approximations in Chapter 8)

$$A = \frac{f(a+h) - f(a)}{h}$$

which is the slope of a secant line, where $h > 0$ is a small spatial step (if $h < 0$, then it is called backward finite difference). Assume $f(x) \in C^2_{[a,a+h]}$. Use Taylor's theorem to determine how the absolute error in the approximation depends on the spatial step h. [**Solution:**] We use Taylor's theorem in the form in terms of h, i.e. Eq. (2.30), with $n+1 = 2$ (note that $f(x) \in C^2_{[a,a+h]}$ is twice differentiable):

$$f(a+h) = f(a) + f'(a)h + \frac{f''(c)}{2!}h^2$$

where $c \in (a, a+h)$. We therefore obtain the dependence of the absolute error on h as

$$|T - A| = \left| \frac{f(a+h) - f(a)}{h} - f'(a) \right| = \frac{|f''(c)|}{2!}h$$

Note that if h is sufficiently small, then $f''(c) \approx f''(a)$ as $c \in (a, a+h)$ and the absolute error is approximately $|f''(a)|h/2$ (a constant multiple of h). ∎

Previously we used the big O notation to present the efficiency of a method. We can also use **the big O notation** to present the **accuracy** of a method. If the absolute error e of the approximation in a method depends on a small parameter h, we say the absolute error is of order $g(h)$, i.e.

$$e = O(g(h)) \tag{2.32}$$

which means that the absolute error e is bounded from above as

$$e \leq Cg(h) \tag{2.33}$$

for some constant $C > 0$ when h is small enough. In the last example above, we have the absolute error

$$|T - A| = \left| \frac{f(a+h) - f(a)}{h} - f'(a) \right| = \frac{|f''(c)|}{2!}h \leq Ch \tag{2.34}$$

where $C = \max_{a \leq x \leq a+h} |f''(x)|/2$. So we can say the absolute error in this example is of $O(h)$.

2.3 Alternating Series Theorem

An alternating series is a series whose terms alternate in signs. It has the form

$$b_0 - b_1 + b_2 - b_3 + \cdots \equiv \sum_{k=0}^{\infty} (-1)^k b_k \quad \text{or} \quad -b_0 + b_1 - b_2 + b_3 - \cdots \equiv \sum_{k=0}^{\infty} (-1)^{k+1} b_k$$

where $b_k > 0$, $k = 0, 1, 2, \ldots$. Looking at the Maclaurin series listed in Section 2.2.2, we recognize that the Maclaurin series of e^x for $x < 0$, $\sin x$ and $\cos x$ for $|x| < \infty$, $1/(1-x)$ for $-1 < x < 0$, $\tan^{-1} x$ for $|x| < 1$, and $\ln(1+x)$ for $0 < x < 1$ are convergent alternating series.

Theorem 2.3 *The alternating series*

$$b_0 - b_1 + b_2 - b_3 + \cdots \equiv \sum_{k=0}^{\infty} (-1)^k b_k \tag{2.35}$$

or

$$-b_0 + b_1 - b_2 + b_3 - \cdots \equiv \sum_{k=0}^{\infty} (-1)^{k+1} b_k \tag{2.36}$$

where $b_k > 0$, $k = 0, 1, 2, \ldots$, is convergent if

$$b_0 > b_1 > b_2 > b_3 > \cdots > 0 \tag{2.37}$$

and

$$\lim_{k \to \infty} b_k = 0 \tag{2.38}$$

Let the sum of the convergent alternating series be S, i.e.

$$S = \sum_{k=0}^{\infty} (-1)^k b_k \quad \text{or} \quad S = \sum_{k=0}^{\infty} (-1)^{k+1} b_k \tag{2.39}$$

Let the partial sum S_n (the truncated series) be

$$S_n = \sum_{k=0}^{n} (-1)^k b_k \quad \text{or} \quad S_n = \sum_{k=0}^{n} (-1)^{k+1} b_k \tag{2.40}$$

Then

$$|S - S_n| \le b_{n+1} \tag{2.41}$$

The Maclaurin series mentioned above satisfy the condition of the alternating series theorem for x in the specified intervals and are therefore convergent (as we have already known from Taylor's theorem).

The alternating series theorem provides an upper bound b_{n+1} for the error in approximating the sum S by the partial sum S_n of a convergent alternating series that satisfies the conditions in the theorem.

Example 2.6 (Approximation by an alternating series) If we use the partial sum of the Maclaurin series of $\cos x$ to approximate $\cos 1$, at which term (included) should we stop in the partial sum to make the absolute error less than 10^{-6}?
[Solution:] The Maclaurin series of $\cos x$ is

$$1 - \frac{x^2}{2!} + \frac{x^4}{4!} - \frac{x^6}{6!} + \cdots \equiv \sum_{k=0}^{\infty} \frac{(-1)^k x^{2k}}{(2k)!}, \quad |x| < \infty$$

which converges to the sum $S(x) = \cos x$. Suppose we stop the partial sum $S_n(x)$ right at the term $(-1)^n x^{2n}/(2n)!$ (included). (Note that $S_n(x)$ is the Taylor polynomial of degree $2n$ for $f(x) = \cos x$ about $a = 0$.) By the alternating series theorem, we have

$$|S(x) - S_n(x)| = |\cos x - S_n(x)| \leq \frac{|x|^{2(n+1)}}{(2(n+1))!}$$

So the absolute error in approximating $\cos 1$ by the partial sum $S_n(1)$ is bounded as

$$|\cos 1 - S_n(1)| \leq \frac{1}{(2n+2)!}$$

Let $1/(2n+2)! < 10^{-6}$, we obtain $n \geq 4$. Note that $S_4(x)$ and $S_4(1)$ are

$$S_4(x) = 1 - \frac{x^2}{2!} + \frac{x^4}{4!} - \frac{x^6}{6!} + \frac{x^8}{8!}, \quad S_4(1) = 1 - \frac{1}{2!} + \frac{1}{4!} - \frac{1}{6!} + \frac{1}{8!}$$

■

2.4   Exercises

Exercise 2.1 Write the polynomial $p_4(x) = 1 + 5x + 4x^2 + 3x^3 + 2x^4$ in the form of nested multiplication. Then evaluate $p_4(2)$ using the nested form (i.e. Horner's method). How many multiplications and additions are used in your evaluation?

Exercise 2.2 How to evaluate $p(x) = 2 + 3x^3 - 4x^6 + 8x^9 - 11x^{12}$ at a given value of x using only 6 multiplications?

Exercise 2.3 Count how many multiplications (in terms of n) are used in the following pseudocode (at the beginning of next page). Write down the number of multiplications in the big O notation. Hint: $\sum_{k=1}^{n} k^2 = n(n+1)(2n+1)/6$.

```
for k from 1 to n − 1 do
    for i from k + 1 to n do
        for j from k + 1 to n do
            Aij ← Aij − AkjAik/Akk
        end for
    end for
end for
```

Exercise 2.4 Find the Taylor polynomial $T_3(x)$ of degree 3 for $f(x) = 1 + 4x + 3x^2 + 2x^3$ about $a = 1$. Show that $T_3(x) = f(x)$ in this case by simplifying $T_3(x)$ or by applying Taylor's theorem.

Exercise 2.5 Find the Taylor polynomial $T_1(x)$ of degree 1 for $f(x) = \sqrt[3]{x^5}$ about $a = 0$. Sketch the graphs of $T_1(x)$ and $f(x)$ in the same plot. Can you find the Taylor polynomial $T_2(x)$ of degree 2 for $f(x)$ about $a = 0$?

Exercise 2.6 Write down the Taylor polynomial $T_4(x)$ of degree 4 for $f(x)$ about a. Verify that $T_4^{(k)}(a) = f^{(k)}(a)$ for $k = 0, 1, 2, 3, 4$. Note that we define the 0-th derivative of a function as the function itself.

Exercise 2.7 Let $T_n(x)$ be the Taylor polynomial of degree n for the polynomial $p_n(x) = c_0 + c_1 x + c_2 x^2 + \cdots + c_n x^n$ about a. Prove that $T_n(x) = p_n(x)$.

Exercise 2.8 Derive the Maclaurin series for $f(x) = \ln(1 - x)$ by two ways: (1) finding $f^{(k)}(0)/k!$, $k = 0, 1, 2, 3, \ldots$; (2) using a substitution in an existing Maclaurin series. State the range of x on which the series converges to $f(x) = \ln(1 - x)$.

Exercise 2.9 Let $\theta = (1/55 \cdots 5)°$ be a small angle in degrees, where the denominator has n fives. (1) Compute $\sin \theta$ for $n = 3, 5, 7, 10$ and write down the results in scientific notation. Do you see something unusual? (2) Why $\sin \theta$ has more and more same decimal digits in the same order as π when n increases? (Hint: What is the truncated degree-1 Maclaurin series of $\sin x$? Is x in the series in degrees or radians?)

Exercise 2.10 Let $T_n(x)$ be the n-th order Taylor polynomial for $f(x) = e^x$ about $a = 0$. (1) Find the expression for $T_n(x)$. (2) If $T_9(x)$ is used to approximate $f(x) = e^x$ for $-2 \leq x \leq 1$, find an upper bound of the error. (3) If $T_n(x)$ is used to approximate $f(x) = e^x$ for $-1 \leq x \leq 1$ with the absolute error at most 10^{-3}, how large should n be?

Exercise 2.11 Let $f(x) = \sin x$, $a = \pi$. (1) Derive the degree-3 Taylor polynomial $T_3(x)$ for $f(x)$ at a. (2) If $f(x)$ is approximated by $T_3(x)$ for $\pi - 1 \leq x \leq \pi + 1$, find an upper bound for the error of this approximation using Taylor's Theorem.

Exercise 2.12 Derive the degree-2 Taylor polynomial $T_2(x)$ for $f(x) = \ln x$ at $a = 1$. If $f(x)$ is approximated by $T_2(x)$ for $0.5 \leq x \leq 1.5$, find an upper bound for the absolute error of this approximation using Taylor's remainder.

Exercise 2.13 Prove Taylor's theorem. (Hint: See remarks for Theorem 2.2.)

Exercise 2.14 Show $2^{n+1}/(n+1)!$ is decreasing with n for $n = 1, 2, 3, \ldots$ using proof by induction.

Exercise 2.15 Use the alternating series theorem to determine the value of n in the Taylor polynomial

$$T_{2n}(x) = \sum_{k=0}^{n} \frac{(-1)^k x^{2k}}{(2k)!}$$

for $\cos x$ about 0 such that $\cos 1$ is approximated by $T_{2n}(1)$ with the absolute error less than 10^{-6}.

Exercise 2.16 Approximate $T = f(a)$ by $A = f(a+h)$, where h is small, and $f(x)$ has continuous first derivative in an interval containing a and $a+h$. (1) Use Taylor's theorem to find an expression of the absolute error $|T - A|$ in terms of h. (2) Write the absolute error in big-O notation. (3) What is the limit of the error as $h \to 0$?

Exercise 2.17 If $f(a)$ is approximated by

$$f(a) \approx \frac{f(a+h) + f(a-h)}{2}$$

where h is a small value, and $f(x)$ has continuous second derivative everywhere, (1) find the absolute error of the approximation in terms of h using Taylor's theorem $f = T_n + R_n$ with $n = 1$; (2) write the absolute error in big-O notation.

Exercise 2.18 $f(x) \in C^2$ near a. Approximate $f'(a)$ as

$$f'(a) \approx \frac{f(a) - f(a-h)}{h}$$

where h is a small positive value. Use Taylor's theorem to show that the error of the approximation is $O(h)$.

2.5   Programming Problems

Problem 2.1 Write the MATLAB m-function

```
function [p] = nest(c,x,a)
```

to implement Horner's method (not the Naive method) for evaluating the degree-n polynomial centered at the number a:

$$p_n(x) = c_0 + c_1(x-a) + c_2(x-a)^2 + \cdots + c_n(x-a)^n.$$

The $n+1$ coefficients $c_0, c_1, \ldots, c_n$ are entries of the input vector c. The values of the independent variable x are passed as a vector.

Test your m-function by evaluating $p_3(x) = 1 + 3(x+1) - 2(x+1)^3$ at $x = 0$.

Problem 2.2 Let $T_n(x)$ be the Taylor polynomial of degree n for $f(x) = \ln(1+x)$ at $a = 0$. Write a MATLAB m-script to use your function `nest` in the previous problem to evaluate $T_4(x)$ and $T_9(x)$ at

$$x = -0.5, -0.49, -0.48, ..., 0.48, 0.49, 0.5.$$

(1) Plot $f(x)$, $T_4(x)$ and $T_9(x)$ in one figure using the MATLAB command `plot`.
(2) Plot the Taylor's remainders (the error terms) $|f(x) - T_4(x)|$ and $|f(x) - T_9(x)|$ in another figure using the MATLAB command `semilogy` (use `help semilogy` to learn what `semilogy` does and how to use it).
(3) Derive an upper bound of $|\ln(1+x) - T_9(x)|$ for $-0.5 \le x \le 0.5$ using Taylor's theorem. Is the absolute error $|\ln(1+x) - T_9(x)|$ in your second plot less than your derived upper bound?

3

Roundoff Errors and Error Propagation

In this chapter, we explains why roundoff errors are inevitable in scientific computing and how roundoff errors are propagated in arithmetic and algorithms. We use examples to illustrate and analyze the stability of algorithms.

Let's start by running the following MATLAB demo code

```
a = 0;
n = 0;
while a~=1 && n<20
    a = a+0.1;
    n = n+1;
end
frpintf('a = %18.16f after addition of 0.1 for %d times\n',a,n)
```

The output of the code is

```
a = 2.000000000000000444 after addition of 0.1 for 20 times
```

We might expect that the while loop would end when a could have reached 1 (to violate the condition a~=1) after $n = 10$ additions of 0.1, but it stops only when n goes to 20 (to violate the condition n<20).

Why is the repeated addition of 0.1 for 10 times not equal to 1 in this demo? The reason is that the innocent looking real number 0.1 cannot be exactly represented by computers. As we will see shortly, a computer can exactly represent only a finite number of real numbers which are called machine numbers. Most likely, a real number (for example 0.1) cannot be represented exactly by a computer and is represented inexactly as a nearby machine number, introducing the so-called roundoff error.

3.1 Numbers

We have seen the following different types of numbers.
- natural numbers: $\mathbb{N} = \{1, 2, 3, \ldots\}$
- integers: $\mathbb{Z} = \{\ldots, -3, -2, -1, 0, 1, 2, 3, \ldots\}$
- rational numbers: $\mathbb{Q} = \{p/q|, p \in \mathbb{Z}, q \in \mathbb{Z}, q \neq 0\}$

DOI: 10.1201/9781003201694-3

25

- real numbers: $\mathbb{R}$ = set of all the numbers on the real number line, including rational numbers (such as 0.1 and $1/3$) and irrational numbers (such as $\sqrt{2}$, π and e).
- complex numbers: $\mathbb{C} = \{a + bi | a \in \mathbb{R}, b \in \mathbb{R}, i = \sqrt{-1}\}$

The above sets of numbers expand from top to bottom: $\mathbb{N} \subset \mathbb{Z} \subset \mathbb{Q} \subset \mathbb{R} \subset \mathbb{C}$.

A computer stores a number in a physical unit such as registers, RAMs, or disk drives which may be regarded as an ordered list of switches. Each switch has 2 statuses: on and off. We assign the on and off statuses the values 1 and 0, respectively, and call such a switch a **bit**. So a computer represents a number as a pattern of ordered bits (for example 01101010). The mapping between a bit pattern and a number is determined by the IEEE (Institute of Electrical and Electronics Engineers, pronounced as I-triple-E) standards. Below we describe how a computer represents integers and real numbers.

3.1.1   Integers

Signed integers are commonly used in computations. An 8-bit (1-byte) signed integer is stored as the bit pattern

$$b_7 b_6 b_5 b_4 b_3 b_2 b_1 b_0$$

which corresponds to the value

$$(-b_7) \times 2^7 + b_6 \times 2^6 + b_5 \times 2^5 + b_4 \times 2^4 + b_3 \times 2^3 + b_2 \times 2^2 + b_1 \times 2^1 + b_0 \times 2^0 \quad (3.1)$$

With 8 bits, only $2^8 = 128$ signed integers $-128, -127, \ldots, 126, 127$ can be represented.

We also use **unsigned integers** in indexing (for example, row and column indices of a matrix) and in the representing real numbers (soon later). An 8-bit unsigned integer is stored as the bit pattern

$$b_7 b_6 b_5 b_4 b_3 b_2 b_1 b_0$$

which corresponds to the value

$$b_7 \times 2^7 + b_6 \times 2^6 + b_5 \times 2^5 + b_4 \times 2^4 + b_3 \times 2^3 + b_2 \times 2^2 + b_1 \times 2^1 + b_0 \times 2^0 \quad (3.2)$$

With 8 bits, only $2^8 = 128$ unsigned integers $0, 1, \ldots, 254, 255$ can be represented.

3.2   Floating-Point Numbers

Real numbers are represented by a computer as floating-point numbers, including 64-bit double precision (DP) type and 32-bit single precision (SP) type (by default, MATLAB use DP floating-point numbers for real number arithmetic). What are floating-point numbers? Let's look at scientific notation first to answer this question.

3.2.1 Scientific Notation and Rounding

We are familiar with decimal numbers. The decimal number system is a base-10 system in which the place values are integer powers of 10 (for example, $234.56 = 2 \times 10^2 + 3 \times 10^1 + 4 \times 10^0 + 5 \times 10^{-1} + 6 \times 10^{-2}$).

Definition 3.1 *The scientific notation of a decimal number T is*

$$T = \sigma \cdot \bar{T} \cdot 10^e \tag{3.3}$$

where
- $\sigma = +1/-1$: *sign of T*
- $1 \leq \bar{T} < 10$: *significand/mantissa of T*
- *e: integer exponent*

The mantissa $\bar{T}$ has the form

$$\bar{T} = d_1.d_2d_3\cdots \tag{3.4}$$

where $d_i \in \{0,1,2,\ldots,9\}$ ($i = 1,2,3,\ldots$) except $d_1 \neq 0$. The leftmost nonzero digit d_1 in the mantissa corresponds to the largest place value and is called the first **significant digit** (the most significant digit), the second digit d_2 from left in the mantissa is called the second significant digit, and so on.

A floating-point representation A of the number T in the decimal system has the same form as scientific notation:

$$A = \sigma \cdot \bar{A} \cdot 10^E \tag{3.5}$$

but the number of digits in the mantissa is limited. If the mantissa $\bar{A}$ of a floating-point representation allows only t decimal digits, then it can be written as

$$\bar{A} = \tilde{d}_1.\tilde{d}_2\tilde{d}_3\cdots\tilde{d}_{t-1}\tilde{d}_t \tag{3.6}$$

It is called floating-point representation because the decimal point is always floating between $\tilde{d}_1$ and $\tilde{d}_2$ with the help of the power 10^E. If the mantissa $\bar{T}$ of the number T in scientific notation has more than t digits:

$$\bar{T} = d_1.d_2d_3\cdots d_t d_{t+1}\cdots \tag{3.7}$$

then its floating-point representation A can be obtained by rounding. The rule for rounding is "round to nearest, ties to even" by comparing the digit d_{t+1} (the first digit to be discarded) with 5 (half of the base value 10) as follows.
- **"Round to nearest"**: If $d_{t+1} < 5$, round $\bar{T}$ down to $\bar{A}$ by simply chopping off all the less significant digits to the right of d_t. If $d_{t+1} > 5$, round $\bar{T}$ up to $\bar{A}$ by discarding all the digits to the right of d_t and adding 1 to the digit d_t, which

may lead to carrying (or even the adjustment of the exponent e to $E = e + 1$). If $d_{t+1} = 5$ but is not the rightmost nonzero digit (i.e. there are other nonzero digits after it), round $\bar{T}$ up.

- **"Ties to even"**: If $d_{t+1} = 5$ and is the rightmost nonzero digit, then when d_t is even, round $\bar{T}$ down; and when d_t is odd, round $\bar{T}$ up. In either case, the digit $\tilde{d}_t$ is even.

Example 3.1 (Rounding) The fixed-point numbers

$$3.1416, -124.63, -43.652, 0.002375, -0.2385$$

can be written in scientific notation as

$$+3.1416 \times 10^0, -1.2463 \times 10^2, -4.3652 \times 10^1, 2.375 \times 10^{-3}, -2.385 \times 10^{-1}$$

respectively. If the mantissa of a floating-point representation (in base 10) can have only 3 significant digits, then the floating-point representations of these numbers are obtained by rounding as

$$+3.14 \times 10^0, -1.25 \times 10^2, -4.37 \times 10^1, 2.38 \times 10^{-3}, -2.38 \times 10^{-1}$$

respectively. ■

So the floating-point representation A can be an approximation of the decimal number T. We say A is a t-digit approximation of T, meaning A has t correct/significant digits. The error induced by rounding is called **roundoff error**. The rounding rule implies that the absolute roundoff error is bounded by 5 times the place value at the digit d_{t+1}:

$$|T - A| \leq 5 \cdot 10^{-t} \cdot 10^e \tag{3.8}$$

Since $|T| \geq 1 \cdot 10^e$, the relative roundoff error is bounded as

$$\frac{|T - A|}{|T|} \leq \frac{5 \cdot 10^{-t} \cdot 10^e}{10^e} = \frac{10^{-(t-1)}}{2} = 5 \cdot 10^{-t} \tag{3.9}$$

If the relative error in an approximation A of a number T does not exceed $5 \cdot 10^{-t}$, we say A has t correct/significant decimal digits.

Example 3.2 (Significant digits) The floating-point representation with a 3-digit mantissa for the number -1.2463×10^2 is -1.25×10^2. The absolute roundoff error in the representation is

$$|(-1.2463 \times 10^2) - (1.25 \times 10^2)| < 0.005 \times 10^2$$

and the relative error is

$$\frac{|(-1.2463 \times 10^2) - (1.25 \times 10^2)|}{|-1.2463 \times 10^2|} < \frac{0.005 \times 10^2}{10^2} = 5 \times 10^{-3}$$

The representation has 3 significant decimal digits. ■

The above description can be extended to a floating-point representation $\mathrm{fl}_\beta(x)$ of a number x in base-β system. In particular, the relative roundoff error is bounded as

$$\frac{|x - \mathrm{fl}_\beta(x)|}{|x|} \le \frac{\beta^{-(t-1)}}{2} \tag{3.10}$$

where t is the number of digits in the mantissa of $\mathrm{fl}_\beta(x)$.

Definition 3.2 *The* **rounding unit** η *of a floating-point number system characterized by the base β and the t-digit mantissa is defined as*

$$\eta = \frac{\beta^{-(t-1)}}{2} \tag{3.11}$$

which is the sharp upper bound for the relative roundoff error in the floating-point representation of a number.

3.2.2 DP Floating-Point Representation

Since a computer represents a real number as a bit (0 or 1) pattern, it naturally uses the binary number system to represent the number as a floating-point number in base 2. If 64 bits are used in the floating-point representation of the number x, the representation is denoted as $\mathrm{fl}_{DP}(x)$, which is called a double precision (DP) floating-point representation or a DP machine number. Soon we will find out that generally

$$x \ne \mathrm{fl}_{DP}(x)$$

The bit pattern of the 64 bits for storing x as $\mathrm{fl}_{DP}(x)$ can be given as

$$\boxed{b_{64}} \; \boxed{b_{63}b_{62}\cdots b_{53}} \; \boxed{b_{52}b_{51}\cdots b_1} \tag{3.12}$$
$$\underbrace{\phantom{b_{64}}}_{s} \quad \underbrace{\phantom{b_{63}b_{62}}}_{b} \quad \underbrace{\phantom{b_{52}b_{51}}}_{f}$$

which is divided into three fields:
- the sign bit s: a 1-bit unsigned integer b_{64}, so s is either 0 or 1.
- the biased exponent b: an 11-bit unsigned integer

$$b = b_{63} \times 2^{10} + b_{62} \times 2^9 + \cdots b_{53} \times 2^0, \quad b \in [0, 2047] \tag{3.13}$$

- the fraction field f: a 52-bit unsigned integer

$$f = b_{52} \times 2^{51} + b_{51} \times 2^{50} + \cdots b_1 \times 2^0, \quad f \in [0, 2^{52} - 1] \tag{3.14}$$

The decimal value $\mathrm{fl}_{DP}(x)$ corresponding to this bit pattern as determined by the IEEE standard is

$$
\mathrm{fl}_{DP}(x) = \begin{cases}
(-1)^s \left(1 + \frac{f}{2^{52}}\right) 2^{b-1023}, & \text{if } 1 \le b \le 2046 \\
(-1)^s \left(\frac{f}{2^{52}}\right) 2^{-1022}, & \text{if } b = 0, \ f \ne 0 \\
\pm 0, & \text{if } b = 0, \ f = 0 \\
\pm \infty, & \text{if } b = 2047, \ f = 0 \\
\mathrm{NaN}, & \text{if } b = 2047, \ f \ne 0
\end{cases} \tag{3.15}
$$

Remark In Eq. (3.15), $e := b - 1023 \in [-1022, 1023]$ is called the unbiased exponent; and $f/2^{52} = (0.b_{52}b_{51} \cdots b_1)_2$ (the reason f is call the fraction field), where the subscript 2 is used to denote a binary number (so $1 + f/2^{52} = (1.b_{52}b_{51} \cdots b_1)_2$ is a 53-digit mantissa in base 2).

Example 3.3 Find the decimal value of the DP floating-point representation of the real number x, the DP number $\mathrm{fl}_{DP}(x)$, whose bit pattern is

| 1 | 10000000101 | 111000010$\cdots$0 |

[**Solution:**] The unsigned integers corresponding to the three fields in the pattern are
- $s = 1$
- $b = 2^{10} + 2^2 + 2^0 = 1029$
- $f = 2^{51} + 2^{50} + 2^{49} + 2^{44}$

So according to Eq. (3.15), we have

$$
\mathrm{fl}_{DP}(x) = (-1)^1 \left(1 + \frac{2^{51} + 2^{50} + 2^{49} + 2^{44}}{2^{52}}\right) 2^{1029-1023} = -120.25
$$

■

Example 3.4 The real number 1 is a DP number with the bit pattern

| 0 | 01111111111 | 00$\cdots$0 |

The next larger DP number adjacent to 1 has the bit pattern

| 0 | 01111111111 | 00$\cdots$01 |

and its value is $1 + 2^{-52}$. ■

We call the distance from 1 to its next larger DP number the DP **machine epsilon** ε_{DP}. As we will see shortly, this value characterizes the precision (the level of relative roundoff error) of the DP floating-point representation.

There are 2^{64} bit patterns with 64 bits, so the DP floating-point representation on a computer provides 2^{64} DP numbers. They are a finite number of discrete points on the real number line, but the real number line is continuous and contains infinitely many numbers/points. If a real number x on the real number line is not one of those discrete DP numbers (i.e. there is no bit pattern whose corresponding value is exactly x), how does the computer represent x and do arithmetic involving x? The computer represents x by $\mathrm{fl}_{DP}(x)$ which is the DP number nearest to x as illustrated in Fig. 3.1, and use $\mathrm{fl}_{DP}(x)$ to do arithmetic. In this case, the true value x is rounded to the approximation $\mathrm{fl}_{DP}(x)$, introducing roundoff error in the approximation $\mathrm{fl}_{DP}(x)$. By Eq. (3.11), the relative roundoff error satisfies

$$\left| \frac{x - \mathrm{fl}_{DP}(x)}{x} \right| \leq \eta = \frac{2^{-(53-1)}}{2} = \frac{\varepsilon_{DP}}{2} = 2^{-53} \approx 1.11 \times 10^{-16} < 5 \times 10^{-16} \quad (3.16)$$

which implies that $\mathrm{fl}_{DP}(x)$, as a representation of x, has at least **16 correct/significant decimal digits**.

FIGURE 3.1
A computer rounds a non-DP number x to $\mathrm{fl}_{DP}(x)$, the DP number nearest to x.

Example 3.5 The real number 1 is a DP number. The next larger DP number adjacent to 1 is $1 + 2^{-52}$. If $x = 1 + 2^{-54}$, what is the value of $\mathrm{fl}_{DP}(x)$?
[Solution:] $x = 1 + 2^{-54}$ is between the consecutive DP numbers 1 and $1 + 2^{-52}$. It is closer to 1 than $1 + 2^{-52}$. So $\mathrm{fl}_{DP}(x) = 1$, i.e. $x = 1 + 2^{-54}$ is rounded to 1. The relative roundoff error is

$$\left| \frac{x - \mathrm{fl}_{DP}(x)}{x} \right| = \frac{2^{-54}}{1 + 2^{-54}} < 2^{-54} < \frac{\varepsilon_{DP}}{2} = 2^{-53}$$

∎

Now we can explain why 0.1 used in the beginning demo is not a DP number. If we convert 0.1 to a binary number, we get

$$0.1 = (0.0001\ \underline{1001}\ \underline{1001} \cdots)_2 = [1 + (0.\underline{1001}\ \underline{1001} \cdots)_2] \cdot 2^{-4}$$

where the pattern 1001 repeats forever. So 0.1 in the floating-point binary form requires infinitely many bits in the fraction field and cannot be a DP number with only 52 fraction bits. In the demo, MATLAB rounds 0.1 to its nearest DP number $\mathrm{fl}_{DP}(0.1)$, which is slightly different from 0.1, and repeatedly adds $\mathrm{fl}_{DP}(0.1)$ (instead of 0.1), so the result is not exactly equal to 1, which is a DP machine number.

From Eq. (3.15), we can figure out the largest and smallest DP numbers, giving the DP numbers a range. If a number resulted from a computation has a magnitude that is outside this range, we say overflow occurs (or the computation overflows). Fatal error or exception in the execution of a program may be caused by overflow. Sometimes we may avoid overflow by properly scaling the numbers.

3.3   Error Propagation

We now know roundoff error is inevitable in computation on a computer. So we need to know how roundoff error affects computational results. Below we first analyze how errors propagate in arithmetic operations.

Let $\hat{x}$ and $\hat{y}$ be approximations of the nonzero true values x and y, respectively (for example $\hat{x} = \text{fl}_{DP}(x)$ and $\hat{y} = \text{fl}_{DP}(y)$). We may write

$$\hat{x} = x(1 + \varepsilon_x), \quad \hat{y} = y(1 + \varepsilon_y) \tag{3.17}$$

Note that $|\varepsilon_x|$ and $|\varepsilon_y|$ are relative errors of the approximations $\hat{x}$ and $\hat{y}$, respectively.

Let's first consider error propagation in multiplication. The true product $x \cdot y$ of x and y is approximated by the approximate product $\hat{x} \cdot \hat{y}$. We have

$$\hat{x} \cdot \hat{y} = x(1 + \varepsilon_x) \cdot y(1 + \varepsilon_y) = x \cdot y(1 + \varepsilon_x + \varepsilon_y + \varepsilon_x \varepsilon_y) \equiv x \cdot y(1 + \varepsilon)$$

where the relative error of the approximate product is $|\varepsilon| = |\varepsilon_x + \varepsilon_y + \varepsilon_x \varepsilon_y|$. So the errors in the multiplicand and the multiplier are propagated to the product. By the triangle inequality,

$$||a| - |b|| \le |a \pm b| \le |a| + |b| \tag{3.18}$$

we have

$$|\varepsilon| \le |\varepsilon_x| + |\varepsilon_y| + |\varepsilon_x \varepsilon_y| \tag{3.19}$$

So if the relative errors $|\varepsilon_x|$ and $|\varepsilon_y|$ of the multiplicand and the multiplier are small, the relative error $|\varepsilon|$ of the approximate product it also small. Multiplication is therefore safe in terms of error propagation.

3.3.1   Catastrophic Cancellation

Now let's consider error propagation in subtraction. The true difference $x - y$ (assuming nonzero) of x and y is approximated by the approximate difference $\hat{x} - \hat{y}$. We have

$$\hat{x} - \hat{y} = x(1 + \varepsilon_x) - y(1 + \varepsilon_y) = (x - y)\left(1 + \frac{x}{x - y}\varepsilon_x - \frac{y}{x - y}\varepsilon_y\right) \equiv (x - y)(1 + \varepsilon)$$

where the relative error of the approximate difference is

$$|\varepsilon| = \left|\varepsilon_x \frac{x}{x - y} - \varepsilon_y \frac{y}{x - y}\right| \le \left|\frac{x}{x - y}\right||\varepsilon_x| + \left|\frac{y}{x - y}\right||\varepsilon_y| \tag{3.20}$$

The upper bound above can be sharp. So if x and y are very close, i.e. $(x - y) \approx 0$, then the relative error can be huge even if ε_x and ε_y are very small because of the division by $(x - y)$ (≈ 0). We call the phenomenon that subtraction of relatively accurate numbers produces relatively inaccurate result **catastrophic cancellation**. Subtraction of close approximate numbers cancels a lot of significant digits and is dangerous.

Example 3.6 (Catastrophic cancellation) $\hat{x} = 1.001$ is an approximation of $x = 1.000$. The relative error of the approximation is 0.1%. $\hat{y} = 0.998$ is an approximation of $y = 0.999$. The relative error of this approximation is approximately 0.1%. The difference $\hat{x} - \hat{y} = 0.003$ approximates (very badly) the true difference $x - y = 0.001$. The relative error of the approximate result is 200%. ∎

We try our best to avoid subtraction of close approximate numbers in our algorithm design to avoid catastrophic cancellation (loss of significant digits). Sometimes we may reformulate the problem to do so.

Example 3.7 (Reformulation) To evaluate $f(x) = \sqrt{x^2 + 1} - 1$ for very small $|x|$, we can reformulate $f(x)$ as $f(x) = x^2/(\sqrt{x^2 + 1} + 1)$. The former expression for $f(x)$ can lead to catastrophic cancellation, while the latter does not. ∎

3.3.2 Algorithm Stability

A numerical algorithm is a step-by-step procedure to find the numerical approximate solution of a mathematical problem. An algorithm is **stable** if small errors introduced in the algorithm stay small. If the small errors get amplified to be out of control, the algorithm is **unstable**. Below we use error propagation analysis to check algorithm stability/instability in an example.

In this example, we consider the sequence of values $\{V_j\}_{j=0}^{\infty}$ defined by the definite integrals

$$V_j = \int_0^1 e^{x-1} x^j \mathrm{d}x, \quad j = 0, 1, 2, \ldots \tag{3.21}$$

The first term V_0 is given by

$$V_0 = \int_0^1 e^{x-1} \mathrm{d}x = 1 - \frac{1}{e} \tag{3.22}$$

The sequence satisfies the inequalities

$$0 < V_{j+1} < V_j < \frac{1}{j}, \quad j = 1, 2, 3, \ldots \tag{3.23}$$

and the forward recurrence relation

$$V_j = 1 - jV_{j-1}, \quad j = 1, 2, 3, \ldots \tag{3.24}$$

In an ideal world with exact arithmetic (no errors), we can start with the true value V_0 and use the recurrence relation to compute the true values V_1, V_2, V_3, ... in this order as follows

$$V_1 = 1 - 1V_0, \quad V_2 = 1 - 2V_1, \quad , V_3 = 1 - 3V_2, \quad \cdots \tag{3.25}$$

In the real world, numbers have roundoff errors, and inexact arithmetic are used. To simplify the error propagation analysis, we only consider the propagation of the error in the initial $\hat{V}_0$ (the approximation of the true value V_0) to the later values $\hat{V}_1$, $\hat{V}_2$, $\hat{V}_3$, ... in the sequence computed from $\hat{V}_0$ by the forward recurrence relation. (We neglect errors due to inexact arithmetic in applying the recurrence relation). Let $\hat{V}_0 = V_0 - \varepsilon_0$, where $\varepsilon_0 = V_0 - \hat{V}_0$ is the error in $\hat{V}_0$ as an approximation of V_0. We have

$$\hat{V}_1 = 1 - 1\hat{V}_0 = 1 - 1(V_0 - \varepsilon_0) = 1 - 1V_0 + \varepsilon_0 = V_1 + 1\varepsilon_0, \tag{3.26}$$
$$\hat{V}_2 = 1 - 2\hat{V}_1 = 1 - 2(V_1 + 1\varepsilon_0) = 1 - 2V_1 - 2 \cdot 1\varepsilon_0 = V_2 - 2 \cdot 1\varepsilon_0, \tag{3.27}$$
$$\hat{V}_3 = 1 - 3\hat{V}_2 = 1 - 3(V_2 - 2 \cdot 1\varepsilon_0) = 1 - 3V_2 + 3 \cdot 2 \cdot 1\varepsilon_0 = V_3 + 3 \cdot 2 \cdot 1\varepsilon_0 \tag{3.28}$$

So $\hat{V}_1$ approximates V_1 with the error $V_1 - \hat{V}_1 = -1\varepsilon_0$, $\hat{V}_2$ approximates V_2 with the error $V_2 - \hat{V}_2 = +2 \cdot 1\varepsilon_0$, and $\hat{V}_3$ approximates V_3 with the error $V_3 - \hat{V}_3 = -3 \cdot 2 \cdot 1\varepsilon_0$. The pattern indicates that $\hat{V}_j$ approximates V_j with the error

$$V_j - \hat{V}_j = (-1)^j j! \varepsilon_0, \quad j = 0, 1, 2, \ldots \tag{3.29}$$

Another way to see the error propagation is to establish the recurrence relation between the error of the current approximation and the error of the previous approximation. We have the recurrence for true values

$$V_j = 1 - jV_{j-1}, \quad j = 1, 2, 3, \ldots \tag{3.30}$$

and the recurrence for approximate values

$$\hat{V}_j = 1 - j\hat{V}_{j-1}, \quad j = 1, 2, 3, \ldots \tag{3.31}$$

Let $e_j = V_j - \hat{V}_j$, $j = 0, 1, 2, \ldots$. By subtracting the above two recurrence relations, we have

$$e_j = -je_{j-1}, \quad j = 1, 2, 3, \ldots \tag{3.32}$$

which is the recurrence relation between errors. By applying this recurrence repeatedly, we obtain

$$e_j = -je_{j-1} = -j(-(j-1)e_{j-2}) = -j(-(j-1))(-(j-2)e_{j-3})$$
$$= \cdots = -j(-(j-1))(-(j-2)) \cdots (-2)(-1)e_0 = (-1)^j j! \varepsilon_0 \tag{3.33}$$

Remark The latter way to do error propagation analysis is used in later chapters to analyze the convergence of an iterative method.

The error $e_j = V_j - \hat{V}_j = (-1)^j j! \varepsilon_0$ tells us that the initial error ε_0 in $\hat{V}_0$ is propagated into $\hat{V}_j$ with the amplification factor $(-1)^j j!$. If j is large, the magnitude of this factor is huge. Recall that the true value $V_j < 1/j$ and the approximate value $\hat{V}_j = V_j - (-1)^j j! \varepsilon_0$. So for large j, $\hat{V}_j \approx -(-1)^j j! \varepsilon_0$ is dominated by the error, and the true value V_j is swamped and lost in the approximation. So the algorithm to compute V_j using the forward recurrence relation is unstable.

However, V_j can be approximated/computed with high accuracy by different algorithms (we say computing V_j numerically is a well-conditioned problem. More on problem conditioning later). For example, we may directly approximate the definite integral in the definition of V_j by quadrature rules (introduced in Chapter 7). We can even reformulate the forward recurrence relation into the backward recurrence relation

$$V_{j-1} = \frac{1-V_j}{j}, \quad j = \cdots, 3, 2, 1 \tag{3.34}$$

to accurately compute V_M for any M. To do so, we can start from a guess value of a later entry V_{M+N}, where N is large enough (say $N \geq 18$), and runs the backward recurrence from V_{M+N} toward V_M. It turns out that the error in the initial guess of V_{M+N} is significantly shrunk by the backward recurrence when it is propagated in V_M.

3.4 Exercises

Exercise 3.1 Find the decimal value of each of the binary numbers (1) 10000001, (2) 11111111, (3) 101.101, (4) 1111.1111, (5) 0.0001$\underline{1100}$ $\underline{1100}$ $\underline{1100}\cdots$

Exercise 3.2 Write in scientific notation each of the decimal numbers (1) -2.718, (2) 2020, (3) 0.0005678, (4) -737.2×10^{-8}, (5) 0.00256×10^{23}

Exercise 3.3 Convert into a binary number each of the decimal numbers (1) 256 (2) 13 (3) 127 (4) 3.75 (5) 0.1

Exercise 3.4 Find the decimal value of the DP number for each of the bit patterns

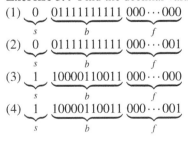

Exercise 3.5 (1) Show that the decimal value of the DP number with the bit pattern

$$\underbrace{1}_{s} \; \underbrace{01111111111}_{b} \; \underbrace{00111\cdots111}_{f}$$

is $(-1.25 + \varepsilon_{DP})$, where ε_{DP} is the DP machine epsilon.
(2) What is the decimal value of the DP number whose bit pattern is

$$\underbrace{1}_{s} \; \underbrace{01111111111}_{b} \; \underbrace{00111\cdots110}_{f}$$

which differs from (1) at the last digit? Write your answer similar as in (1) in terms of ε_{DP}.
(3) Use the fact that the two DP numbers in (1) and (2) are consecutive to find $\mathrm{fl}_{DP}(-1.25 + 1.25\varepsilon_{DP})$.
(4) Verify that the relative roundoff error to represent the real number $(-1.25 + 1.25\varepsilon_{DP})$ as a DP number $\mathrm{fl}_{DP}(-1.25 + 1.25\varepsilon_{DP})$ is less than $\varepsilon_{DP}/2$.

Exercise 3.6 Are DP numbers uniformly distributed on the real number line?

Exercise 3.7 What is the DP machine epsilon ε_{DP}? What is its exact value (in terms of a power of 2)? What constant in MATLAB corresponds to it?

Exercise 3.8 What is the upper bound of the relative roundoff error $|x - \mathrm{fl}_{DP}(x)|/|x|$ for a nonzero real value x? At least how many correct decimal digits does the DP representation of a real number have?

Exercise 3.9 2^{52} and $2^{52} + 1$ are two consecutive DP numbers. So what are the values of $\mathrm{fl}_{DP}(2^{52} + 0.1)$ and $\mathrm{fl}_{DP}(2^{52} + 0.7)$? What is the relative roundoff error in the DP floating-point representation of the number $2^{52} + 0.2$?

Exercise 3.10 $(x-1)^{10} = x^{10} - 10x^9 + 45x^8 - 120x^7 + 210x^6 - 252x^5 + 210x^4 - 120x^3 + 45x^2 - 10x + 1$. Fig. 3.2 shows the plots of the factorized form (at the left-hand side of the equation) and the expanded form (at the right-hand side) using 201 points with x between 0.95 and 1.05. Why is the former smooth while the latter noisy? (Hint: Error propagation in arithmetic and catastrophic cancellation.)

Exercise 3.11 $V_j = \int_0^1 e^{x-1} x^j \mathrm{d}x, \quad j = 0, 1, 2, \cdots$. Show that
(1) $V_j = 1 - jV_{j-1}, \quad j = 1, 2, 3, \cdots$ (Hint: Integration by parts)
(2) $0 < V_{j+1} < V_j < \frac{1}{j}, \quad j = 1, 2, 3, \cdots$ (Hint: $e^{x-1} < 1$ for $x \in (0, 1)$)

Exercise 3.12 Conduct error propagation analysis for the arithmetic operation division.

Exercise 3.13 Reformulate each of the following formulas to avoid catastrophic cancellation.
(1) $\ln x - \ln y$, when $x \approx y$
(2) $\sqrt{1+x} - \sqrt{1-x}$, when $|x|$ is small

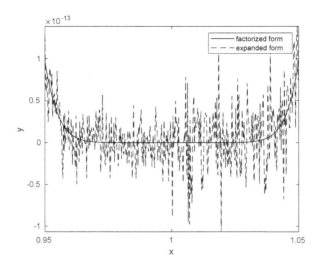

FIGURE 3.2
Plots of the two different forms of the same polynomial $y = (x-1)^{10}$.

(3) $\dfrac{-b+\sqrt{b^2-4ac}}{2a}$ with $b > 0$, when $ac \approx 0$

(4) $\dfrac{-b-\sqrt{b^2-4ac}}{2a}$ with $b < 0$, when $ac \approx 0$

(5) $\sqrt[3]{1+x} - \sqrt[3]{1-x}$, when $|x|$ is small

(6) $\sin(a+x) - \sin(a-x)$, when $|x|$ is small

Exercise 3.14 Conduct error propagation analysis for the backward recurrence relation
$$V_{j-1} = \frac{1-V_j}{j}, \quad j = M+N, M+N-1, \ldots, M+1, M$$

3.5 Programming Problems

Problem 3.1 Write a MATLAB m-script to compute the backward difference approximation $A = \dfrac{f(a) - f(a-h)}{h}$ of the derivative $T = f'(a)$ for $f(x) = \sin(x)$ and $a = \pi/4$ using each value of h in the sequence 2^{-n} ($n = 1, 2, 3, \ldots, 52$).

(1) For each value of h, compute the absolute relative error $r = |A-T|/|T|$ and its upper bound $R = c_1 h + c_2 \dfrac{1}{h}$, where the constants c_1 and c_2 are defined below (keep reading).

(2) Plot in the same figure r versus h with open circles and R versus h with a solid line using the MATLAB plotting command `loglog` (use `help loglog` to learn what `loglog` does and how to use it.)

(3) Qualitatively describe how r and R behave as the increment h goes from its largest value to its smallest value.

(4) Based on your understanding of the mathematical analysis below, give an explanation to your observation.

(5) Find the value of h at which R is minimum using calculus or algebra, and check if your derived results are consistent with MATLAB results.

Below is an analysis of the two sources of the relative error r. Make sure you understand it to explain your results.

For a sufficiently smooth function $f(x)$, the backward difference estimate of the derivative $f'(a)$ is

$$\delta^- f(a) = \frac{f(a) - f(a-h)}{h} \tag{3.35}$$

where the small $h > 0$ is called the step size. Taylor's theorem tells us

$$f(a-h) = f(a) - f'(a)h + \frac{f''(\xi)}{2}h^2 \tag{3.36}$$

where ξ is between a and $a-h$, and ξ is close to a for small values of h. So for very small values of h, we have

$$|\delta^- f(a) - f'(a)| \approx \frac{|f''(a)|}{2}h \tag{3.37}$$

which is the absolute error in approximating the derivative $f'(a)$ by the backward difference $\delta^- f(a)$. This error is the error due to mathematical approximation. As $h \to 0$, the error approaches 0, and $\delta^- f(a) \to f'(a)$, a familiar fact from Calculus.

When we compute the mathematical approximation $\delta^- f(a)$ on a computer, there is another source of error, the roundoff error. Suppose the only roundoff errors that occur are caused by rounding the exact values of $f(a)$ and $f(a-h)$ to their double precision (DP) floating-point versions $\text{fl}_{DP}(f(a))$ and $\text{fl}_{DP}(f(a-h))$ (We choose h to be DP floating-point machine numbers with no rounding errors). We may write

$$\text{fl}_{DP}(f(a)) = f(a) - \mu_1 f(a), \quad \text{fl}_{DP}(f(a-h)) = f(a-h) - \mu_2 f(a-h) \tag{3.38}$$

where $|\mu_1|$ and $|\mu_2|$ are relative errors in the rounding, and they are bounded by

$$|\mu_1| \le \frac{\varepsilon_{DP}}{2}, \quad |\mu_2| \le \frac{\varepsilon_{DP}}{2} \tag{3.39}$$

where $\varepsilon_{DP} = 2^{-52}$ is the DP machine epsilon.

So there are two sources of errors in approximating $f'(a)$ above. One is due to the *mathematical approximation*, i.e. approximation of $f'(a)$ by the backward difference estimate $\delta^- f(a)$, as shown in Eq. (3.37). The other is due to the DP *floating-point representations* of numbers in computing, as indicated by Eq. (3.38), which implies that $\delta^- f(a)$ is not computed exactly.

Let $\delta^-_{DP} f(a)$ denote the computed approximate value of the mathematical approximation $\delta^- f(a)$. With Eqs. (3.38) and (3.37), we have

$$
\begin{aligned}
\delta^-_{DP} f(a) &= \frac{fl_{DP}(f(a)) - fl_{DP}(f(a-h))}{h} \\
&= \frac{(f(a) - \mu_1 f(a)) - (f(a-h) - \mu_2 f(a-h))}{h} \\
&= \delta^- f(a) + \frac{\mu_2 f(a-h) - \mu_1 f(a)}{h} \\
&\approx f'(a) - \frac{f''(a)}{2} h + \frac{\mu_2 f(a) - \mu_1 f(a)}{h}
\end{aligned}
\tag{3.40}
$$

which gives the absolute relative error r in approximating $f'(a)$ by $\delta^-_{DP} f(a)$ as

$$
\begin{aligned}
r &= \left| \frac{f'(a) - \delta^-_{DP} f(a)}{f'(a)} \right| \\
&\approx \left| h \frac{f''(a)}{2f'(a)} + \frac{1}{h} \frac{-\mu_2 f(a) + \mu_1 f(a)}{f'(a)} \right| \\
&\leq h \left| \frac{f''(a)}{2f'(a)} \right| + \frac{1}{h} \left| \frac{-\mu_2 f(a) + \mu_1 f(a)}{f'(a)} \right| \\
&\leq h \left| \frac{f''(a)}{2f'(a)} \right| + \frac{1}{h} \frac{|f(a)|(|\mu_2| + |\mu_1|)}{|f'(a)|} \\
&\leq h \left| \frac{f''(a)}{2f'(a)} \right| + \frac{1}{h} \left| \frac{f(a)\varepsilon_{DP}}{f'(a)} \right|
\end{aligned}
\tag{3.41}
$$

where the last inequality is due to the inequality (3.39). Define

$$
c_1 = \left| \frac{f''(a)}{2f'(a)} \right|, \quad c_2 = \left| \frac{f(a)\varepsilon_{DP}}{f'(a)} \right|, \quad R = c_1 h + c_2 \frac{1}{h}
\tag{3.42}
$$

We have therefore obtained the upper bound R of the absolute relative error r, i.e. $r \leq R$.

Problem 3.2

$$
p(x) = 1 + (x+1) + (x+1)^2 + \ldots + (x+1)^{50}
$$

is a finite geometric series with the common ratio $(x+1)$, so it can be written as

$$
p(x) = \frac{(x+1)^{51} - 1}{x}
$$

(1) Evaluate both forms of $p(x)$ at $x = 0.00001$ (you can use the m-function nest.m for Problem 2.1 in Chapter 2 to evaluate the first form). (2) Which form gives the more accurate result. Why?

Problem 3.3 Let $p_n(x)$ be the Taylor polynomial of degree n for $f(x) = e^x$ at $a = 0$. Use the m-function nest.m for Problem 2.1 in Chapter 2 to evaluate $p_{30}(x)$ at -11.1 as an approximation of $e^{-11.1}$. (1) Calculate the relative error of your result by regarding exp(-11.1) as the true value. Explain why the error is large. (2) You can approximate $e^{-11.1}$ by $1/p_{30}(11.1)$. Calculate the relative error of this new approximation and explain why it gives much better result.

Problem 3.4 The sequence of values $\{V_k\}_{k=0}^{\infty}$ defined by the definite integrals

$$V_k = \int_0^1 \frac{x^k}{x+10}dx, \quad k = 0,1,2,\ldots \tag{3.43}$$

satisfies the recurrence relation

$$V_k = \frac{1}{k} - 10V_{k-1}, \quad k = 1,2,\ldots \tag{3.44}$$

and the inequalities

$$\frac{1}{11}\frac{1}{k+1} < V_k < \frac{1}{10}\frac{1}{k+1}, \quad k = 0,1,2,\ldots \tag{3.45}$$

Write a MATLAB program to calculate the values of $V_1, V_2, \ldots, V_{20}$ using the recurrence relation (3.44) starting from

$$V_0 = \int_0^1 \frac{1}{x+10}dx = \ln(11/10) \tag{3.46}$$

Your program must meanwhile compute the values of $V_k/10^k$ for $k = 0,1,2,\ldots,20$.
   (1) Display your results in a table in which each row contains the values of k, V_k, $1/(10(k+1))$ and $V_k/10^k$ ($k = 0,1,\ldots,20$).
   (2) How do the computed values V_k change as k increases?
   (3) Explain why do they violate the inequalities (3.45) when k is large by conducting an error propagation analysis.
   Now modify the formula (3.46) in the previous problem to

$$V_0 = \int_0^1 \frac{1}{x+10}dx \approx \ln(11/10) + \varepsilon \tag{3.47}$$

and re-run your program with $\varepsilon = 0.17, 0.0023$ and 0.000036.
   (4) What value does $|V_k/10^k|$ approach with increasing k for the each value of ε?

Problem 3.5 To obtain an accurate approximation of V_j, we can instead run the recurrence in the previous problem backwards as

$$V_{k-1} = \frac{1}{10}\left(\frac{1}{k} - V_k\right), \quad k = M, M-1, \ldots, 1 \tag{3.48}$$

If we want accurate approximations of the values V_k, we can start with $M = k + 20$ and $V_M = 0$ and compute the values V_k by the backward recurrence relation. Write another MATLAB program to calculate the value of V_0 using the backward recurrence starting with $V_{20} = 0$.

(1) Display the relative error in the computed value V_0 using the value of V_0 calculated from the formula (3.46) as the true value.

(2) Explain why the value V_0 computed from the backward recurrence relation (3.48) is accurate by conducting an error propagation analysis.

4

Direct Methods for Linear Systems

In this chapter, we presents some basic direct methods for solving linear systems. We start with simply-solved triangular systems, proceed to Gaussian elimination and LU decomposition, and end at the conditioning of linear systems.

Many computational problems in science and engineering end up with solving large linear systems. Below is a simple illustration of the need to solve a linear system. We want to determine the current through each resistor in the electronic circuit in Fig. 4.1, which is formed by a battery and three resistors. Kirchhoff's circuit laws state that the algebraic sum of the currents entering a node and the algebraic sum of the voltage drops around a loop are zero at any instant. Consider the node K and the loops 1 and 2 in Fig. 4.1, we have the following linear equations for the three currents.

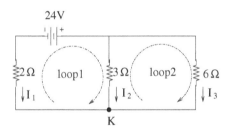

FIGURE 4.1
An electronic circuit.

$$I_1 + I_2 + I_3 = 0 \qquad (4.1)$$
$$-2I_1 - 24 + 3I_2 = 0 \qquad (4.2)$$
$$-3I_2 + 6I_3 = 0 \qquad (4.3)$$

So we have a linear system of three linear equations in three unknowns. If we use x_1, x_2 and x_3 (instead of I_1, I_2 and I_3) as the unknowns, the linear system can be rewritten in the standard form as

$$\begin{cases} x_1 & +x_2 & +x_3 & = & 0 \\ -2x_1 & +3x_2 & & = & 24 \\ & -3x_2 & +6x_3 & = & 0 \end{cases} \qquad (4.4)$$

To find the three currents, we need to solve this linear system.

DOI: 10.1201/9781003201694-4

43

4.1　Matrices and Vectors

The linear system in the motivating example

$$\begin{cases} x_1 & +x_2 & +x_3 & = & 0 \\ -2x_1 & +3x_2 & & = & 24 \\ & -3x_2 & +6x_3 & = & 0 \end{cases} \tag{4.5}$$

can be written as a matrix equation

$$A\mathbf{x} = \mathbf{b}$$

where A is the coefficient matrix consisting of all the coefficients of the unknowns, $\mathbf{x}$ is the vector of unknowns consisting of all the unknowns, and $\mathbf{b}$ is the right-hand-side vector consisting of all the right-hand sides. They are

$$A = \begin{bmatrix} 1 & 1 & 1 \\ -2 & 3 & 0 \\ 0 & -3 & 6 \end{bmatrix}, \quad \mathbf{x} = \begin{bmatrix} x_1 \\ x_2 \\ x_3 \end{bmatrix}, \quad \mathbf{b} = \begin{bmatrix} 0 \\ 24 \\ 0 \end{bmatrix}$$

The i-th row of A, denoted as $\mathrm{Row}_i(A)$, contains the coefficients of all the unknowns in $\mathbf{x}$ in the i-th linear equation; and the j-th column of A, denoted as $\mathrm{Col}_j(A)$, contains the coefficients of x_j in all the linear equations. Note that the left-hand side of the i-th equation is the product of $\mathrm{Row}_i(A)$ (the transpose of a column vector) and $\mathbf{x}$, which is called an inner product. For example, the left-hand of the second equation is

$$\begin{bmatrix} -2 & 3 & 0 \end{bmatrix} \begin{bmatrix} x_1 \\ x_2 \\ x_3 \end{bmatrix} = -2x_1 + 3x_2 + 0x_3 = -2x_1 + 3x_2$$

The linear system is fully described by its augmented matrix $[A\,|\,\mathbf{b}]$:

$$[A\,|\,\mathbf{b}] = \begin{bmatrix} 1 & 1 & 1 & 0 \\ -2 & 3 & 0 & 24 \\ 0 & -3 & 6 & 0 \end{bmatrix}$$

To distinguish the coefficients from the right-hand sides, we can use a vertical line to separate them.

　　In general, we consider a real linear system $A\mathbf{x} = \mathbf{b}$ with m linear equations in n unknowns, where $A \in \mathbb{R}^{m \times n}$, $\mathbf{x} \in \mathbb{R}^n$ and $\mathbf{b} \in \mathbb{R}^m$. If $m = n$, the linear system is a square linear system of order n. Below we focus on such linear systems. If $m > n$ (more equations than unknowns), the linear system is over-determined and can be consistent (with one or infinitely many solutions) or inconsistent (with no solution), and we will find its so-called least squares solution in Chapter 11. The augmented matrix of a linear system of order n can be denoted as

$$[A\,|\,\mathbf{b}] = \begin{bmatrix} a_{11} & a_{12} & \cdots & a_{1n} & b_1 \\ a_{21} & a_{22} & \cdots & a_{2n} & b_2 \\ \vdots & \vdots & & \vdots & \vdots \\ a_{n1} & a_{n2} & \cdots & a_{nn} & b_n \end{bmatrix} \tag{4.6}$$

where the entry a_{ij} at the i-th row and j-th column is the coefficient for x_j in the equation i.

4.2 Triangular Systems

An upper triangular system of order n, $Ax = \mathbf{b}$, has the form

$$\begin{bmatrix} a_{11} & a_{12} & \cdots & a_{1i} & \cdots & a_{1n} \\ & a_{22} & \cdots & a_{2i} & \cdots & a_{2n} \\ & & \ddots & \vdots & & \vdots \\ & & & a_{ii} & \cdots & a_{in} \\ & & & & \ddots & \vdots \\ & & & & & a_{nn} \end{bmatrix} \begin{bmatrix} x_1 \\ x_2 \\ \vdots \\ x_i \\ \vdots \\ x_n \end{bmatrix} = \begin{bmatrix} b_1 \\ b_2 \\ \vdots \\ b_i \\ \vdots \\ b_n \end{bmatrix} \tag{4.7}$$

where the entries below the main diagonal of the coefficient matrix A are zero and are left blank, and A is called an upper triangular matrix. We assume A is invertible/nonsingular, so $\det A = a_{11}a_{22}\cdots a_{nn}(= \prod_{i=1}^{n} a_{ii}) \neq 0$, i.e. all the diagonal entries are nonzero, and the system has a unique solution.

This upper triangular linear system can be simply solved using **backward substitution**: solving the equations backward from the last equation to the first for the unknowns $x_n, x_{n-1}, \dots, x_1$ in turn by substituting previously solved unknowns into the current equation. At the step to solve the i-th equation ($i = n, n-1, \dots, 1$)

$$a_{ii}\boxed{x_i} + a_{i,i+1}x_{i+1} + \cdots + a_{in}x_n = b_i \tag{4.8}$$

the only unknown is x_i as $x_{i+1}, \dots, x_n$ have been solved in previous steps. So we have

$$x_i = \frac{b_i - (a_{i,i+1}x_{i+1} + \cdots + a_{in}x_n)}{a_{ii}} \tag{4.9}$$

where $(a_{i,i+1}x_{i+1} + \cdots + a_{in}x_n)$ is the inner product

$$\begin{bmatrix} a_{i,i+1} & \cdots & a_{in} \end{bmatrix} \begin{bmatrix} x_{i+1} \\ \vdots \\ x_n \end{bmatrix} = \sum_{j=i+1}^{n} a_{ij}x_j \tag{4.10}$$

Algorithm 2 Backward substitution for upper triangular system $A\mathbf{x} = \mathbf{b}$

---

for i from n down to 1 **do**
   $p \leftarrow b_i$
   for j from $i+1$ to n **do**
      $p \leftarrow p - A_{ij}x_j$
   end for
   $x_i \leftarrow p/A_{ii}$
end for

---

with the convention $\sum_{j=n+1}^{n}(\cdot) = 0$ for the first step when $i = n$. See the pseudocode for backward substitution.

In backward substitution above, the number of divisions is n, and the number of multiplications is the same as the number of additions/subtractions, which is

$$\sum_{i=1}^{n}\sum_{j=i+1}^{n} 1 = \sum_{i=1}^{n}(n-i) = (n-1) + (n-2) + \cdots + (n-n) = \frac{n(n-1)}{2} \qquad (4.11)$$

So the number of arithmetic operations in backward substitution is $O(n^2)$.

A lower triangular system of order n, $A\mathbf{x} = \mathbf{b}$, has the form

$$\begin{bmatrix} a_{11} & & & & & \\ a_{21} & a_{22} & & & & \\ \vdots & \vdots & \ddots & & & \\ a_{i1} & a_{i2} & \cdots & a_{ii} & & \\ \vdots & \vdots & & \vdots & \ddots & \\ a_{n1} & a_{n2} & \cdots & a_{ni} & \cdots & a_{nn} \end{bmatrix} \begin{bmatrix} x_1 \\ x_2 \\ \vdots \\ x_i \\ \vdots \\ x_n \end{bmatrix} = \begin{bmatrix} b_1 \\ b_2 \\ \vdots \\ b_i \\ \vdots \\ b_n \end{bmatrix} \qquad (4.12)$$

Similarly, this system can be solved by **forward substitution**. The number of arithmetic operations is also $O(n^2)$.

4.3 GE and A=LU

The idea of Gaussian Elimination (GE) to solve a linear system is to transform the linear system to an equivalent upper triangular system. The transformation must not change the solution so that the beginning and ending systems are equivalent (have the same solution).

In linear algebra, we have seen three **elementary row operations** on a matrix. They are

- **interchange**: swap two rows
- **scaling**: multiply a row by a nonzero number
- **replacement**: subtract a row by a multiple of another row

It can be easily seen that an elementary row operation on the augmented matrix of a linear system corresponds to swapping, scaling or combining equations of the linear system, leading to an equivalent linear system with the same solution. So we can use the elementary row operations to transform the original augmented matrix (original linear system) such that the transformed augmented matrix (equivalent linear system) has an upper triangular coefficient matrix.

Below we use an example to demonstrate how we can do the transformation systematically so that we can readily extend the process to the general case and develop the general algorithm. In this example, we use only replacements (no interchanges and scaling), and the process is called **naive Gaussian Elimination** without pivoting (**pivoting** is a special interchange operation). We call it **GEnoP**.

Example 4.1 (GEnoP and backward substitution) Solve the linear system $Ax = \mathbf{b}$ whose augmented matrix is

$$[A \,|\, \mathbf{b}] = \begin{bmatrix} \boxed{6} & -2 & 2 & 4 & 16 \\ 12 & \boxed{-8} & 6 & 10 & 26 \\ 3 & -13 & \boxed{9} & 3 & -19 \\ -6 & 4 & 1 & \boxed{-18} & -34 \end{bmatrix}$$

by GEnoP, which transforms the linear system to an equivalent upper triangular system with only replacements.

[Solution:] Our goal is to eliminate entries under the boxed entries at the main diagonal in the above augmented matrix, which can be achieved in 3 steps for this system of order 4.

Step 1: We start from the matrix

$$[A \,|\, \mathbf{b}] = \begin{bmatrix} \boxed{6} & -2 & 2 & 4 & 16 \\ 12 & -8 & 6 & 10 & 26 \\ 3 & -13 & 9 & 3 & -19 \\ -6 & 4 & 1 & -18 & -34 \end{bmatrix}$$

We eliminate the entries (in gray) under the first boxed diagonal entry in the first step. We replace a row that has a to-be-eliminated gray entry by subtracting from this row a multiple of the row that has the boxed diagonal entry. We therefore have the following three row replacements.

- Row$_2$ replaced by Row$_2 - m_{21}$Row$_1$, where $m_{21} = a_{21}/a_{11} = 12/\boxed{6}$ is called a multiplier
- Row$_3$ replaced by Row$_3 - m_{31}$Row$_1$, where $m_{31} = a_{31}/a_{11} = 3/\boxed{6}$
- Row$_4$ replaced by Row$_4 - m_{41}$Row$_1$, where $m_{41} = a_{41}/a_{11} = -6/\boxed{6}$

The result after the replacements is

$$\begin{bmatrix} \boxed{6} & -2 & 2 & 4 & 16 \\ & -4 & 2 & 2 & -6 \\ & -12 & 8 & 1 & -27 \\ & 2 & 3 & -14 & -18 \end{bmatrix}$$

which is denoted as $[A^{(1)} \mid \mathbf{b}^{(1)}]$, where the zero entries under the first boxed diagonal entry are left blank.

Step 2: We start from the matrix

$$[A^{(1)} \mid \mathbf{b}^{(1)}] = \begin{bmatrix} 6 & -2 & 2 & 4 & \bigm| & 16 \\ & \boxed{-4} & 2 & 2 & \bigm| & -6 \\ & -12 & 8 & 1 & \bigm| & -27 \\ & 2 & 3 & -14 & \bigm| & -18 \end{bmatrix}$$

We eliminate the entries (in gray) under the second boxed diagonal entry in the second step. We replace a row that has a to-be-eliminated gray entry by subtracting from this row a multiple of the row that has the boxed diagonal entry. We therefore have the following two row replacements.

- Row$_3$ replaced by Row$_3 - m_{32}$Row$_2$, where $m_{32} = a_{32}^{(1)}/a_{22}^{(1)} = -12/(\boxed{-4})$
- Row$_4$ replaced by Row$_4 - m_{42}$Row$_2$, where $m_{42} = a_{42}^{(1)}/a_{22}^{(1)} = 2/(\boxed{-4})$

The result after the replacements is

$$\begin{bmatrix} 6 & -2 & 2 & 4 & \bigm| & 16 \\ & \boxed{-4} & 2 & 2 & \bigm| & -6 \\ & & 2 & -5 & \bigm| & -9 \\ & & 4 & -13 & \bigm| & -21 \end{bmatrix}$$

which is denoted as $[A^{(2)} \mid \mathbf{b}^{(2)}]$.

Step 3: We start from the matrix

$$[A^{(2)} \mid \mathbf{b}^{(2)}] = \begin{bmatrix} 6 & -2 & 2 & 4 & \bigm| & 16 \\ & -4 & 2 & 2 & \bigm| & -6 \\ & & \boxed{2} & -5 & \bigm| & -9 \\ & & 4 & -13 & \bigm| & -21 \end{bmatrix}$$

We eliminate the entry (in gray) under the third boxed diagonal entry in the third step. We replace the row that has a to-be-eliminated gray entry by subtracting from this row a multiple of the row that has the boxed diagonal entry. We therefore have the following one row replacement.

- Row$_4$ replaced by Row$_4 - m_{43}$Row$_3$, where $m_{43} = a_{43}^{(2)}/a_{33}^{(2)} = 4/\boxed{2}$

The result after this replacement is

$$\begin{bmatrix} 6 & -2 & 2 & 4 & \bigm| & 16 \\ & -4 & 2 & 2 & \bigm| & -6 \\ & & \boxed{2} & -5 & \bigm| & -9 \\ & & & -3 & \bigm| & -3 \end{bmatrix}$$

which is denoted as $[A^{(3)} \mid \mathbf{b}^{(3)}]$.

So we have successfully transformed the original linear system of the augmented matrix $[A \,|\, \mathbf{b}]$ to an equivalent upper triangular linear system of the augmented matrix $[A^{(3)} \,|\, \mathbf{b}^{(3)}]$, which is

$$
\begin{bmatrix}
\boxed{6} & -2 & 2 & 4 & 16 \\
 & \boxed{-4} & 2 & 2 & -6 \\
 & & \boxed{2} & -5 & -9 \\
 & & & \boxed{-3} & -3
\end{bmatrix}
$$

Applying backward substitution, we have

$$x_4 = -3/(-3) = 1$$

$$x_3 = \frac{-9 - (-5x_4)}{2} = \frac{-9 - (-5)(1)}{2} = -2$$

$$x_2 = \frac{-6 - (2x_3 + 2x_4)}{-4} = \frac{-6 - 2(-2) - 2(1)}{2} = 1$$

$$x_1 = \frac{16 - (-2x_2 + 2x_3 + 4x_4)}{6} = \frac{16 - (-2)(1) - 2(-2) - 4(1)}{2} = 3$$

that is

$$
\mathbf{x} = \begin{bmatrix} x_1 \\ x_2 \\ x_3 \\ x_4 \end{bmatrix} = \begin{bmatrix} 3 \\ 1 \\ -2 \\ 1 \end{bmatrix}
$$

■

Remark GEnoP may break down if a diagonal entry used for elimination is zero, as the calculation of the multiplier then involves dividing by zero. We will fix this breakdown later with the pivoting (swapping rows) strategy.

GEnoP for the general case is similar to the above example. Consider a linear system of order n with the following augmented matrix:

$$
[A \,|\, \mathbf{b}] = \begin{bmatrix}
\boxed{a_{11}} & a_{12} & a_{13} & \cdots & a_{1n} & b_1 \\
a_{21} & a_{22} & a_{23} & \cdots & a_{2n} & b_2 \\
a_{31} & a_{32} & a_{33} & \cdots & a_{3n} & b_2 \\
\vdots & \vdots & \vdots & & \vdots & \vdots \\
a_{n1} & a_{n2} & a_{n3} & \cdots & a_{nn} & b_n
\end{bmatrix}
\tag{4.13}
$$

We can solve the system by transforming it to an upper triangular system in $(n-1)$ steps (assuming GEnoP does not break down).

Step 1: Compute the multipliers

$$m_{i1} = a_{i1}/a_{11} \quad (i = 2, 3, \ldots, n) \tag{4.14}$$

to eliminate the entries under the diagonal entry a_{11} by row replacements. Entries a_{ij} and b_i in the submatrix below and to the right of the diagonal entry a_{11} are updated to

$$a_{ij}^{(1)} = a_{ij} - m_{i1}a_{1j}, \quad b_i^{(1)} = b_i - m_{i1}b_1 \quad (i = 2, 3, \ldots, n; \; j = 2, 3, \ldots, n) \tag{4.15}$$

After this step, the matrix becomes

$$[A^{(1)} \mid \mathbf{b}^{(1)}] = \begin{bmatrix} a_{11} & a_{12} & a_{13} & \cdots & a_{1n} & b_1 \\ & \boxed{a_{22}^{(1)}} & a_{23}^{(1)} & \cdots & a_{2n}^{(1)} & b_2^{(1)} \\ & a_{32}^{(1)} & a_{33}^{(1)} & \cdots & a_{3n}^{(1)} & b_3^{(1)} \\ & \vdots & \vdots & & \vdots & \vdots \\ & a_{n2}^{(1)} & a_{n3}^{(1)} & \cdots & a_{nn}^{(1)} & b_n^{(1)} \end{bmatrix} \tag{4.16}$$

Step 2: Step 2 applies the same process as in Step 1 to the lower-right submatrix consisting of the entries with the superscript (1). Compute the multipliers

$$m_{i2} = a_{i2}^{(1)}/a_{22}^{(1)} \quad (i = 3, \ldots, n) \tag{4.17}$$

to eliminate the entries under the diagonal entry $a_{22}^{(1)}$ by row replacements. Entries $a_{ij}^{(1)}$ and $b_i^{(1)}$ in the submatrix below and to the right of the diagonal entry $a_{22}^{(1)}$ are updated to

$$a_{ij}^{(2)} = a_{ij}^{(1)} - m_{i2}a_{2j}^{(1)}, \quad b_i^{(2)} = b_i^{(1)} - m_{i2}b_2^{(1)} \quad (i = 3, \ldots, n; \; j = 3, \ldots, n) \tag{4.18}$$

After this step, the matrix becomes

$$[A^{(2)} \mid \mathbf{b}^{(2)}] = \begin{bmatrix} a_{11} & a_{12} & a_{13} & \cdots & a_{1n} & b_1 \\ & a_{22}^{(1)} & a_{23}^{(1)} & \cdots & a_{2n}^{(1)} & b_2^{(1)} \\ & & \boxed{a_{33}^{(2)}} & \cdots & a_{3n}^{(2)} & b_3^{(2)} \\ & & \vdots & & \vdots & \vdots \\ & & a_{n3}^{(2)} & \cdots & a_{nn}^{(2)} & b_n^{(2)} \end{bmatrix} \tag{4.19}$$

$\vdots$

Step k: After Step $(k-1)$, the matrix becomes

$$[A^{(k-1)} \mid \mathbf{b}^{(k-1)}] = \begin{bmatrix} a_{11} & \cdots & a_{1k} & \cdots & a_{1n} & b_1 \\ & \ddots & \vdots & & \vdots & \vdots \\ & & \boxed{a_{kk}^{(k-1)}} & \cdots & a_{kn}^{(k-1)} & b_k^{(k-1)} \\ & & \vdots & & \vdots & \vdots \\ & & a_{nk}^{(k-1)} & \cdots & a_{nn}^{(k-1)} & b_n^{(k-1)} \end{bmatrix} \tag{4.20}$$

In Step k, compute the multipliers

$$m_{ik} = a_{ik}^{(k-1)}/a_{kk}^{(k-1)} \quad (i = k+1,\ldots,n) \quad (4.21)$$

to eliminate the entries under the diagonal entry $a_{kk}^{(k-1)}$ by row replacements, and update entries $a_{ij}^{(k-1)}$ and $b_i^{(k-1)}$ in the submatrix below and to the right of the diagonal entry $a_{kk}^{(k-1)}$ to

$$a_{ij}^{(k)} = a_{ij}^{(k-1)} - m_{ik}a_{kj}^{(k-1)}, \quad b_i^{(k)} = b_i^{(k-1)} - m_{ik}b_k^{(k-1)} \quad (4.22)$$

$$(i = k+1,\ldots,n; \; j = k+1,\ldots,n)$$

$\vdots$

After $(n-1)$ steps, the original linear systems is transformed to an upper triangular linear system.

By putting Step k in a loop (with the loop variable k), we have the following pseudocode of GEnoP.

Algorithm 3 GEnoP for a square system $A\mathbf{x} = \mathbf{b}$

---

for k from 1 to $n-1$ **do**
    for i from $k+1$ to n **do**
        $A_{ik} \leftarrow A_{ik}/A_{kk}$
        for j from $k+1$ to n **do**
            $A_{ij} \leftarrow A_{ij} - A_{ik}A_{kj}$
        end for
        $b_i \leftarrow b_i - A_{ik}b_k$
    end for
end for

Note that the above pseudocode does not fill zeros under the diagonal. Instead, the multipliers m_{ik} are stored to replace A_{ik} under the diagonal. So the final updated matrix A consists of the upper triangular matrix in its upper portion and all the multipliers in its lower portion, which gives the so-called LU decomposition (see below) of the original matrix A. The number of arithmetic operations in GEnoP is of $O(n^3)$ (see Exercises 2.3 and 4.9). We can pass the updated matrix A and the updated vector $\mathbf{b}$ to backward substitution to find the solution of the linear system. The backward substitution does not use entries under the diagonal of A and therefore does not care if they are zero or not.

4.3.1 Elementary Matrices

Each of the three elementary row operations (interchange, scaling and replacement) on a $m \times n$ matrix A can be achieved by left-multiplying this matrix A by a so-called elementary matrix. The elementary matrix is obtained by performing the same elementary row operation on the $m \times m$ identity matrix. Below is a demonstration for row replacement.

Let the matrix resulted from a row interchange on the matrix A be the matrix A_1:

$$A = \begin{bmatrix} 1 & 2 \\ 3 & 4 \\ 5 & 6 \end{bmatrix} \xrightarrow[\text{Row}_3 = \text{Row}_3 + (-5)\text{Row}_1]{\text{replacement}} A_1 = \begin{bmatrix} 1 & 2 \\ 3 & 4 \\ 0 & -4 \end{bmatrix}$$

Then $A_1 = E_1 A$, where E_1 is the elementary matrix obtained by performing the same interchange on an identity matrix as:

$$I = \begin{bmatrix} 1 & 0 & 0 \\ 0 & 1 & 0 \\ 0 & 0 & 1 \end{bmatrix} \xrightarrow[\text{Row}_3 = \text{Row}_3 + (-5)\text{Row}_1]{\text{replacement}} E_1 = \begin{bmatrix} 1 & 0 & 0 \\ 0 & 1 & 0 \\ -5 & 0 & 1 \end{bmatrix}$$

It can be easily checked that $A_1 = E_1 A$, i.e.

$$A_1 = \begin{bmatrix} 1 & 2 \\ 3 & 4 \\ 0 & -4 \end{bmatrix} = \begin{bmatrix} 1 & 0 & 0 \\ 0 & 1 & 0 \\ -5 & 0 & 1 \end{bmatrix} \begin{bmatrix} 1 & 2 \\ 3 & 4 \\ 5 & 6 \end{bmatrix} = E_1 A$$

Note that an elementary matrix is invertible because an elementary row operation can be reversed. Its inverse is simply the elementary row operation corresponding to the reverse row operation. Below is a demonstration.

$$A_1 = \begin{bmatrix} 1 & 2 \\ 3 & 4 \\ 0 & -4 \end{bmatrix} \xrightarrow[\text{Row}_3 = \text{Row}_3 + (+5)\text{Row}_1]{\text{reverse replacement}} A = \begin{bmatrix} 1 & 2 \\ 3 & 4 \\ 5 & 6 \end{bmatrix}$$

Then $A = E_{-1} A_1 = E_{-1} E_1 A$ (which is true for any A). So the elementary matrix $E_{-1} = E_1^{-1}$, which can be obtained as

$$I = \begin{bmatrix} 1 & 0 & 0 \\ 0 & 1 & 0 \\ 0 & 0 & 1 \end{bmatrix} \xrightarrow[\text{Row}_3 = \text{Row}_3 + (+5)\text{Row}_1]{\text{reverse replacement}} E_1^{-1} = \begin{bmatrix} 1 & 0 & 0 \\ 0 & 1 & 0 \\ +5 & 0 & 1 \end{bmatrix}$$

4.3.2   A=LU

If we apply GEnoP to the matrix A, the matrix A is transformed to a sequence of matrices as:

$$A \rightarrow A^{(1)} \rightarrow A^{(2)} \rightarrow \cdots \rightarrow A^{(n-1)} \tag{4.23}$$

where $A^{(n-1)}$ is an upper triangular matrix, and is denoted as U.

We can now represent each elimination of an entry (a row replacement) in GEnoP as a left-multiplication by an elementary matrix for the row replacement. For example,

$$A^{(1)} = (E_n \cdots (E_3(E_2 A)) \cdots) \tag{4.24}$$

where $E_2, E_3, \ldots, E_n$ are respectively the elementary matrices for row replacements to eliminate $a_{21}, a_{31}, \ldots, a_{n1}$. They are

$$E_2 = \begin{bmatrix} 1 & 0 & 0 & \cdots & 0 \\ -m_{21} & 1 & 0 & \cdots & 0 \\ 0 & 0 & 1 & \cdots & 0 \\ \vdots & \vdots & \vdots & & \vdots \\ 0 & 0 & 0 & \cdots & 1 \end{bmatrix}, \tag{4.25}$$

$$E_3 = \begin{bmatrix} 1 & 0 & 0 & \cdots & 0 \\ 0 & 1 & 0 & \cdots & 0 \\ -m_{31} & 0 & 1 & \cdots & 0 \\ \vdots & \vdots & \vdots & & \vdots \\ 0 & 0 & 0 & \cdots & 1 \end{bmatrix}, \tag{4.26}$$

$$E_n = \begin{bmatrix} 1 & 0 & 0 & \cdots & 0 \\ 0 & 1 & 0 & \cdots & 0 \\ 0 & 0 & 1 & \cdots & 0 \\ \vdots & \vdots & \vdots & & \vdots \\ -m_{n1} & 0 & 0 & \cdots & 1 \end{bmatrix} \tag{4.27}$$

We can write

$$A^{(1)} = E^{(1)}A \tag{4.28}$$

where $E^{(1)}$ is the product $E_n \cdots E_3 E_2$. It can be easily seen (by considering left-multiplication of an elementary matrix as an elementary row operation) that this product is

$$E^{(1)} = E_n \cdots E_3 E_2 = \begin{bmatrix} 1 & 0 & 0 & \cdots & 0 \\ -m_{21} & 1 & 0 & \cdots & 0 \\ -m_{31} & 0 & 1 & \cdots & 0 \\ \vdots & \vdots & \vdots & & \vdots \\ -m_{n1} & 0 & 0 & \cdots & 1 \end{bmatrix} \tag{4.29}$$

Similarly, we can write

$$A^{(2)} = E^{(2)}A^{(1)}, \, A^{(3)} = E^{(3)}A^{(2)}, \, \ldots, \, A^{(n-1)} = E^{(n-1)}A^{(n-2)} \tag{4.30}$$

Finally we have

$$U = A^{(n-1)} = E^{(n-1)} \cdots E^{(3)} E^{(2)} E^{(1)} A \tag{4.31}$$

Since each elementary matrix is invertible, the product of elementary matrices is invertible. We thus have

$$A = [E^{(1)}]^{-1}[E^{(2)}]^{-1}[E^{(3)}]^{-1} \cdots [E^{(n-1)}]^{-1} U := LU \tag{4.32}$$

Now we want to find L. It can be easily shown that $[E^{(1)}]^{-1} = E_2^{-1}E_3^{-1}\cdots E_n^{-1}$ is computed as

$$
\begin{bmatrix}
1 & 0 & 0 & \cdots & 0 \\
m_{21} & 1 & 0 & \cdots & 0 \\
0 & 0 & 1 & \cdots & 0 \\
\vdots & \vdots & \vdots & & \vdots \\
0 & 0 & 0 & \cdots & 1
\end{bmatrix}
\begin{bmatrix}
1 & 0 & 0 & \cdots & 0 \\
0 & 1 & 0 & \cdots & 0 \\
m_{31} & 0 & 1 & \cdots & 0 \\
\vdots & \vdots & \vdots & & \vdots \\
0 & 0 & 0 & \cdots & 1
\end{bmatrix}
\cdots
\begin{bmatrix}
1 & 0 & 0 & \cdots & 0 \\
0 & 1 & 0 & \cdots & 0 \\
0 & 0 & 1 & \cdots & 0 \\
\vdots & \vdots & \vdots & & \vdots \\
m_{n1} & 0 & 0 & \cdots & 1
\end{bmatrix}
\tag{4.33}
$$

which gives

$$
[E^{(1)}]^{-1} =
\begin{bmatrix}
1 & 0 & 0 & \cdots & 0 \\
m_{21} & 1 & 0 & \cdots & 0 \\
m_{31} & 0 & 1 & \cdots & 0 \\
\vdots & \vdots & \vdots & & \vdots \\
m_{n1} & 0 & 0 & \cdots & 1
\end{bmatrix}
\tag{4.34}
$$

Similarly, we have

$$
[E^{(2)}]^{-1} =
\begin{bmatrix}
1 & 0 & 0 & \cdots & 0 \\
0 & 1 & 0 & \cdots & 0 \\
0 & m_{32} & 1 & \cdots & 0 \\
\vdots & \vdots & \vdots & & \vdots \\
0 & m_{n2} & 0 & \cdots & 1
\end{bmatrix}
, [E^{(n-1)}]^{-1} =
\begin{bmatrix}
1 & 0 & \cdots & 0 & 0 \\
0 & 1 & \cdots & 0 & 0 \\
\vdots & \vdots & & \vdots & \vdots \\
0 & 0 & \cdots & 1 & 0 \\
0 & 0 & \cdots & m_{n,n-1} & 1
\end{bmatrix}
\tag{4.35}
$$

It can be easily shown that $L = [E^{(1)}]^{-1}[E^{(2)}]^{-1}[E^{(3)}]^{-1}\cdots[E^{(n-1)}]^{-1}$ is computed as

$$
\begin{bmatrix}
1 & 0 & 0 & \cdots & 0 \\
m_{21} & 1 & 0 & \cdots & 0 \\
m_{31} & 0 & 1 & \cdots & 0 \\
\vdots & \vdots & \vdots & & \vdots \\
m_{n1} & 0 & 0 & \cdots & 1
\end{bmatrix}
\begin{bmatrix}
1 & 0 & 0 & \cdots & 0 \\
0 & 1 & 0 & \cdots & 0 \\
0 & m_{32} & 1 & \cdots & 0 \\
\vdots & \vdots & \vdots & & \vdots \\
0 & m_{n2} & 0 & \cdots & 1
\end{bmatrix}
\cdots
\begin{bmatrix}
1 & 0 & \cdots & 0 & 0 \\
0 & 1 & \cdots & 0 & 0 \\
\vdots & \vdots & & \vdots & \vdots \\
0 & 0 & \cdots & 1 & 0 \\
0 & 0 & \cdots & m_{n,n-1} & 1
\end{bmatrix}
\tag{4.36}
$$

which gives a lower triangular matrix

$$
L =
\begin{bmatrix}
1 & 0 & \cdots & 0 & 0 \\
m_{21} & 1 & \cdots & 0 & 0 \\
m_{31} & m_{32} & \cdots & 0 & 0 \\
\vdots & \vdots & & \vdots & \vdots \\
m_{n-1,1} & m_{n-1,2} & \cdots & 1 & 0 \\
m_{n1} & m_{n2} & \cdots & m_{n,n-1} & 1
\end{bmatrix}
\tag{4.37}
$$

Finally we obtain the so-called LU decomposition of A as

$$
A = LU
$$

where

$$
L =
\begin{bmatrix}
1 & 0 & \cdots & 0 & 0 \\
m_{21} & 1 & \cdots & 0 & 0 \\
m_{31} & m_{32} & \cdots & 0 & 0 \\
\vdots & \vdots & & \vdots & \vdots \\
m_{n-1,1} & m_{n-1,2} & \cdots & 1 & 0 \\
m_{n1} & m_{n2} & \cdots & m_{n,n-1} & 1
\end{bmatrix}
\tag{4.38}
$$

$$
U =
\begin{bmatrix}
a_{11} & a_{12} & a_{13} & \cdots & a_{1,n-1} & a_{1n} \\
 & a_{22}^{(1)} & a_{23}^{(1)} & \cdots & a_{2,n-1}^{(1)} & a_{2n}^{(1)} \\
 & & a_{33}^{(2)} & \cdots & a_{3,n-1}^{(2)} & a_{3n}^{(2)} \\
 & & & \ddots & \vdots & \vdots \\
 & & & & a_{n-1,n-1}^{(n-2)} & a_{n-1,n}^{(n-2)} \\
 & & & & & a_{nn}^{(n-1)}
\end{bmatrix}
\tag{4.39}
$$

Note that the pseudocode of GEnoP returns an updated matrix A which stores L and U as

$$
\begin{bmatrix}
a_{11} & a_{12} & a_{13} & \cdots & a_{1,n-1} & a_{1n} \\
m_{21} & a_{22}^{(1)} & a_{23}^{(1)} & \cdots & a_{2,n-1}^{(1)} & a_{2n}^{(1)} \\
m_{31} & m_{32} & a_{33}^{(2)} & \cdots & a_{3,n-1}^{(2)} & a_{3n}^{(2)} \\
\vdots & \vdots & \vdots & \ddots & \vdots & \vdots \\
m_{n-1,1} & m_{n-1,2} & m_{n-1,3} & \cdots & a_{n-1,n-1}^{(n-2)} & a_{n-1,n}^{(n-2)} \\
m_{n-1,1} & m_{n2} & m_{n3} & \cdots & m_{n,n-1} & a_{nn}^{(n-1)}
\end{bmatrix}
\tag{4.40}
$$

4.3.3 Solving $Ax = b$ by A=LU

Why is the **LU decomposition** of a matrix A useful? The answer is that we can efficiently solve multiple linear systems with the same coefficient matrix A but different right-hand sides. Here is how. Suppose we need to solve the following m linear systems

$$
A\mathbf{x} = \mathbf{b}_i, \quad i = 1, 2, \ldots, m
\tag{4.41}
$$

We use GEnoP to find the LU decomposition of A as $A = LU$ in $O(n^3)$ arithmetic operations. Since

$$
A\mathbf{x} = \mathbf{b}_i \iff LU\mathbf{x} = \mathbf{b}_i
\tag{4.42}
$$

we can solve each linear system using the LU decomposition by first solving for $\mathbf{y} = U\mathbf{x}$ from

$$
L\mathbf{y} = \mathbf{b}_i
\tag{4.43}
$$

using forward substitution in $O(n^2)$ arithmetic operations, and then solving for $\mathbf{x}$ from

$$
U\mathbf{x} = \mathbf{y}
\tag{4.44}
$$

using backward substitution in $O(n^2)$ arithmetic operations. The total cost in the process is of $O(n^3 + mn^2)$, as compared with $O(mn^3 + mn^2)$ if using GEnoP separately to solve for each linear system.

Example 4.2 (LU decomposition) Use the LU decomposition to solve the linear system $Ax = \mathbf{b}$, where

$$[A \,|\, \mathbf{b}] = \left[\begin{array}{cccc|c} \boxed{6} & -2 & 2 & 4 & 16 \\ 12 & \boxed{-8} & 6 & 10 & 26 \\ 3 & -13 & \boxed{9} & 3 & -19 \\ -6 & 4 & 1 & \boxed{-18} & -34 \end{array}\right]$$

[Solution:] From Example 4.1, we have the LU decomposition $A = LU$, where

$$L = \begin{bmatrix} 1 & & & \\ m_{21} & 1 & & \\ m_{31} & m_{32} & 1 & \\ m_{41} & m_{42} & m_{43} & 1 \end{bmatrix} = \begin{bmatrix} 1 & & & \\ 2 & 1 & & \\ 0.5 & 3 & 1 & \\ -1 & -0.5 & 2 & 1 \end{bmatrix}$$

$$U = A^{(3)} = \begin{bmatrix} 6 & -2 & 2 & 4 \\ & -4 & 2 & 2 \\ & & 2 & -5 \\ & & & -3 \end{bmatrix}$$

By $Ax = \mathbf{b} \iff LUx = \mathbf{b}$, we define $\mathbf{y} = U\mathbf{x}$ and first solve for $\mathbf{y}$ from $L\mathbf{y} = \mathbf{b}$:

$$\begin{bmatrix} 1 & & & \\ 2 & 1 & & \\ 0.5 & 3 & 1 & \\ -1 & -0.5 & 2 & 1 \end{bmatrix} \begin{bmatrix} y_1 \\ y_2 \\ y_3 \\ y_4 \end{bmatrix} = \begin{bmatrix} 16 \\ 26 \\ -19 \\ -34 \end{bmatrix}$$

using forward substitution as

$$y_1 = 16$$
$$y_2 = 26 - 2y_1 = -6$$
$$y_3 = -19 - 0.5y_1 - 3y_2 = -9$$
$$y_4 = -34 - (-1)y_1 - (-0.5)y_2 - 2y_3 = -3$$

Note that $\mathbf{y}$ is just $\mathbf{b}^{(3)}$ after the last step of GEnoP in Example 4.1. We then solve for $\mathbf{x}$ from $U\mathbf{x} = \mathbf{y}$ using backward substitution, which is exactly the same as in Example 4.1. The final solution is $\mathbf{x} = [3, 1, -2, 1]^T$. ∎

4.4   GEPP and PA=LU

Let's start this session with two examples. In the first example, GEnoP breaks down for a linear system that has a unique solution. In the second example, GEnoP in inexact arithmetic gives very inaccurate solution of a linear system. We point out simple remedies to fix the breakdown and inaccuracy problems.

Example 4.3 (GEnoP breakdown) Use GEnoP to solve the linear system $Ax = b$:

$$
\begin{bmatrix} 1 & 1 & 1 \\ 1 & 1 & 2 \\ 1 & 2 & 2 \end{bmatrix}
\begin{bmatrix} x_1 \\ x_2 \\ x_3 \end{bmatrix}
=
\begin{bmatrix} 3 \\ 4 \\ 5 \end{bmatrix}
$$

Note that this linear system is nonsingular and has a unique solution $\mathbf{x} = [1,1,1]^T$ (as shown at the end of this example).

[Solution:] We apply GEnoP to this linear system in the following steps.

Step 1: We start from the augmented matrix

$$
[A \,|\, \mathbf{b}] =
\begin{bmatrix} \boxed{1} & 1 & 1 & 3 \\ 1 & 1 & 2 & 4 \\ 1 & 2 & 2 & 5 \end{bmatrix}
$$

We use the following two row replacements for elimination.

- Row_2 replaced by $\text{Row}_2 - m_{21}\text{Row}_1$, where the multiplier $m_{21} = a_{21}/a_{11} = \boxed{1}/1$.
- Row_3 replaced by $\text{Row}_3 - m_{31}\text{Row}_1$, where the multiplier $m_{31} = a_{31}/a_{11} = \boxed{1}/1$.

The result after the replacements is

$$
[A^{(1)} \,|\, \mathbf{b}^{(1)}] =
\begin{bmatrix} \boxed{1} & 1 & 1 & 3 \\ & 0 & 1 & 1 \\ & 1 & 1 & 2 \end{bmatrix}
$$

which is denoted as $[A^{(1)} \,|\, \mathbf{b}^{(1)}]$.

Step 2: In GEnoP, we want to continue elimination with the following row replacement on the matrix $[A^{(1)} \,|\, \mathbf{b}^{(1)}]$.

- Row_3 replaced by $\text{Row}_3 - m_{32}\text{Row}_2$, where the multiplier $m_{32} = a_{32}^{(1)}/a_{22}^{(1)} = 1/0$, which involves dividing by zero.

We have the following observations and comments.

Breakdown: The multiplier $m_{32} = a_{32}^{(1)}/a_{22}^{(1)} = 1/0$ is calculated by dividing zero and is not defined. So GEnoP breaks down at this step even though the linear system is solvable and has a unique solution.

Remedy: From linear algebra, we know the second column is a pivot column with the pivot position at the diagonal (the 2nd row and 2nd column) because there is a nonzero entry under the diagonal. A simple remedy we learned from linear algebra is to use a row interchange to bring a nonzero entry (a pivot) underneath (not atop) to the pivot position. So we do the following interchange

- interchange Row_2 with Row_3

The result after the interchange is

$$\begin{bmatrix} \boxed{1} & 1 & 1 & | & 3 \\ & \boxed{1} & 1 & | & 2 \\ & & 1 & | & 1 \end{bmatrix}$$

We can then proceed elimination by row replacements using multipliers calculated by diving the nonzero pivot. In this example, no further elimination is needed as the matrix after the interchange is already upper triangular. We denote the final upper triangular matrix as $[A^{(2)} \,|\, \mathbf{b}^{(2)}]$. We can then solve the linear system using backward substitution to obtain the solution $\mathbf{x} = [1,1,1]^T$. ∎

Example 4.4 (GEnoP inaccuracy) Solve the following linear system $A\mathbf{x} = \mathbf{b}$ using GEnoP with 4-significant-decimal-digit arithmetic (i.e. each decimal number is represented by only 4 significant decimal digits).

$$\begin{bmatrix} 2.5 \times 10^{-5} & 1 \\ 1 & 1 \end{bmatrix} \begin{bmatrix} x_1 \\ x_2 \end{bmatrix} = \begin{bmatrix} 1 \\ 1 \end{bmatrix}$$

Note that if a number has more than 4 significant digits, it is rounded to a 4-significant-digit number with relative roundoff error less than 5×10^{-4}. So the 4-significant-decimal-digit arithmetic is generally inexact. Also note that the exact solution of the linear system is $\mathbf{x} = [40000/39999, 39998/39999]^T \approx [1,1]^T$.

[**Solution:**] We apply GEnoP in the following step.

Step 1: We compute the multiplier and replace entries using 4-significant-decimal-digit arithmetic as follows.

- The multiplier $m_{21} = 1.000/(2.500 \times 10^{-5}) = 4.000 \times 10^4$.
- New entry $a_{22}^{(1)} = a_{22} - m_{21}a_{12} = 1.000 - 1.000 \times (4.000 \times 10^4) = -39999 \approx -4.000 \times 10^4$
- New entry $b_2^{(1)} = b_2 - m_{21}b_1 = 1.000 - 1.000 \times (4.000 \times 10^4) = -39999 \approx -4.000 \times 10^4$

The linear system after the elimination with inexact arithmetic becomes

$$\begin{bmatrix} 2.500 \times 10^{-5} & 1.000 \\ & -4.000 \times 10^4 \end{bmatrix} \begin{bmatrix} x_1 \\ x_2 \end{bmatrix} = \begin{bmatrix} 1.000 \\ -4.000 \times 10^4 \end{bmatrix}$$

The solution (inexact) is $\mathbf{x} = [0, 1.000]^T$.

We have the following observations and comments.

Swamping: The approximate solution obtained by GEnoP in inexact arithmetic differs dramatically from the exact/true solution. What is wrong? Notice that the multiplier m_{21} above is huge due to the division by a very small pivot ($a_{11} = 2.5 \times 10^{-5}$ here). When computing the new entries $a_{22}^{(1)}$ and $b_2^{(1)}$ with rounding, the old small entries a_{22} and b_2 are swamped by the large numbers from the multiplication by the huge multiplier and do not play roles in the new entries (as they disappear after rounding).

Catastrophic cancellation: We do not have catastrophic cancellation in this example, but keep in mind that a very small pivot value may come from subtraction of two close values in inexact arithmetic and bear a large relative error which leads to a very inaccurate multiplier to negatively affect new entries computed with the multiplier.

Remedy: So it is clearly a bad idea to use a very small pivot value to compute multipliers in inexact arithmetic. We would like to use row interchange to bring a large entry as the pivot to the pivot position. In this example, we interchange the two rows/equations first to get

$$\begin{bmatrix} \boxed{1} & 1 \\ 2.5 \times 10^{-5} & 1 \end{bmatrix} \begin{bmatrix} x_1 \\ x_2 \end{bmatrix} = \begin{bmatrix} 1 \\ 1 \end{bmatrix}$$

We then proceed with elimination using the 4-significant-decimal-digit arithmetic as follows

- The multiplier $m_{21} = 2.5 \times 10^{-5}/1.000 = 2.500 \times 10^{-5}$
- New Entry $a_{22}^{(1)} = a_{22} - m_{21}a_{12} = 1.000 - 1.000 \times (2.500 \times 10^{-4}) = 0.99975 \approx 1.000$
- New entry $b_2^{(1)} = b_2 - m_{21}b_1 = 1.000 - 1.000 \times (2.500 \times 10^{-4}) = 0.99975 \approx 1.000$

The linear system after the elimination becomes

$$\begin{bmatrix} 1.000 & 1.000 \\ & 1.000 \end{bmatrix} \begin{bmatrix} x_1 \\ x_2 \end{bmatrix} = \begin{bmatrix} 1.000 \\ 1.000 \end{bmatrix}$$

which can be solved by backward substitution to give the approximate solution $\mathbf{x} = [1.000, 1.000]^T$, which is very close to the exact/true solution. ∎

4.4.1 GEPP

We want to introduce pivoting strategies in Gaussian elimination. The **partial pivoting** strategy is to bring an entry with the largest amplitude under the pivot position in the same pivot column to the pivot position (at the diagonal) by using a row interchange. The Gaussian elimination with partial pivoting is abbreviated as **GEPP**. Suppose after GEPP step $(k-1)$, the augmented matrix is

$$[A^{(k-1)} \,|\, \mathbf{b}^{(k-1)}] = \begin{bmatrix} a_{11} & \cdots & a_{1k} & \cdots & a_{1n} & b_1 \\ & \ddots & \vdots & & \vdots & \vdots \\ & & \boxed{a_{kk}^{(k-1)}} & \cdots & a_{kn}^{(k-1)} & b_k^{(k-1)} \\ & & \vdots & & \vdots & \vdots \\ & & a_{nk}^{(k-1)} & \cdots & a_{nn}^{(k-1)} & b_n^{(k-1)} \end{bmatrix} \quad (4.45)$$

In GEPP step k, we do a row interchange (if needed) before we proceed with row replacements for elimination as follows

- **Partial Pivoting:** Find the first largest value among $|a_{kk}^{(k-1)}|$, $|a_{k+1,k}^{(k-1)}|$, ..., $|a_{nk}^{(k-1)}|$. Denote the row number where the largest value locates as m, i.e. $|a_{mk}^{(k-1)}| = \max_{k \le i \le n}|a_{ik}^{(k-1)}|$. If $m \ne k$, swap Row_k with Row_m to bring the entry $a_{mk}^{(k-1)}$ (the pivot with the largest amplitude under the pivot position) to the pivot position (at the diagonal).
- **Gaussian Elimination:** Compute the multipliers m_{ik} ($i = k, k+1, \ldots, n$) using the new pivot (note that $|m_{ik}| \le 1$ now) and do row replacements toward elimination as in GEnoP.

Example 4.5 (GEPP) Solve $Ax = b$ using GEPP in exact arithmetic, where

$$[A \,|\, \mathbf{b}] = \begin{bmatrix} 1 & 2 & -1 & 0 \\ 2 & -1 & 1 & 7 \\ -3 & 1 & 2 & 3 \end{bmatrix}$$

[Solution:] We initialize the vector $\mathbf{p} = [1, 2, 3]^T$ to track all the row interchanges in GEPP. It is updated whenever a row interchange occurs such that the i-th ($i = 1, 2, 3$ here) entry p_i in the vector $\mathbf{p}$ indicates that i-th row of the current augmented matrix comes from the row p_i of the initial augmented matrix $[A \,|\, \mathbf{b}]$.

Step 1: We start with

$$[A \,|\, \mathbf{b}] = \begin{bmatrix} 1 & 2 & -1 & 0 \\ 2 & -1 & 1 & 7 \\ -3 & 1 & 2 & 3 \end{bmatrix}, \quad \mathbf{p} = \begin{bmatrix} 1 \\ 2 \\ 3 \end{bmatrix}$$

We do partial pivoting by swapping Row_1 and Row_3 of both the augmented matrix and the tracking vector $\mathbf{p}$. We have an updated augmented matrix and the vector $\mathbf{p}$:

$$\begin{bmatrix} \boxed{-3} & 1 & 2 & 3 \\ 2 & -1 & 1 & 7 \\ 1 & 2 & -1 & 0 \end{bmatrix}, \quad \begin{bmatrix} 3 \\ 2 \\ 1 \end{bmatrix}$$

Note that the first entry $p_1 = 3$ in the tracking vector $\mathbf{p}$ indicates that new Row_1 of the updated augmented matrix comes from Row_3 of the initial augmented matrix, and the third entry $p_3 = 1$ indicates that new Row_3 from old Row_1. We then do row replacements for elimination as usual. We store the corresponding multiplier (instead of 0) at the location of an eliminated entry. We thus have

$$\begin{bmatrix} \boxed{-3} & 1 & 2 & 3 \\ -2/3 & -1/3 & 7/3 & 9 \\ -1/3 & 7/3 & -1/3 & 1 \end{bmatrix}, \quad \begin{bmatrix} 3 \\ 2 \\ 1 \end{bmatrix}$$

Step 2: We start with

$$\begin{bmatrix} -3 & 1 & 2 & 3 \\ -2/3 & -1/3 & 7/3 & 9 \\ -1/3 & 7/3 & -1/3 & 1 \end{bmatrix}, \quad \begin{bmatrix} 3 \\ 2 \\ 1 \end{bmatrix}$$

We do partial pivoting by swapping Row$_2$ and Row$_3$ of the above matrix (multipliers are swapped together too) and the tracking vector **p**. We have

$$
\begin{bmatrix}
-3 & 1 & 2 & 3 \\
-1/3 & \boxed{7/3} & -1/3 & 1 \\
-2/3 & -1/3 & 7/3 & 9
\end{bmatrix}, \qquad
\begin{bmatrix}
3 \\
1 \\
2
\end{bmatrix}
$$

We then do row replacements and store multipliers at the locations of eliminated entries as usual. We finally obtain the following upper triangular system (with multipliers stored under the diagonal)

$$
\begin{bmatrix}
\boxed{-3} & 1 & 2 & 3 \\
-1/3 & \boxed{7/3} & -1/3 & 1 \\
-2/3 & -1/7 & \boxed{16/7} & 64/7
\end{bmatrix}, \qquad
\begin{bmatrix}
3 \\
1 \\
2
\end{bmatrix}
$$

We can now perform backward substitution to find the solution $\mathbf{x} = [2,1,4]^T$. ∎

Below gives the pseudocode of GEPP. Compared with the pseudocode of GEnoP, it adds permutation tracking and partial pivoting.

Note that GEPP does not break down (due to division by zero) in exact arithmetic if the square coefficient matrix is invertible (as each diagonal position is a pivot position). The GEPP algorithm in inexact arithmetic generally is numerically stable to produce accurate solutions, but there are pathetic examples where it is unstable to produce inaccurate solutions.

4.4.2   PA=LU

In the previous example, Example 4.5, the final matrix and the permutation tracking vector after GEPP are

$$
\begin{bmatrix}
\boxed{-3} & 1 & 2 & 3 \\
-1/3 & \boxed{7/3} & -1/3 & 1 \\
-2/3 & -1/7 & \boxed{16/7} & 64/7
\end{bmatrix}, \qquad
\begin{bmatrix}
3 \\
1 \\
2
\end{bmatrix}
$$

From this final matrix, we have the upper triangular matrix U

$$
U = \begin{bmatrix}
\boxed{-3} & 1 & 2 \\
 & \boxed{7/3} & -1/3 \\
 & & \boxed{16/7}
\end{bmatrix}
$$

which comes from the row reduction to the initial coefficient matrix

$$
A = \begin{bmatrix}
1 & 2 & -1 \\
2 & -1 & 1 \\
-3 & 1 & 2
\end{bmatrix}
$$

Algorithm 4 GEPP for a square system $A\mathbf{x} = \mathbf{b}$

---

for i from 1 to n **do**
  $p_i \leftarrow i$
end for
for k from 1 to $n-1$ **do**
  $pivot \leftarrow |A_{kk}|$
  $m = k$
  for i from $k+1$ to n **do**
    if $pivot < |A_{ik}|$ **then**
      $pivot \leftarrow |A_{ik}|$
      $m = i$
    end if
  end for
  $itmp \leftarrow p_k$
  $p_k \leftarrow p_m$
  $p_m \leftarrow itmp$
  $btmp \leftarrow b_k$
  $b_k \leftarrow b_m$
  $b_m \leftarrow btmp$
  for j from 1 to n **do**
    $atmp \leftarrow A_{kj}$
    $A_{kj} \leftarrow A_{mj}$
    $A_{mj} \leftarrow atmp$
  end for
  for i from $k+1$ to n **do**
    $A_{ik} \leftarrow A_{ik}/A_{kk}$
    for j from $k+1$ to n **do**
      $A_{ij} \leftarrow A_{ij} - A_{ik}A_{kj}$
    end for
    $b_i \leftarrow b_i - A_{ik}b_k$
  end for
end for

According to the last permutation tracking vector $\mathbf{p} = [3,1,2]^T$, the 1st row of U comes from the 3rd row of A, the 2nd row from the 1st and the 1st from the 2nd. From the final matrix after GEPP, we also have the lower triangular matrix L

$$L = \begin{bmatrix} 1 & & \\ -1/3 & 1 & \\ -2/3 & -1/7 & 1 \end{bmatrix}$$

which stores the multipliers. The multipliers in each row of L (say the multipliers $-2/3$ and $-1/7$ in the 3rd row) are used for elimination to obtain the corresponding row of U (say the 3rd row with the entries 0, 0 and $16/7$) since we also swap stored multipliers in the partial pivoting process. It is expected (as GEnoP leads to the LU decomposition) the product LU gives a matrix which has the rows of A in the order specified by $\mathbf{p} = [3,1,2]^T$. The 1st row of LU should be the 3rd row of A, the 2nd be the 1st of A and the 3rd be the 2nd, as verified below

$$LU = \begin{bmatrix} 1 & & \\ -1/3 & 1 & \\ -2/3 & -1/7 & 1 \end{bmatrix} \begin{bmatrix} \boxed{-3} & 1 & 2 \\ & \boxed{7/3} & -1/3 \\ & & \boxed{16/7} \end{bmatrix} = \begin{bmatrix} -3 & 1 & 2 \\ 1 & 2 & -1 \\ 2 & -1 & 1 \end{bmatrix}$$

We can write $LU = PA$, where P is called a **permutation matrix**, which is formed by swapping rows of the identity matrix. In this case,

$$PA = \begin{bmatrix} -3 & 1 & 2 \\ 1 & 2 & -1 \\ 2 & -1 & 1 \end{bmatrix}, \quad P = \begin{bmatrix} 0 & 0 & 1 \\ 1 & 0 & 0 \\ 0 & 1 & 0 \end{bmatrix}$$

The 1st row of P is the 3rd row of the 3×3 identity matrix, 2nd row of P is the 1st row of I and 3rd row of P is the 2nd row of I. Let $\mathbf{p} = [p_1, p_2, p_3]^T = [3,1,2]^T$, we can form P as

$$P = \begin{bmatrix} \text{Row}_{p_1}(I) \\ \text{Row}_{p_2}(I) \\ \text{Row}_{p_3}(I) \end{bmatrix}$$

It can be easily shown that how PA swaps the rows of A is exactly the same as how the rows of I are swapped to obtain P.

In conclusion, GEPP leads to the so called PA=LU decomposition.

Remark Another way to make sense of the PA=LU decomposition is to swap the rows of A first to obtain PA and then apply GEnoP on PA. The diagonal entry sitting at each pivot position of PA in the GEnoP process is already a pivot. GEnoP on PA leads to the LU decomposition of PA.

We can also show that GEPP leads to the PA=LU decomposition using elementary matrices. Suppose the $n \times n$ matrix A is transformed to a sequence of matrices by the $(n-1)$ GEPP steps as

$$A \to A^{(1)} \to A^{(2)} \to \cdots \to A^{(n-1)} \tag{4.46}$$

where $A^{(n-1)}$ is the final upper triangular matrix U. As $A^{(k)} = E^{(k)}A^{(k-1)}$ in GEnoP, we now have $A^{(k)} = E^{(k)}P_k A^{(k-1)}$ after the k-th GEPP step, where P_k is the elementary matrix that is applied to $A^{(k-1)}$ first to interchange k-th row of $A^{(k-1)}$ with a row of the pivot under the k-th row, and $E^{(k)}$ is the product of elementary matrices for row replacements applied to $P_k A^{(k-1)}$ toward elimination. So the final upper triangular matrix U can be written as

$$U = A^{(n-1)} = E^{(n-1)}P_{n-1}E^{(n-2)}P_{n-2}\cdots E^{(2)}P_2 E^{(1)}P_1 A \tag{4.47}$$

We have $A^{(1)} = E^{(1)}P_1 A$ after the 1st GEPP step. As in GEnoP, the matrix $E^{(1)}$ has the form

$$E^{(1)} = \begin{bmatrix} 1 & 0 & 0 & \cdots & 0 \\ \boxed{*} & 1 & 0 & \cdots & 0 \\ * & 0 & 1 & \cdots & 0 \\ \vdots & \vdots & \vdots & & \vdots \\ * & 0 & 0 & \cdots & 1 \end{bmatrix} \tag{4.48}$$

where $*$ represents a multiplier (whose magnitude $|*| \leq 1$). We now show that

$$P_2 E^{(1)} = E^{(1)\mathrm{i}}P_2 \implies P_2 A^{(1)} = P_2 E^{(1)}P_1 A = E^{(1)\mathrm{i}}P_2 P_1 A \tag{4.49}$$

where $E^{(1)\mathrm{i}}$ comes from $E^{(1)}$ and has the same form as $E^{(1)}$. The matrix P_2 is the elementary matrix for the interchange of the 2nd row of a matrix with the m-th row ($m \geq 2$) in the 2nd GEPP step. Let's demonstrate the case $m = 3$ (the case $m \neq 3$ is similar). Left-multiplying $E^{(1)}$ by P_2 interchanges the 2nd and 3rd rows of $E^{(1)}$:

$$P_2 = \begin{bmatrix} 1 & 0 & 0 & \cdots & 0 \\ 0 & 0 & 1 & \cdots & 0 \\ 0 & 1 & 0 & \cdots & 0 \\ \vdots & \vdots & \vdots & & \vdots \\ 0 & 0 & 0 & \cdots & 1 \end{bmatrix}, \quad P_2 E^{(1)} = \begin{bmatrix} 1 & 0 & 0 & \cdots & 0 \\ * & 0 & 1 & \cdots & 0 \\ \boxed{*} & 1 & 0 & \cdots & 0 \\ \vdots & \vdots & \vdots & & \vdots \\ * & 0 & 0 & \cdots & 1 \end{bmatrix} \tag{4.50}$$

We have $P_2 E^{(1)} = E^{(1)\mathrm{i}}P_2$, where

$$E^{(1)\mathrm{i}}P_2 = \begin{bmatrix} 1 & 0 & 0 & \cdots & 0 \\ * & 1 & 0 & \cdots & 0 \\ \boxed{*} & 0 & 1 & \cdots & 0 \\ \vdots & \vdots & \vdots & & \vdots \\ * & 0 & 0 & \cdots & 1 \end{bmatrix}\begin{bmatrix} 1 & 0 & 0 & \cdots & 0 \\ 0 & 0 & 1 & \cdots & 0 \\ 0 & 1 & 0 & \cdots & 0 \\ \vdots & \vdots & \vdots & & \vdots \\ 0 & 0 & 0 & \cdots & 1 \end{bmatrix} \tag{4.51}$$

Note that P_2 is now right-multiplied to $E^{(1)_1}$ to interchange the 2nd and 3rd columns of $E^{(1)_1}$, where $E^{(1)_1}$ is generated from $E^{(1)}$ by swapping only multipliers in the 2nd and 3rd rows of $E^{(1)}$. The subscript 1 of (1) in $E^{(1)_1}$ is used to indicate that one elementary permutation matrix (here P_2) is moved across $E^{(1)}$ from the left to the right of $E^{(1)}$ (here $P_2E^{(1)} = E^{(1)_1}P_2$), which changes $E^{(1)}$ to $E^{(1)_1}$. So we have

$$P_2E^{(1)} = E^{(1)_1}P_2, \quad A^{(2)} = E^{(2)}P_2E^{(1)}P_1A = E^{(2)}E^{(1)_1}P_2P_1A \tag{4.52}$$

Similarly, we can show that

$$P_iE^{(j)} = E^{(j)_{i-j}}P_i \tag{4.53}$$

where the subscript $i - j > 0$ indicates that $i - j$ elementary matrices $P_{j+1}, P_{j+2}, \ldots, P_i$ have moved across $E^{(j)}$, which changes $E^{(j)}$ to $E^{(j)_{i-j}}$, where $E^{(j)_{i-j}}$ has a similar form as $E^{(j)}$. By moving all the elementary permutation matrices to the right across all the elimination matrices, we end up with

$$
\begin{aligned}
U = A^{(n-1)} &= E^{(n-1)}P_{n-1}E^{(n-2)}P_{n-2}\cdots E^{(2)}P_2E^{(1)}P_1A \\
&= E^{(n-1)}E^{(n-2)_1}\cdots E^{(2)_{n-3}}E^{(1)_{n-2}}P_{n-1}P_{n-2}\cdots P_2P_1A \quad (4.54)
\end{aligned}
$$

So we finally have

$$P_{n-1}P_{n-2}\cdots P_2P_1A = [E^{(1)_{n-2}}]^{-1}[E^{(2)_{n-3}}]^{-1}\cdots[E^{(n-2)_1}]^{-1}[E^{(n-1)}]^{-1}U \tag{4.55}$$

which is the PA=LU decomposition with

$$P = P_{n-1}P_{n-2}\cdots P_2P_1 \tag{4.56}$$

$$L = [E^{(1)_{n-2}}]^{-1}[E^{(2)_{n-3}}]^{-1}\cdots[E^{(n-2)_1}]^{-1}[E^{(n-1)}]^{-1} \tag{4.57}$$

$$U = A^{(n-1)} \tag{4.58}$$

where L is a lower triangular matrix with multipliers under the main diagonal.

4.4.3 Solving $Ax = b$ by PA=LU

With the PA=LU decomposition, we can efficiently solve the following m linear systems

$$Ax = b_i, \quad i = 1, 2, \ldots, m \tag{4.59}$$

We use GEPP to find the PA=LU decomposition in $O(n^3)$ arithmetic operations. Since

$$Ax = b_i \iff PAx = Pb_i \iff LUx = Pb_i \tag{4.60}$$

we can solve each linear system using the PA=LU decomposition by first solving for $y = Ux$ from

$$Ly = Pb_i \tag{4.61}$$

using forward substitution in $O(n^2)$ arithmetic operations, and then solving for x from

$$Ux = y \tag{4.62}$$

Direct Methods for Linear Systems

using backward substitution in $O(n^2)$ arithmetic operations. The total cost in the process is of $O(n^3 + mn^2)$, as compared with $O(mn^3 + mn^2)$ if using GEPP separately to solve for each linear system.

It should be noted that the pseudocode of GEPP already returns the PA=LU decomposition, where L and U are in the updated matrix A, and the permutation matrix P can be built from the permutation vector $\mathbf{p}$.

Example 4.6 (PA=LU) In the previous example, Example 4.5, the linear system $A\mathbf{x} = \mathbf{b}$, where

$$[A \mid \mathbf{b}] = \begin{bmatrix} 1 & 2 & -1 & 0 \\ 2 & -1 & 1 & 7 \\ -3 & 1 & 2 & 3 \end{bmatrix}$$

is solved by GEPP, which also leads to the decomposition $PA = LU$, where

$$P = \begin{bmatrix} 0 & 0 & 1 \\ 1 & 0 & 0 \\ 0 & 1 & 0 \end{bmatrix}, L = \begin{bmatrix} 1 & & \\ -1/3 & 1 & \\ -2/3 & -1/7 & 1 \end{bmatrix}, U = \begin{bmatrix} \boxed{-3} & 1 & 2 \\ & \boxed{7/3} & -1/3 \\ & & \boxed{16/7} \end{bmatrix}$$

We now solve the linear system $A\mathbf{x} = \mathbf{b}$ using the given PA=LU decomposition. We have

$$A\mathbf{x} = \mathbf{b} \iff PA\mathbf{x} = P\mathbf{b} \iff LU\mathbf{x} = P\mathbf{b}$$

where $P\mathbf{b} = [3,0,7]^T$, denoted as $\mathbf{b}^{(2)} = [b_1^{(2)}, b_2^{(2)}, b_3^{(2)}]$ (to be consistent with the notations in GEPP after step 2), can also be formed using the permutation vector $\mathbf{p} = [p_1, p_2, p_3] = [3,1,2]^T$ returned from GEPP as

$$b_1^{(2)} = b_{p_1} = b_3, \quad b_2^{(2)} = b_{p_2} = b_1, \quad b_3^{(2)} = b_{p_3} = b_2$$

where $[b_1, b_2, b_3]^T = \mathbf{b}$. We first solve for $\mathbf{y} = U\mathbf{x}$ from $L\mathbf{y} = \mathbf{b}^{(2)}$:

$$\begin{bmatrix} 1 & & \\ -1/3 & 1 & \\ -2/3 & -1/7 & 1 \end{bmatrix} \mathbf{y} = \begin{bmatrix} 3 \\ 0 \\ 7 \end{bmatrix}$$

using forward substitution and get $\mathbf{y} = [3, 1, 64/7]^T$. We then solve for $\mathbf{x}$ from $U\mathbf{x} = \mathbf{y}$:

$$\begin{bmatrix} \boxed{-3} & 1 & 2 \\ & \boxed{7/3} & -1/3 \\ & & \boxed{16/7} \end{bmatrix} \mathbf{x} = \begin{bmatrix} 3 \\ 1 \\ 64/7 \end{bmatrix}$$

using backward substitution and get $\mathbf{x} = [2, 1, 4]^T$. ∎

4.5 Tridiagonal Systems

The linear system $T\mathbf{x} = \mathbf{d}$ with

$$T = \begin{bmatrix} b_1 & c_1 & & & & \\ a_2 & b_2 & c_2 & & & \\ & a_3 & b_3 & c_3 & & \\ & & \ddots & \ddots & \ddots & \\ & & & a_{n-1} & b_{n-1} & c_{n-1} \\ & & & & a_n & b_n \end{bmatrix}, \quad \mathbf{d} = \begin{bmatrix} d_1 \\ d_2 \\ d_3 \\ \vdots \\ d_{n-1} \\ d_n \end{bmatrix} \qquad (4.63)$$

is a tridiagonal system as the nonzero entries in T appear only on the subdiagonal $(a_2, a_3, \ldots, a_n)$, the main diagonal $(b_1, b_2, \ldots, b_n)$ and the superdiagonal $(c_1, c_2, \ldots, c_{n-1})$. Such linear systems are common in many applications (see Problem 4.6 and finding cubic splines in Chapter 6). A matrix is called a **sparse matrix** if it has small percentage of nonzeros entries. The tridiagonal matrix T is a sparse matrix if $n \gg 3$. Obviously we do not need to store the full matrix T with a lot of zeros. Instead, we can store only the three vectors $\mathbf{a} = [0, a_2, a_3, \ldots, a_n]^T$, $\mathbf{b} = [b_1, b_2, \ldots, b_n]^T$ and $\mathbf{c} = [c_1, c_2, \ldots, c_{n-1}, 0]^T$ for the three diagonals (note that $\mathbf{a}$ and $\mathbf{c}$ are padded with zeros to make them have the same number of entries as $\mathbf{b}$ with correct entry indices).

The tridiagonal matrix T is often **weakly diagonally dominant** as

$$|b_1| > |c_1| > 0, \quad |b_i| \geq |a_i| + |c_i| \; (i = 2, 3, \ldots, n-1), \quad |b_n| > |a_n| > 0 \quad (4.64)$$

When applied to such matrix T, the naive Gaussian elimination GEnoP does not break down and leads to the LU decomposition $T = LU$, where L and U take the following forms

$$L = \begin{bmatrix} 1 & & & & & \\ \alpha_2 & 1 & & & & \\ & \alpha_3 & 1 & & & \\ & & \ddots & \ddots & & \\ & & & \alpha_{n-1} & 1 & \\ & & & & \alpha_n & 1 \end{bmatrix} \qquad (4.65)$$

$$U = \begin{bmatrix} \beta_1 & c_1 & & & & \\ & \beta_2 & c_2 & & & \\ & & \beta_3 & c_3 & & \\ & & & \ddots & \ddots & \\ & & & & \beta_{n-1} & c_{n-1} \\ & & & & & \beta_n \end{bmatrix} \qquad (4.66)$$

We can find L and U by matching the i-th $(i = 1, 2, \ldots, n)$ row of the product LU with the i-th row of T down from the top ($i = 1$) to the bottom ($i = n$). By doing so, we have

$$\beta_1 = b_1 \tag{4.67}$$
$$\alpha_i \beta_{i-1} = a_i, \quad \alpha_i c_{i-1} + \beta_i = b_i \quad (i = 2, 3, \ldots, n) \tag{4.68}$$

from which we can in turn solve for β_1 and the pairs α_i and β_i for $i = 2, 3, \ldots, n$. The pseudocode of this method is

---

Algorithm 5 LU decomposition of a triangular matrix

$\beta_1 \leftarrow b_1$
for i from 2 to n **do**
$\quad \alpha_i \leftarrow a_i / \beta_{i-1}$
$\quad \beta_i \leftarrow b_i - \alpha_i c_{i-1}$
end for

Remark This method/algorithm is called Doolittle's method/algorithm. It can also be used to find the LU decomposition of a general dense matrix (if the LU decomposition exists).

With the LU decomposition, the linear system $T\mathbf{x} = \mathbf{d}$ can be solved as follows. Since

$$T\mathbf{x} = \mathbf{d} \iff LU\mathbf{x} = \mathbf{d} \tag{4.69}$$

we can first solve for $\mathbf{y} = U\mathbf{x}$ from $L\mathbf{y} = \mathbf{d}$ using forward substitution. The corresponding pseudocode is

---

Algorithm 6 Forward substitution part for solving a triangular system

$y_1 \leftarrow b_1$
for i from 2 to n **do**
$\quad y_i \leftarrow d_i - \alpha_i y_{i-1}$
end for

We then solve for $\mathbf{x}$ from $U\mathbf{x} = \mathbf{y}$ using backward substitution. The corresponding pseudocode is given below

---

Algorithm 7 Backward substitution part for solving a triangular system

$x_n \leftarrow y_n / \beta_n$
for i from $n - 1$ down to 1 **do**
$\quad x_i \leftarrow (y_i - c_i x_{i+1}) / \beta_i$
end for

4.6 Conditioning of Linear Systems

Let's consider the following example.

Example 4.7 (Ill-conditioned) The linear system $Ax = b$:

$$\begin{bmatrix} 1 & 1 \\ 1.0001 & 1 \end{bmatrix} \begin{bmatrix} x_1 \\ x_2 \end{bmatrix} = \begin{bmatrix} 2 \\ 2.0001 \end{bmatrix}$$

can be solved in exact arithmetic to give the exact solution $\mathbf{x} = [x_1, x_2]^T = [1, 1]^T$.

If the entry 2.0001 in the right-hand-side vector $\mathbf{b}$ is slightly perturbed to 2.0002 (in reality, perturbations may come from roundoff errors or other types of errors), we end up with a nearby linear system $Ax' = b'$:

$$\begin{bmatrix} 1 & 1 \\ 1.0001 & 1 \end{bmatrix} \begin{bmatrix} x_1' \\ x_2' \end{bmatrix} = \begin{bmatrix} 2 \\ 2.0002 \end{bmatrix}$$

which can solved in exact arithmetic to give the exact solution $\mathbf{x}' = [x_1', x_2']^T = [2, 0]^T$, which differ dramatically from the exact solution of the original linear system $Ax = b$ even though the perturbation to only one entry is very small relative to the magnitude of the entry ($|2.0002 - 2.0001|/|2.0001| \approx 0.005\%$). ■

What is interesting with this example is the linear system $Ax = b$ itself is very sensitive to small perturbations regardless of what algorithms are used to solve it (in this example, we even do not specify what methods are used, as we directly give the exact solutions). This sensitivity (large change to the solution caused by small change to the linear system) can be understood graphically as in Fig. 4.2. The dramatic difference in the two solutions are not caused by error amplification in an unstable algorithm. Instead, it is due to the totally different exact solutions of the two close linear systems. It can also be shown that small perturbations to entries of A or entries of both A and $\mathbf{b}$ can lead to dramatic change of the solution of $Ax = b$. We say this linear system $Ax = b$ is **ill-conditioned**.

Remark We can also regard $Ax = b$ as the linear system coming from a perturbation to $Ax' = b'$, and we can also conclude that $Ax' = b'$ is ill-conditioned.

Now let's consider a different linear system in the next example.

Example 4.8 (Well-conditioned) The linear system $Ax = b$:

$$\begin{bmatrix} 1 & 0 \\ 0 & 1 \end{bmatrix} \begin{bmatrix} x_1 \\ x_2 \end{bmatrix} = \begin{bmatrix} 1 \\ 1 \end{bmatrix}$$

has exact solution $\mathbf{x} = [x_1, x_2]^T = [1, 1]^T$.

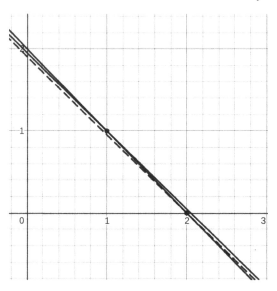

FIGURE 4.2

The two lines (1) $x + y = 2$ and (2) $x + 1.05y = 2.05$ have close slopes and intersect at $(1, 1)$. A small perturbation to the line (2) toward the line (3) $x + 1.05y = 2$ (dashed line) alter the intersection point dramatically to $(2, 0)$.

If one entry 0 in the coefficient matrix A is slightly perturbed to 0.0001, we end up with a nearby linear system $A'\mathbf{x}' = \mathbf{b}$:

$$\begin{bmatrix} 1 & 0 \\ 0.0001 & 1 \end{bmatrix} \begin{bmatrix} x'_1 \\ x'_2 \end{bmatrix} = \begin{bmatrix} 1 \\ 1 \end{bmatrix}$$

which has the exact solution $\mathbf{x}' = [x'_1, x'_2]^T = [1, 0.9999]^T$, which is very close to the exact solution of the original linear system $A\mathbf{x} = \mathbf{b}$. ∎

In contrast, the linear system $A\mathbf{x} = \mathbf{b}$ in this example is actually not sensitive to **any** small perturbations, and we say it is **well-conditioned**.

Remark In Chapter 3, we compute entries in the sequence $\{V_j = \int_0^1 e^{x-1} x^j dx\}_{j=0}^{\infty}$. The algorithm of forward recurrence is unstable and leads to inaccurate results, while the algorithm of backward recurrence is stable and leads to accurate results. In this case, the problem itself is well-conditioned, but the algorithms used to solve the problem can be stable or unstable.

We should distinguish conditioning of a problem and stability of an algorithm to solve the problem. Conditioning is the property of a problem, while the stability is the property of an algorithm. It is very challenging to solve an ill-conditioned problem accurately when perturbations/errors are inevitable no matter what algorithms are used. The difference between conditioning and stability is illustrated in Fig. 4.3.

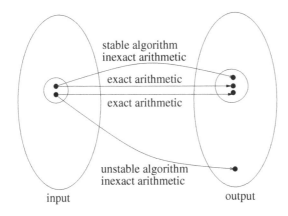

well-conditioned

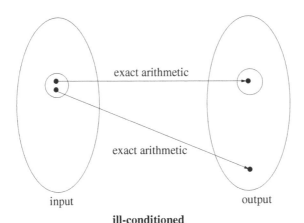

ill-conditioned

FIGURE 4.3
A problem can be regarded as finding output from input (for example, solving $A\mathbf{x} = \mathbf{b}$ is finding the output $\mathbf{x}$ from the input A and $\mathbf{b}$). The illustrations show how conditioning of the problem and stability of algorithms used to solve the problem affects results. The small circles around input or output represent small neighborhoods.

Remark A square Vandermonde matrix V of order $(n+1)$

$$V = \begin{bmatrix} 1 & x_0 & x_0^2 & \cdots & x_0^n \\ 1 & x_1 & x_1^2 & \cdots & x_1^n \\ 1 & x_2 & x_2^2 & \cdots & x_2^n \\ \vdots & \vdots & \vdots & & \vdots \\ 1 & x_n & x_n^2 & \cdots & x_n^n \end{bmatrix} \qquad (4.70)$$

can be formed by entry-wise powers of the vector $[x_0, x_1, x_2, \ldots, x_n]^T$, where the entries in the vector are all distinct. The Vandermonde linear system $V\mathbf{x} = \mathbf{f}$ become more and more badly ill-conditioned as n increases. The Vandermonde linear system appears in interpolation (Chapter 6) and numerical differentiation (Chapter 8).

4.6.1 Vector and Matrix Norms

The magnitude of a number (say -1) can be measured by the absolute value of the number (say $|-1| = 1$). The magnitude of the error in an approximation A of a number T is the absolute error $|T - A|$.

The solution of a linear system is a vector containing more than one numbers.

Definition 4.1 *Suppose* $\mathbf{x}_T$ *is the exact solution of a linear system* $A\mathbf{x} = \mathbf{b}$ *and* $\mathbf{x}_A$ *is an approximate solution. We define the error vector* $\mathbf{e}$ *as*

$$\mathbf{e} = \mathbf{x}_T - \mathbf{x}_A$$

We define the residual vector for $\mathbf{x}_A$ *as*

$$\mathbf{r} = \mathbf{b} - A\mathbf{x}_A$$

Remark The error vector and residual vector for the exact solution $\mathbf{x}_T$ are zero vectors. The error vector $\mathbf{e}$ and residual vector $\mathbf{r}$ for the approximate solution $\mathbf{x}_A$ satisfy

$$A\mathbf{e} = \mathbf{r} \qquad (4.71)$$

How can we measure the "magnitude" of a vector (say the error vector) that contains more than one number? We use different norms to measure a vector or even a matrix, which lead to different perspectives of the "magnitude".

Example 4.9 (Magnitudes) A rectangular box has three dimensions a, b and c. We may measure the "magnitude" of the box by its diagonal length $\sqrt{a^2 + b^2 + c^2}$, its maximum dimension $\max\{|a|, |b|, |c|\}$, or the sum of the three dimensions $|a| + |b| + |c|$. We may also use the volume abc. In USPS, the girth is used, which is the sum of length and transverse perimeter. ∎

Definition 4.2 *The 2-norm, 1-norm and ∞-norm of a vector* $\mathbf{x} = [x_1, x_2, \ldots, x_n]^T$ *in* $\mathbb{R}^n$ *are defined respectively as*

$$\|\mathbf{x}\|_2 = \sqrt{x_1^2 + x_2^2 + \cdots + x_n^2}$$

$$\|\mathbf{x}\|_1 = |x_1| + |x_2| + \cdots + |x_n|$$

$$\|\mathbf{x}\|_\infty = \max\{|x_1|, |x_2|, \ldots, |x_n|\}$$

These norms belong to the p-norm $\|\mathbf{x}\|_p$ *of the vector* $\mathbf{x}$ *with* $p = 2$, $p = 1$ *and* $p = \infty$, *respectively.*

Remark Let c be a scalar (number). It can be shown that the p-norm of the scalar multiple $c\mathbf{x}$ is related to the p-norm of $\mathbf{x}$ as

$$\|c\mathbf{x}\|_p = |c| \|\mathbf{x}\|_p \tag{4.72}$$

Example 4.10 (Vector norms) The vector $\mathbf{x} = [0.3, -0.4]^T$ can be represented as a point in the $\mathbb{R}^2$ plane. Its 2-norm $\|\mathbf{x}\|_2 = \sqrt{(0.3)^2 + (-0.4)^2} = 0.5$ is the distance from the point to the origin. Its 1-norm $\|\mathbf{x}\|_1 = |0.3| + |-0.4| = 0.7$ is the sum of the distances from the point to the coordinate axes. Its ∞-norm $\|\mathbf{x}\|_\infty = \max\{|0.3|, |-0.4|\} = 0.4$ is the maximum of the distances from the point to the coordinate axes. ∎

Shown in Fig. 4.4 are graphs of the sets of vectors (represented as points) in the $\mathbb{R}^2$ plane with unit norms.

Now we introduce norms for a matrix A. When solving the linear system $A\mathbf{x} = \mathbf{b}$, we can regard A as the standard matrix of a matrix (linear) transformation $\mathbf{v} \mapsto A\mathbf{v}$, and we want to find the vector $\mathbf{x}$ such that its image under the transformation is the vector $\mathbf{b}$. The matrix transformation $\mathbf{v} \mapsto A\mathbf{v}$ transforms the vector $\mathbf{v}$ of the "magnitude" $\|\mathbf{v}\|_p$ (the p-norm of $\mathbf{v}$, where $p = 2$, 1, ∞ or other values for other norms) to the image $A\mathbf{v}$ of the "magnitude" $\|A\mathbf{v}\|_p$. The "magnitude" amplification factor of this

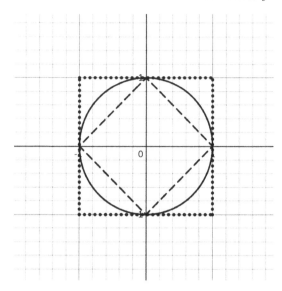

FIGURE 4.4
Graphs of the sets of vectors with $\|\mathbf{x}\|_2 = 1$ (solid), $\|\mathbf{x}\|_1 = 1$ (dashed) and $\|\mathbf{x}\|_\infty = 1$ (dotted).

transformation is simply $\|A\mathbf{v}\|_p / \|\mathbf{v}\|_p$. We define an induced norm (operator norm) $\|A\|_p$ of the matrix A as the maximum amplification factor as the following

---

Definition 4.3

$$\|A\|_p = \max_{\mathbf{v} \neq \mathbf{0}} \frac{\|A\mathbf{v}\|_p}{\|\mathbf{v}\|_p}$$

Remark   Since $1/\|\mathbf{v}\|_p$ is a positive scalar, we can write

$$\frac{\|A\mathbf{v}\|_p}{\|\mathbf{v}\|_p} = \left\| A \frac{\mathbf{v}}{\|\mathbf{v}\|_p} \right\| \tag{4.73}$$

where $\mathbf{v}/\|\mathbf{v}\|_p$ is a unit vector from the normalization of $\mathbf{v}$. Let $\mathbf{u} = \mathbf{v}/\|\mathbf{v}\|_p$. We can equivalently define the induced matrix norm as

$$\|A\|_p = \max_{\|\mathbf{u}\|_p = 1} \|A\mathbf{u}\|_p \tag{4.74}$$

Remark The square of the 2-norm of A is

$$\|A\|_2^2 = \max_{\|\mathbf{u}\|_2=1} \|A\mathbf{u}\|_2^2 = \max_{\|\mathbf{u}\|_2=1} (A\mathbf{u})^T (A\mathbf{u}) = \max_{\|\mathbf{u}\|_2=1} \mathbf{u}^T (A^T A)\mathbf{u} \qquad (4.75)$$

which is to maximize the quadratic form with the symmetric positive definite (SPD) matrix $A^T A$ (if A has linearly independent columns) under the constrain $\|\mathbf{u}\|_2 = 1$ (finding $\|A\|_2^2$ is called a constrained optimization problem). From linear algebra, we know the SPD matrix $A^T A$ is orthogonally diagonalizable as $A^T A = PDP^T$, where P is an orthogonal matrix with $P^{-1} = P^T$ and D is a diagonal matrix with positive eigenvalues $\lambda_1 \geq \lambda_2 \geq \cdots \geq \lambda_n > 0$ of $A^T A$ on the diagonal, which are the squares of the singular values $\sigma_1 \geq \sigma_2 \geq \cdots \geq \sigma_n > 0$ of A. So we have

$$\|A\|_2 = \sigma_1 = \sqrt{\lambda_1} \qquad (4.76)$$

i.e. the 2-norm of the matrix A is the largest singular value of A.

Remark Let A be an $n \times n$ matrix with the entry a_{ij} at i-th row and j-th column. It can be proved using the definition of a matrix norm that

$$\|A\|_1 = \max_{1 \leq j \leq n} \sum_{i=1}^{n} |a_{ij}| \qquad (4.77)$$

$$\|A\|_\infty = \max_{1 \leq i \leq n} \sum_{j=1}^{n} |a_{ij}| \qquad (4.78)$$

Below is a very useful theorem about a vector norm and its induced matrix norm.

Theorem 4.1

$$\|A\mathbf{x}\|_p \leq \|A\|_p \|\mathbf{x}\|_p \qquad (4.79)$$

Proof *If $\mathbf{x} = \mathbf{0}$, then $\|A\mathbf{x}\|_p = \|A\|_p \|\mathbf{x}\|_p = 0$. For a given nonzero vector $\mathbf{x}$, we have*

$$\frac{\|A\mathbf{x}\|_p}{\|\mathbf{x}\|_p} \leq \max_{\mathbf{v} \neq \mathbf{0}} \frac{\|A\mathbf{v}\|_p}{\|\mathbf{v}\|_p} = \|A\|_p \qquad (4.80)$$

which is simply the statement that an amplification factor is less than or equal to the maximum possible amplification factor. So we have

$$\frac{\|A\mathbf{x}\|_p}{\|\mathbf{x}\|_p} \leq \|A\|_p \Rightarrow \|A\mathbf{x}\|_p \leq \|A\|_p \|\mathbf{x}\|_p \qquad (4.81)$$

■

> **Remark** Note that $\|A\mathbf{x}\|_p$ can be equal to $\|A\|_p\|\mathbf{x}\|_p$ for particular $\mathbf{x}$ which leads to the maximum amplification factor.

4.6.2 Condition Numbers

We can now measure the conditioning of a linear system using vector and matrix norms. We analyze the conditioning of $A\mathbf{x} = \mathbf{b}$ in a simple case below, where A is a square invertible matrix so that the linear system has a unique solution.

In this case, we perturb only the vector $\mathbf{b}$ to the new vector $\mathbf{b}'$ with the perturbation $\delta\mathbf{b}$, which is $\mathbf{b}' - \mathbf{b}$. We denote the exact solution of $A\mathbf{x} = \mathbf{b}$ as $\mathbf{x}$, the exact solution of the new linear system (after the perturbation) $A\mathbf{x} = \mathbf{b}'$ as $\mathbf{x}'$, and the change of the solution due to the perturbation as $\delta\mathbf{x}$, which is $\mathbf{x}' - \mathbf{x}$.

Using the vector p-norm, the "magnitude" of the perturbation $\delta\mathbf{b}$ relative to the "magnitude" of the unperturbed vector $\mathbf{b}$ is the fraction $\|\delta\mathbf{b}\|_p/\|\mathbf{b}\|_p$. Similarly, the "magnitude" of the change of the solution $\delta\mathbf{x}$ relative to the "magnitude" of the original solution $\mathbf{x}$ is the fraction $\|\delta\mathbf{x}\|_p/\|\mathbf{x}\|_p$. So the ratio of these two fractions (i.e. the ratio of change in output to change in input) can measure the conditioning (see Fig. 4.3 for the illustration of the conditioning of a problem). We can establish the following upper bound for this ratio:

$$\frac{\|\delta\mathbf{x}\|_p/\|\mathbf{x}\|_p}{\|\delta\mathbf{b}\|_p/\|\mathbf{b}\|_p} \leq \|A\|_p\|A^{-1}\|_p \tag{4.82}$$

Clearly the larger the ratio is, the worse the conditioning is. The upper bound $\|A\|_p\|A^{-1}\|_p$ is a sharp one that can be reached in worst situations. So it is called the condition number of A (or the linear system $A\mathbf{x} = \mathbf{b}$).

> **Definition 4.4** *The condition number of the matrix A (in p-norm) is*
>
> $$\kappa_p(A) = \|A\|_p\|A^{-1}\|_p$$

Therefore if the condition number $\kappa_p(A) = \|A\|_p\|A^{-1}\|_p$ is small, the linear system $A\mathbf{x} = \mathbf{b}$ is not overly sensitive to perturbations and well-conditioned. Otherwise if $\kappa(A)$ is very large, the linear system is potentially very sensitive to perturbations and ill-conditioned.

Remark Similar analysis can be carried out when perturbations are applied to A or both A and **b**. It turns out that the conclusion is the same, that is the conditioning of the linear system $A\mathbf{x} = \mathbf{b}$ is measured by the condition number $\kappa_p(A) = \|A\|_p \|A^{-1}\|_p$.

We now prove that

$$\frac{\|\delta\mathbf{x}\|_p/\|\mathbf{x}\|_p}{\|\delta\mathbf{b}\|_p/\|\mathbf{b}\|_p} \leq \|A\|_p \|A^{-1}\|_p = \kappa_p(A) \tag{4.83}$$

Proof *By definition,* $A\mathbf{x} = \mathbf{b}$, $A\mathbf{x}' = \mathbf{b}'$, $\mathbf{x}' = \mathbf{x} + \delta\mathbf{x}$ *and* $\mathbf{b}' = \mathbf{b} + \delta\mathbf{b}$, *so we have*

$$A(\mathbf{x} + \delta\mathbf{x}) = \mathbf{b} + \delta\mathbf{b}, \quad A\delta\mathbf{x} = \delta\mathbf{b}, \quad \delta\mathbf{x} = A^{-1}\delta\mathbf{b} \tag{4.84}$$

$$\|\delta\mathbf{x}\|_p = \|A^{-1}\delta\mathbf{b}\|_p \leq \|A^{-1}\|_p \|\delta\mathbf{b}\|_p \tag{4.85}$$

where the last inequality above is due to Theorem 4.1. We then have

$$\left(\frac{\|\delta\mathbf{x}\|_p}{\|\mathbf{x}\|_p}\right) \leq \frac{\|A^{-1}\|_p \|\delta\mathbf{b}\|_p}{\|\mathbf{x}\|_p} = \frac{\|A^{-1}\|_p \|\mathbf{b}\|_p}{\|\mathbf{x}\|_p} \left(\frac{\|\delta\mathbf{b}\|_p}{\|\mathbf{b}\|_p}\right) \tag{4.86}$$

Note that

$$\mathbf{b} = A\mathbf{x}, \quad \|\mathbf{b}\|_p = \|A\mathbf{x}\|_p \leq \|A\|_p \|\mathbf{x}\|_p \tag{4.87}$$

So

$$\left(\frac{\|\delta\mathbf{x}\|_p}{\|\mathbf{x}\|_p}\right) \leq \frac{\|A^{-1}\|_p \|\mathbf{b}\|_p}{\|\mathbf{x}\|_p} \left(\frac{\|\delta\mathbf{b}\|_p}{\|\mathbf{b}\|_p}\right) \leq \frac{\|A^{-1}\|_p \|A\|_p \|\mathbf{x}\|_p}{\|\mathbf{x}\|_p} \left(\frac{\|\delta\mathbf{b}\|_p}{\|\mathbf{b}\|_p}\right) \tag{4.88}$$

which leads to

$$\frac{\|\delta\mathbf{x}\|_p/\|\mathbf{x}\|_p}{\|\delta\mathbf{b}\|_p/\|\mathbf{b}\|_p} \leq \|A\|_p \|A^{-1}\|_p = \kappa_p(A) \tag{4.89}$$

■

4.6.3 Error and Residual Vectors

We need to distinguish errors and residuals.

Suppose $\mathbf{x}_T$ is the exact solution of a linear system $A\mathbf{x} = \mathbf{b}$ and $\mathbf{x}_A$ is an approximate solution. The error vector $\mathbf{e} = \mathbf{x}_T - \mathbf{x}_A$ is the difference of the approximate solution $\mathbf{x}_A$ and the true solution $\mathbf{x}_T$, while the residual vector $\mathbf{r} = \mathbf{b} - A\mathbf{x}_A$ is the difference of the left-hand side $A\mathbf{x}_A$ and the right-hand side $\mathbf{b}$ of the linear system when the approximate solution $\mathbf{x}_A$ is plugged in.

So the accuracy of the approximate solution $\mathbf{x}_A$ is indicated by the norm of the error vector $\|\mathbf{e}\|_p$, while how well the approximate solution $\mathbf{x}_A$ satisfies the linear system $A\mathbf{x} = \mathbf{b}$ is indicated by the norm of the residual vector $\|\mathbf{r}\|_p$.

Example 4.11 The linear system $Ax = \mathbf{b}$:

$$\begin{bmatrix} 1 & 1 \\ 1.0001 & 1 \end{bmatrix} \begin{bmatrix} x_1 \\ x_2 \end{bmatrix} = \begin{bmatrix} 2 \\ 2.0001 \end{bmatrix}$$

has the exact solution $\mathbf{x}_T = [1, 1]^T$. Let $\mathbf{x}_A = [2, 0]^T$ be an approximate solution of this linear system. From Example 4.7, we know that $A\mathbf{x}_A = \mathbf{b}'$:

$$\begin{bmatrix} 1 & 1 \\ 1.0001 & 1 \end{bmatrix} \begin{bmatrix} 2 \\ 0 \end{bmatrix} = \begin{bmatrix} 2 \\ 2.0002 \end{bmatrix}$$

Clearly $\mathbf{x}_A$ is a bad approximation as $\|\mathbf{e}\|_2 = \sqrt{(1-2)^2 + (1-0)^2} = \sqrt{2}$ is large. However the residual vector for $\mathbf{x}_A$ is $\mathbf{r} = \mathbf{b} - A\mathbf{x}_A = \mathbf{b} - \mathbf{b}' = [0, -0.0001]^T$ and has the very small norm $\|\mathbf{r}\|_2 = \sqrt{0^2 + 0.0001^2} = 0.0001$. So this bad approximation almost satisfies the original linear system well.

Noticing that $\mathbf{b}' = \mathbf{b} + (-\mathbf{r})$, we can regard $\mathbf{r}$ as a perturbation to $\mathbf{b}$ of the linear system $A\mathbf{x} = \mathbf{b}$ to produce the new linear system $A\mathbf{x}' = \mathbf{b}' = \mathbf{b} - \mathbf{r}$. The linear system $A\mathbf{x} = \mathbf{b}$ is ill-conditioned, so the small perturbation $\mathbf{r} = [0, 0.0001]^T$ leads to dramatic change of the solution from $\mathbf{x}_T = [1, 1]^T$ to $\mathbf{x}_A = [2, 0]^T$ (large error $\mathbf{e} = [-1, 1]^T$ if $\mathbf{x}_A$ is regarded as an approximation of $\mathbf{x}_T$). ■

> **Remark** For a well-conditioned linear system $A\mathbf{x} = \mathbf{b}$ with the true solution $\mathbf{x}_T$, if a small residual $\mathbf{r}$ for an approximate solution $\mathbf{x}_A$ is regarded as a small perturbation to the linear system, the induced change of the solution (from $\mathbf{x}_T$ to $\mathbf{x}_A$) of the linear system must also be small. That is the error $\mathbf{e} = \mathbf{x}_T - \mathbf{x}_A$ is small.

In practice, we cannot compute the error of an approximate solution as we do not know the true solution, while we can compute the residual for the approximate solution. However, we should be careful with the use of the residual. For an ill-conditioned linear system, we cannot use the residual to measure the accuracy of the solution. We can only if a linear system is well-conditioned.

4.7   Software

In MATLAB, the backslash operator can be used to solve a linear system $A\mathbf{x} = \mathbf{b}$ simply as x = A\b. The backslash operator has very simple syntax, but it is very powerful and involves numerous lines of codes behind by choosing a suitable algorithm according to the structure and property of the coefficient matrix A.

In Python, the `numpy.linalg.solve(A,b)` method in the numpy library can be used to solve the linear system `Ax = b`. The similar method `scipy.linalg.solve()` is also available in the library `scipy`.

LAPACK is a freely-available software package available from `netlib.org` that provides routines for solving systems of linear equations as well as other linear algebra problems (such as least-squares problems, eigenvalue problems and singular value problems). LAPACK routines are written so that as much as possible of the computation is performed by calls to the Basic Linear Algebra Subprograms (BLAS). LAPACK is written in the computer language Fortran 90.

4.8 Exercises

Exercise 4.1 Consider the linear system $A\mathbf{x} = \mathbf{b}$ with

$$A = \begin{bmatrix} 2 & 1 & -1 & 3 \\ -2 & 0 & 0 & 0 \\ 4 & 1 & -2 & 6 \\ -6 & -1 & 2 & -3 \end{bmatrix}, \quad \mathbf{b} = \begin{bmatrix} 13 \\ -2 \\ 24 \\ -14 \end{bmatrix}$$

(1) Use GEnoP to solve the linear system. (2) Find the LU decomposition of A. (3) Use your LU decomposition to solve the linear system again.

Exercise 4.2 Find the LU factorization of the matrix A using GEnoP.

$$A = \begin{bmatrix} 4 & 2 & 1 \\ -4 & -6 & 1 \\ 8 & 16 & -3 \end{bmatrix}$$

Exercise 4.3 The LU factorization of the matrix A is given as

$$A = LU = \begin{bmatrix} 1 & 0 & 0 \\ -2 & 1 & 0 \\ -1 & 1 & 1 \end{bmatrix} \begin{bmatrix} -3 & 2 & -1 \\ 0 & -2 & 5 \\ 0 & 0 & -2 \end{bmatrix}$$

Use the LU factorization $A = LU$ to solve $A\mathbf{x} = \mathbf{b}$ for

$$\mathbf{b} = \begin{bmatrix} -1 \\ -7 \\ -6 \end{bmatrix}$$

Exercise 4.4 Let A be an $n \times n$ symmetric matrix, i.e. $A^T = A$ and its entries satisfy $a_{ij} = a_{ji}$ ($i = 1, 2, \ldots, n$ and $j = 1, 2, \ldots, n$). Suppose A is transformed by the 1st step of GEnoP to the following matrix

$$A = \begin{bmatrix} a_{11} & a_{12} & \cdots & a_{1n} \\ 0 & & & \\ \vdots & & \hat{A} & \\ 0 & & & \end{bmatrix}$$

where $\hat{A}$ is an $(n-1) \times (n-1)$ submatrix. Show that $\hat{A}$ is also symmetric.

Exercise 4.5 Use GEPP to solve $A\mathbf{x} = \mathbf{b}$, where

$$A = \begin{bmatrix} 1 & 3 & -4 \\ 0 & -1 & 5 \\ 2 & 0 & 4 \end{bmatrix}, \quad \mathbf{b} = \begin{bmatrix} 0 \\ 4 \\ 6 \end{bmatrix}$$

Exercise 4.6 Consider the linear system $A\mathbf{x} = \mathbf{b}$ with

$$A = \begin{bmatrix} 0 & 4 & 1 \\ 1 & 1 & 3 \\ 2 & -2 & 1 \end{bmatrix}, \quad \mathbf{b} = \begin{bmatrix} 9 \\ 6 \\ -1 \end{bmatrix}$$

(1) Use GEPP to solve the linear system. (2) Find P, L and U in the PA=LU decomposition for A. (3) Use your PA=LU decomposition to solve the linear system again.

Exercise 4.7 The linear system $A\mathbf{x} = \mathbf{b}$ has

$$A = \begin{bmatrix} 2 & 2 & -4 \\ 1 & 1 & 5 \\ 1 & 3 & 6 \end{bmatrix}, \quad \mathbf{b} = \begin{bmatrix} 10 \\ -2 \\ -5 \end{bmatrix}$$

(1) Use GEPP to solve the linear system. (2) Find P, L and U in the PA=LU decomposition for A. (3) Use your PA=LU decomposition to solve the linear system again.

Exercise 4.8 When A is a symmetric positive definite (SPD) matrix (that is A satisfies $A^T = A$ and $\mathbf{x}^T A \mathbf{x} > 0$ for any $\mathbf{x} \neq \mathbf{0}$), it can be factored in the form

$$A = R^T R$$

where R is an upper triangular matrix that has nonzero diagonal entries, usually taken to be positive. This factorization is called the Cholesky decomposition. Find R in the Cholesky decomposition $A = R^T R$ by the Doolittle method for

$$A = \begin{bmatrix} 1 & -1 \\ -1 & 5 \end{bmatrix}$$

Exercise 4.9 Count how many multiplications (in terms of n) are used in the following pseudocode (at the beginning of the next page). Write down the number of multiplications in the big O notation. Hint: $\sum_{k=1}^{n} k^2 = n(n+1)(2n+1)/6$.

Exercise 4.10 Count the number of arithmetic operations to find the LU decomposition of an $n \times n$ diagonally dominant tridiagonal matrix T. How many arithmetic operations are needed to solve $T\mathbf{x} = \mathbf{b}$ using the known LU decomposition?

Exercise 4.11 The permutation matrix P is given as

$$P = \begin{bmatrix} 0 & 0 & 1 \\ 1 & 0 & 0 \\ 0 & 1 & 0 \end{bmatrix}$$

---

for k from 1 to $n-1$ **do**
  for i from $k+1$ to n **do**
    for j from $k+1$ to n **do**
      $A_{ij} \leftarrow A_{ij} - A_{kj}A_{ik}/A_{kk}$
    end for
  end for
end for

Compute PA and AP for the following A

$$A = \begin{bmatrix} 1 & 2 & 3 \\ 4 & 5 & 6 \\ 7 & 8 & 9 \end{bmatrix}$$

Describe the effects of multiplying A by P to the left and right. Check $PP^T = I$. What is the inverse of P? Is P an orthogonal matrix?

Exercise 4.12 Use the LU decomposition to solve the tridiagonal system

$$
\begin{aligned}
2x_1 + x_2 &= 3 \\
-x_1 + 2x_2 + x_3 &= 2 \\
-x_2 + 2x_3 + x_4 &= 2 \\
-x_3 + 2x_4 + x_5 &= 2 \\
-x_4 + 2x_5 &= 1
\end{aligned}
$$

Exercise 4.13 Consider the problem to evaluate $f(x) = (x-1)^{1/3}$ at x_0. If we perturb the input x_0 to $x_0 + h$ by a small h, the output moves from $f(x_0)$ to $f(x_0 + h)$. The ratio r of the output change to the input change is

$$r = \left| \frac{f(x_0 + h) - f(x_0)}{h} \right|$$

(1) What is the limit of r as $h \to 0$? (2) Denote the limit as $R = \lim_{h \to 0} r$. How is R related to the conditioning of the problem? (3) Is the evaluation of $f(x)$ at $x_0 = 1$ ill-conditioned or well-conditioned? (4) How about at $x_0 = 2$?

Exercise 4.14 Given

$$A = \begin{bmatrix} 1000 & 999 \\ 999 & 998 \end{bmatrix}, \quad b = \begin{bmatrix} 1999 \\ 1997 \end{bmatrix}$$

(1) Check $\mathbf{x} = [1,1]^T$ is the exact solution of $A\mathbf{x} = \mathbf{b}$. Let $\hat{\mathbf{x}} = [20.97, -18.99]^T$ be an approximate solution. (2) Compute the error $\mathbf{e} = \mathbf{x} = \hat{\mathbf{x}}$ and the residual $\mathbf{r} = \mathbf{b} - A\hat{\mathbf{x}}$. (3) Compute $\|\mathbf{e}\|_2$ and $\|\mathbf{r}\|_2$. (4) Comment on your results in terms of the conditioning of the linear system $A\mathbf{x} = \mathbf{b}$.

Exercise 4.15 Let I be an $n \times n$ identity matrix. What is the value of $\|I\|_2$? What is the condition number of I in 2-norm?

Exercise 4.16 Let D be an $n \times n$ invertible diagonal matrix. What is the value of $\|D\|_2$? What is the condition number of D in 2-norm?

4.9   Programming Problems

Problem 4.1 Write a MATLAB m-function BackSubstitution.m to solve upper triangular linear systems using backward substitution. Write an m-script to test your m-function by comparing your results with those obtained from the MATLAB backslash operator on randomly generated upper triangular linear systems (using the MATLAB built-in functions rand and triu) of size 15, 25 and 35. You can use the MATLAB built-in function norm to compute the 2-norm of each vector of difference between your solution and a MATLAB solution.

Problem 4.2 Write a MATLAB m-function ForwardSubstitution.m to solve lower triangular linear systems using forward substitution. Write an m-script to test your m-function by comparing your results with those obtained from the MATLAB backslash operator on randomly generated lower triangular linear systems (using the MATLAB built-in functions rand and tril) of size 15, 25 and 35. You can use the MATLAB built-in function norm to compute the 2-norm of each vector of difference between your solution and a MATLAB solution.

Problem 4.3 Write a MATLAB m-function GEnoP.m to find the LU decomposition of a square matrix using Gaussian elimination without partial pivoting. Then use the LU decomposition to solve a linear system by calling your ForwardSubstitution.m and BackwardSubstitution.m in Problems 4.1 and 4.2. Test your codes by comparing your results with those obtained from the MATLAB backslash operator (x = A\ b in MATLAB finds the solution of Ax=b) on randomly generated linear systems of size 15 25, and 35. You can use the MATLAB built-in function norm to compute the 2-norm of each vector of difference between your solution and a MATLAB solution.

Problem 4.4 Write a MATLAB m-function GEPP.m to find the PA=LU decomposition of a square matrix using Gaussian elimination with partial pivoting. Then use the PA=LU decomposition to solve a linear system by calling your ForwardSubstitution.m and BackwardSubstitution.m in Problems 4.1 and 4.2. Test your codes by comparing your results with those obtained from the MATLAB backslash operator (x = A\ b in MATLAB finds the solution of Ax=b) on randomly generated linear systems of size 15, 25 and 35. You can use the MATLAB built-in function norm to compute the 2-norm of each vector of difference between your solution and a MATLAB solution.

Problem 4.5 Let's consider a linear system $V_n \mathbf{x} = \mathbf{b}$. The coefficient matrix V_n is called an $n \times n$ Vandermonde matrix. A Vandermonde matrix is generated in MATLAB by the command `vander` using a vector with distinct entries (use `help vander` to see the introduction of a Vandermonde matrix in MATLAB). Set the right-hand side $\mathbf{b}$ so that the exact solution to the linear system is

$$x_1 = x_3 = x_5 = \cdots = -2, \quad x_2 = x_4 = x_6 = \cdots = 2.$$

For $n = 7, 9, 11, 13, 15$, find a computed (approximate) solution $\hat{\mathbf{x}}$ of the linear system using the MATLAB backslash operator ($\mathbf{x}$ = A\ b in MATLAB finds the solution of Ax=b); and use `fprintf` to print a table such that each row of the table includes the value of n, the magnitude of the maximum component of the error $\mathbf{x} - \hat{\mathbf{x}}$ and the magnitude of the maximum component of the residual $\mathbf{r} = \mathbf{b} - A\hat{\mathbf{x}}$ for this value of n. Discuss the behaviors of the magnitudes of the errors and residuals as the value of n increases. What is the probable cause of the behaviors?

Problem 4.6 The temperature of a thin straight rod is modeled by the following so-called two-point boundary value problem (BVP)

$$u''(x) = f(x), \quad x \in [0, 1] \tag{4.90}$$
$$u(0) = u(1) = 0 \tag{4.91}$$

where $u(x)$ is the temperature distribution along the rod that starts at $x = 0$ and ends at $x = 1$, and $f(x)$ is a given function that models the heating effect on the rod. The temperature of the two ends of the rod are fixed to zero.

In this problem, we solve the BVP (4.90–4.91) numerically to find its numerical solutions. To do so, we break up the interval $[0, 1]$ in m subintervals of length $h = 1/m$ and solve the temperature at the points $x_i = ih$, $i = 0, 1, \ldots, m$. We approximate the derivatives in (4.90) by the so-called centered finite difference as

$$u''(x_i) = \frac{u(x_{i+1}) - 2u(x_i) + u(x_{i-1})}{h^2} + O(h^2)$$

Thus Eq. (4.90) can be approximated (or say discretized) as

$$\frac{u_{i+1} - 2u_i + u_{i-1}}{h^2} = f(x_i), \quad i = 1, \ldots, m-1$$

where u_i is the approximation of $u(x_i)$. Since the temperature at the two ends is given in (4.91), we have $u_0 = u_m = 0$ and only need to consider the temperature at the interior points. So there are $m - 1$ unknowns u_i, $i = 1, 2, \ldots, m-1$, and there are $m - 1$ equations. In matrix form, the system of the equations can be written as $T\mathbf{u} = \mathbf{f}$, where

$$T = \begin{bmatrix} -2 & 1 & & & \\ 1 & -2 & 1 & & \\ & \ddots & \ddots & \ddots & \\ & & 1 & -2 & 1 \\ & & & 1 & -2 \end{bmatrix}, \quad \mathbf{u} = \begin{bmatrix} u_1 \\ u_2 \\ \vdots \\ u_{m-2} \\ u_{m-1} \end{bmatrix}, \quad \mathbf{f} = h^2 \begin{bmatrix} f(x_1) \\ f(x_2) \\ \vdots \\ f(x_{m-2}) \\ f(x_{m-1}) \end{bmatrix}$$

Since the matrix T is tridiagonal, this tridiagonal system can be solved very efficiently by doing the follows.

(1) Write three MATLAB functions.

The first function

```
function [alpha,beta] = LUofTriDiag(a,b,c)
```

returns α's and β's in the LU factorization (as vectors `alpha` and `beta`) of the following general tridiagonal matrix

$$A = \begin{bmatrix} b_1 & c_1 & & & \\ a_2 & b_2 & c_2 & & \\ & \ddots & \ddots & \ddots & \\ & & a_{n-1} & b_{n-1} & c_{n-1} \\ & & & a_n & b_n \end{bmatrix}$$

which is decomposed as $A = LU$, where L and U are

$$L = \begin{bmatrix} 1 & & & & \\ \alpha_2 & 1 & & & \\ & \ddots & \ddots & \ddots & \\ & & \alpha_{n-1} & 1 & \\ & & & \alpha_n & 1 \end{bmatrix}, \quad U = \begin{bmatrix} \beta_1 & c_1 & & & \\ & \beta_2 & c_2 & & \\ & & \ddots & \ddots & \ddots \\ & & & \beta_{n-1} & c_{n-1} \\ & & & & \beta_n \end{bmatrix}$$

Note that a's, b's and c's in this matrix are passed to your function as input vectors a, b and c.

The second function solves a linear system $L\mathbf{y} = \mathbf{d}$.

The third function solves a linear system $U\mathbf{x} = \mathbf{y}$.

(2) Call your three functions to solve the BVP with $f(x) = \pi^2 \cos(\pi x)$. It is easy to verify that $u(x) = \sin(\pi x)$ is the exact solution. Call your MATLAB functions to find numerical solutions for $m = 10, 20, 40$ and 80. Plot the four numerical solutions as well as the exact solution in one figure. Find the maximum absolute difference

$$\max_i |u(x_i) - u_i|$$

between the numerical solution and the exact solution for each value of m, and then plot them versus the values of $h = 1/m$ with both axes in the log scale in a second figure. What do you observe?

5

Root Finding for Nonlinear Equations

In this chapter, we describe and analyze some fundamental methods for root finding and unconstrained optimization. We highlight Newton's method for root finding and gradient descent for optimization. We analyze order of convergence for Newton's method and secant method.

Let's consider a finance problem that many of us may need to solve at some stage of life. Someone bought a home with most money loaned from a mortgage company charging 4% annual interest rate and requiring equal monthly payment. When he still had 20 years to pay off the loan, he wanted to refinance his loan balance 200,000$ with the term of 20 years. The refinancing fee was 3,000. If 3000$ was invested in a US index mutual fund, an annual 8% return was expected on average. At what refinancing interest rate should he go ahead to do refinancing? (Neglect any investment return due to the difference between monthly payments before and after refinancing.)

We first find the break-even refinance interest rate, denoted as x, at which the refinancing brings no gain and no loss. If the real refinancing interest is lower than x, he should do refinancing to save money.

Denote the monthly payment of the original loan before the refinancing as m_1. The loan would be paid off after 240 months. We have

$$\left\{ \cdots \left[\left(200000 \left(1 + \frac{4\%}{12}\right) - m_1 \right) \underbrace{\left(1 + \frac{4\%}{12}\right)}_{2nd} - m_1 \right] \cdots \right\} \underbrace{\left(1 + \frac{4\%}{12}\right)}_{240th} - m_1 = 0$$

$$\underbrace{\phantom{200000\left(1+\frac{4\%}{12}\right)}}_{1st}$$

(5.1)

which can be reformulated as

$$200000 \left(1 + \frac{4\%}{12}\right)^{240} - m_1 \sum_{k=0}^{239} \left(1 + \frac{4\%}{12}\right)^k = 0 \tag{5.2}$$

So the total payment for the original loan is

$$240 m_1 = 240 \frac{200000 \left(1 + \frac{4\%}{12}\right)^{240}}{\sum_{k=0}^{239} \left(1 + \frac{4\%}{12}\right)^k} \tag{5.3}$$

DOI: 10.1201/9781003201694-5

Denote the monthly payment of the new loan after the refinancing as m_2. Similarly, we obtain the total payment for the new loan as

$$240m_2 = 240\frac{200000\left(1+\frac{x}{12}\right)^{240}}{\sum_{k=0}^{239}\left(1+\frac{x}{12}\right)^k} \tag{5.4}$$

The total return from the investment of the refinancing fee 3000\$ in 20 years is

$$3000\left(1+\frac{8\%}{12}\right)^{20} - 3000 \tag{5.5}$$

To break even the original loan with the refinancing, we require

$$240m_1 = 240m_2 + 3000 + \left[3000\left(1+\frac{8\%}{12}\right)^{20} - 3000\right] \tag{5.6}$$

which is an equation for the break-even interest rate x. The equation for x can be re-written as

$$240\frac{200000\left(1+\frac{4\%}{12}\right)^{240}}{\sum_{k=0}^{239}\left(1+\frac{4\%}{12}\right)^k} = 240\frac{200000\left(1+\frac{x}{12}\right)^{240}}{\sum_{k=0}^{239}\left(1+\frac{x}{12}\right)^k} + 3000\left(1+\frac{8\%}{12}\right)^{20} \tag{5.7}$$

This equation is a nonlinear equation in x that cannot be solved analytically in closed form. In this chapter, we introduce some computational methods to approximate the solutions of such nonlinear equations.

5.1   Roots and Fixed Points

An equation in one unknown variable x with zero right-hand side has the form

$$f(x) = 0 \tag{5.8}$$

A **root** (or solution) of the equation is a number r that satisfies $f(r) = 0$. The number r is thus a **zero** of the function $f(x)$.

Example 5.1 The equation $\ln x = 0$ has one root $r = 1$. The equation $e^x = 0$ has no roots. The equation $x^2 + 1 = 0$ has no real roots. ■

Example 5.2 The quadratic equation $ax^2 + bx + c = 0$, $a \neq 0$ has two roots: $r_1 = \frac{-b+\sqrt{b^2-4ac}}{2a}$, $r_2 = \frac{-b-\sqrt{b^2-4ac}}{2a}$. The two roots are distinct real numbers if the discriminant $b^2 - 4ac > 0$, the same real number if $b^2 - 4ac = 0$, and complex conjugate numbers if $b^2 - 4ac < 0$. ■

Example 5.3 A polynomial equation is an equation of the form

$$c_n x^n + c_{n-1} x^{n-1} + \cdots + c_2 x^2 + c_1 x + c_0 = 0$$

where the function $f(x)$ in Eq. (5.8) is a polynomial $p_n(x)$ of degree $n \geq 1$. By the fundamental theorem of algebra (Theorem 2.1), the polynomial equation $p_n(x) = 0$ has n and only n roots (counting multiplicity) $r_1, r_2, \ldots, r_n$. The **Abel-Ruffini theorem** (also known as Abel's impossibility theorem) states that not all solutions of general polynomial equations of degree five or higher can be obtained by starting with the coefficients and rational constants, and repeatedly forming sums, differences, products, quotients and radicals of previously obtained numbers for a finite number of times. ■

If r is a root of the equation $f(x) = 0$, then r is a zero of the function $f(x)$, and the graph of the function $f(x)$ intersects the x axis at $x = r$, as illustrated in Fig. 5.1. So the geometric interpretation of the root r is the x-intercept of the curve $y = f(x)$.

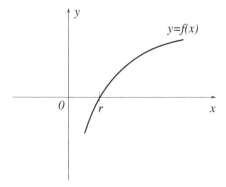

FIGURE 5.1
The root r of $f(x) = 0$ is an x-intercept of the curve $y = f(x)$.

A root r of the equation $f(x) = 0$ (a zero r of the function $f(x)$) has **multiplicity** m ($m \geq 1$) if

$$f(r) = f'(r) = \cdots = f^{(m-1)}(r) = 0, \quad f^{(m)}(r) \neq 0 \qquad (5.9)$$

and it can be shown that

$$f(x) = (x - r)^m g(x), \quad g(r) \neq 0 \qquad (5.10)$$

For $m = 1$, we have $f(r) = 0$ but $f'(r) \neq 0$ (the tangent line to the curve $y = f(x)$ at r is not the horizontal x-axis), and r is called a **simple** root (zero). The root (zero) in Fig. 5.1 is a simple root (zero). For $m \geq 2$, we have $f(r) = 0$ and at least $f'(r) = 0$ (the tangent line to the curve $y = f(x)$ at r is the x-axis), and r is called a **multiple or repeated** root (zero). The larger the multiplicity m of r is, the flatter the curve $y = f(x)$ looks near r.

Example 5.4 Consider the equation $\sin^2 x = 0$. π is a root. Let $f(x) = \sin^2 x$. Then $f(\pi) = f'(\pi) = 0$, but $f''(\pi) \neq 0$. So π is a multiple root of multiplicity 2 (i.e. a root repeated twice, and also called a double root). ∎

Example 5.5 Consider the polynomial in the factorized form

$$(x-1)^{10}$$

and the expanded form

$$x^{10} - 10x^9 + 45x^8 - 120x^7 + 210x^6 - 252x^5 + 210x^4 - 120x^3 + 45x^2 - 10x + 1$$

Fig. 5.2 shows the MATLAB plots of the two forms using 201 points with x between 0.95 and 1.05. The equation $(x-1)^{10} = 0$ (or $x^{10} - 10x^9 + 45x^8 - 120x^7 + 210x^6 - 252x^5 + 210x^4 - 120x^3 + 45x^2 - 10x + 1 = 0$) has the root $r = 1$ with multiplicity 10. It can be expected that it is much more difficult to find such a multiple root with high multiplicity than a simple root. ∎

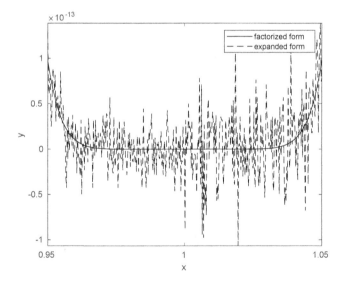

FIGURE 5.2
Plots of two different forms of the polynomial $(x-1)^{10}$.

The root-finding equation $f(x) = 0$ can be reformulated (in various ways) to the form $x = g(x)$, and vice versa. For example $x^2 - 2 = 0$ can be reformulated to $x = 0.5(x + 2/x)$ (and vice versa). A solution x^* of equation $x = g(x)$, i.e. a number x^* that satisfies $x^* = g(x^*)$, is called a **fixed point** of the equation $x = g(x)$. The fixed point x^* can be interpreted as the x-coordinate of the intersection of the line $y = x$ and the curve $y = g(x)$, as illustrated in Fig. 5.3.

Under certain conditions, the intermediate value theorem (IVT) may tell us the existence of a root over an interval.

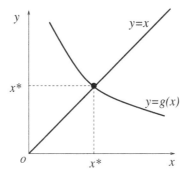

FIGURE 5.3
The fixed point x^* of $x = g(x)$ is the x- (or y-) coordinate of the intersection of the curves $y = x$ and $y = g(x)$.

Theorem 5.1 (Intermediate Value Theorem (IVT)) *If (x) is continuous on $[a,b]$ (i.e. $f \in C_{[a,b]}$), and N is a number between $f(a)$ and $f(b)$, then there must exist at least a number c in (a,b) such that $f(c) = N$ (i.e. $f(x)$ takes all the values between $f(a)$ and $f(b)$ for x between a and b).*

The IVT is illustrated in Fig. 5.4.

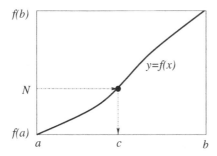

FIGURE 5.4
The illustration of the IVT.

If $f(x)$ is continuous on $[a,b]$ and $f(a)f(b) < 0$, i.e. 0 is between $f(a)$ and $f(b)$, then by the IVT, there exists at least a number $r \in (a,b)$ such that $f(r) = 0$, that is the equation $f(x) = 0$ must have at least a root r in (a,b). The bisection method in the next section is a root finding method based on the IVT to narrow down intervals in which a root lies.

Example 5.6 Show $x - e^{-x} = 0$ has one and only one root.

Solution: Let $f(x) = x - e^{-x}$. We first show the existence of a root.

- $f(x)$ is continuous over any interval.
- $f(0) = -1 < 0$, $f(1) = 1 - 1/e > 0$, so $f(0)f(1) < 0$ (i.e. 0 is between $f(0)$ and $f(1)$).

By the IVT, $f(x) = 0$ must have at least one root r_1 in $(0,1)$.

We then show the uniqueness of the root. To show the root r_1 in $(0,1)$ is the only root of the equation, we use proof by contradiction. Let's make the assumption that $f(x) = 0$ has at least one other root r_2 ($\neq r_1$). Note that $f(x)$ is differentiable everywhere and $f(r_1) = f(r_2) (= 0)$, then by **Rolle's theorem**, there must be a number c between r_1 and r_2 such that $f'(c) = 0$. However, $f'(c) = 1 + e^{-c} > 1$ can never be 0. The contradiction implies that the assumption is not correct, and $f(x) = 0$ cannot have other roots. So $f(x)$ has only one root, the root r_1 in $(0,1)$. ∎

5.2   The Bisection Method

The idea of the bisection method is to repeatedly narrow down an interval in which a root lies according to the IVT.

Let's consider the equation $f(x) = 0$. If $f(x)$ is continuous on an initial interval $[a,b]$ and $f(a)f(b) < 0$, then the interval contains at least one root r of the equation $f(x) = 0$ (and the interval is called a bracket for the root) which can be approximated by the bisection method. As illustrated in Fig. 5.5, the steps of the bisection method in a computer program are given below.

1. Approximate the root r by the middle point $c = (a+b)/2$. Clearly

$$|r - c| \leq (b - a)/2 = b - c \tag{5.11}$$

   i.e. $b - c$ is an **upper bound** of the error in the approximation.

2. If $b - c < \tau$, where τ is the maximum error tolerated by a user, and is called the **error tolerance**, stop the method. Otherwise continue to the next step.

3. If $f(a)$ and $f(c)$ have different signs, the root r is in $[a,c]$, and we assign c to b such that r is in the new $[a,b]$. Otherwise, $f(c)$ and $f(b)$ must have different signs, the root r is in $[c,b]$, and we assign c to a such that r is again in the new $[a,b]$.

4. Go back to Step 1 to repeat the process.

Example 5.7 If the bisection method is programmed to find the zero r of the function $f(x)$ whose graph is shown in Fig. 5.6, the values of a, b, c (current approximation of r), the sign of $f(c)$, and $b - c$ (upper bound for the error of the current approximation) in each of the first four bisections are listed in Table 5.1. ∎

The pseudocode of the bisection method is given below.

In Chapter 4, we solve a linear system of equations $Ax = b$ by Gaussian elimination. Gaussian elimination is a finite sequence of operations that can deliver exact solution in the absence of errors. We therefore call it a **direct method**.

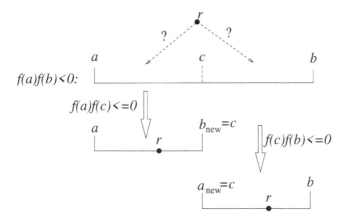

FIGURE 5.5
The illustration of the bisection method, where $c = (a+b)/2$ is the middle point of the interval $[a,b]$ that bisects the interval into two halves.

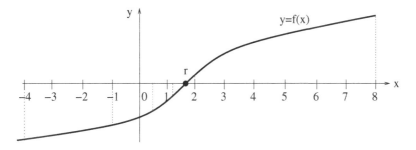

FIGURE 5.6
An example to illustrate the bisection method.

In contrast, the bisection method is an **iterative method**. An iterative method is a procedure that generate an infinite sequence of approximations of the solution of a problem, and a current approximation is generated from the previous ones. Each generation of a new approximation is called an **iteration**. In the bisection method, the sequence of approximations of a root are the middle points, and each iteration is a bisection to generate a new middle point.

Since the iterative procedure can go on and on to generate an infinite sequence, a **stopping criterion** needs to be in place to terminate the procedure. In general, we terminate the procedure when the error of the current approximation is within the given error tolerance (or the allowed maximum number of iterations is reached). Since we do not know the true error (as we do not know the true solution), we derive either an **error estimate** or an **error upper bound** so that we can terminate the procedure when the error estimate or upper bound is within the error tolerance. In the above pesudocode of the bisection method, the upper bound for the error of the current

Bisection Step	a	b	c	sign($f(c)$)	$b-c$
1	-4	8	2	$+$	6
2	-4	2	-1	$-$	3
3	-1	2	0.5	$-$	1.5
4	0.5	2	1.25	$-$	0.75

TABLE 5.1
An example of bisection steps in a computer program.

Algorithm 8 The Bisection Method for solving $f(x) = 0$

$f_a \leftarrow f(a)$
$f_b \leftarrow f(b)$
if $f_a \cdot f_b > 0$ **then**
    display "f(a) and f(b) must have different signs!"
    return
end if
while $b - c > \tau$ **do**
$c \leftarrow (a+b)/2$
$f_c \leftarrow f(c)$
if $f_a \cdot f_c < 0$ **then**
    $b \leftarrow c$
    $f_b \leftarrow f_c$
else
    $a \leftarrow c$
    $f_a \leftarrow f_c$
end if
end while

---

approximation (middle point) is $b - c$, half of the current interval length. We terminate the iterative process when the upper bound is within the error tolerance τ.

The termination criterion can be satisfied if the iterative method is **convergent**, that is the sequence of approximations converges to the true value/solution. The speed of convergence is indicated by convergence order (the order of convergence).

Definition 5.1 (Convergence order) *Suppose   a   sequence $\{x_k\}_{k=0}^{\infty}$ = $\{x_0, x_1, x_2, \ldots\}$ converges   to   the   number r,   i.e. $\lim\limits_{k \to \infty} x_k = r$.   We   say   the order of convergence is p ($p \geq 1$) if*

$$\lim_{k \to \infty} \frac{|r - x_{k+1}|}{|r - x_k|^p} = C, \tag{5.12}$$

where $C > 0$ is a constant and $C < 1$ for $p = 1$.

We say a sequence is

- **linearly convergent** if $p = 1$ (has linear convergence), i.e. $\lim\limits_{k\to\infty} \dfrac{|r - x_{k+1}|}{|r - x_k|} = C < 1$
- **quadratically convergent** if $p = 2$ (has quadratic convergence), i.e. $\lim\limits_{k\to\infty} \dfrac{|r - x_{k+1}|}{|r - x_k|^2} = C$
- **superlinearly convergent** if $1 < p < 2$ (has superlinear convergence).

Example 5.8 The sequence $x_k = 2^{-k}$, $k = 0, 1, 2, \ldots$ converges linearly because $\lim\limits_{k\to\infty} x_k = 0$ and $|0 - x_{k+1}| = 0.5|0 - x_k|$.

The sequence $x_k = 2^{-2^k}$, $k = 0, 1, 2, \ldots$ converges quadratically because $\lim\limits_{k\to\infty} x_k = 0$ and $|0 - x_{k+1}| = |0 - x_k|^2$. ∎

Theorem 5.2 (The Bisection Method Theorem) *If the bisection method is applied to find a root of the equation $f(x) = 0$ in the initial interval $[a, b]$, where $f(x) \in C_{[a,b]}$ and $f(a)f(b) < 0$, then the upper bound of the error in the n-th middle point c_n as approximation of the root r that the method converges to satisfies*

$$|r - c_n| \leq \frac{b - a}{2^n} \tag{5.13}$$

Note that the true error sometimes can even go up after a particular iteration, but it goes down eventually because of the bisection method theorem. This theorem tells us that the upper bound for the error is halved after each iteration. So we may say that the bisection method is overall linearly convergent.

Example 5.9 How many middle points (or bisections) are needed to find a root r by the bisection method such that the error is within 10^{-6} if the initial interval $[a, b] = [-1, 2]$? (Assuming the conditions for the bisection method are satisfied and the method converges to the root r.)

Solution: Let c_n be the n-th middle point. By the bisection method theorem

$$|r - c_n| \leq \frac{b - a}{2^n}$$

So $|r - c_n| \leq 10^{-6}$ if

$$\frac{b - a}{2^n} = \frac{2 - (-1)}{2^n} \leq 10^{-6}$$

which gives $n \geq \log_2(3 \times 10^6) \approx 21.5$. So we need at least 22 middle points to guarantee the satisfaction of the error tolerance. ∎

In summary, the bisection method is an iterative method with guaranteed overall linear convergence if the conditions for the method are satisfied, and the number n of iterations/midpoints can be predicted for a given error tolerance τ as

$$|r - c_n| \le \frac{b-a}{2^n} \le \tau \Rightarrow n \ge \left\lceil \log_2 \frac{b-a}{\tau} \right\rceil \tag{5.14}$$

where $[a,b]$ is the initial interval, and $\lceil x \rceil$ is the ceiling function which returns the smallest integer $N \ge x$ (e.g. $\lceil 21.5 \rceil = 22$).

5.3   Newton's Method

Newton's method, also called Newton-Raphson method, is the most important root-finding method. It is an iterative method that produces a sequence of approximations for a root r of the equation $f(x) = 0$. In each Newton's iteration, a new approximation x_{k+1} is generated from a current approximation x_k ($k = 0, 1, 2, \ldots$), and x_0 is a given initial guess of r to start the procedure.

The idea underlying Newton's method is to approximate the curve $y = f(x)$ by the tangent line $y = L_k(x)$ (i.e. linearization or degree-1 Taylor polynomial) of $f(x)$ at the current x_k. Accordingly, the hard root find problem $f(x) = 0$ is approximated by the easy root finding problem $L_k(x) = 0$, which is solved to produce the next approximation x_{k+1}. This idea is illustrated in Fig. 5.7. Note that when x_k is sufficiently close to r, $L_k(x)$ approximates $f(x)$ well near r and Newton's method can converge to r very fast. Later we will look at the restrictions on x_0 for convergence and the order of convergence. Below we derive **Newton's iteration** formula (a recurrence relation) that gives x_{k+1} using x_k and show an example.

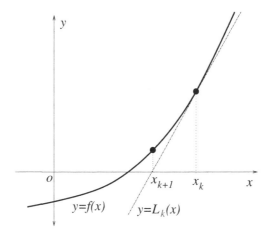

FIGURE 5.7
Illustration of Newton's method.

The tangent line $y = L_k(x)$ (Taylor polynomial of degree 1) to the curve $y = f(x)$ at x_k is

$$y = f(x_k) + f'(x_k)(x - x_k) \tag{5.15}$$

The root of $L_k(x) = 0$ is x_{k+1}. We thus have Newton's iteration formula (provided that $f'(x_k) \neq 0$)

$$x_{k+1} = x_k - \frac{f(x_k)}{f'(x_k)}, \quad k = 0, 1, 2, \ldots \tag{5.16}$$

We can terminate Newton's iteration when the error of the current approximation is within the given error tolerance (if the method converges) or the allowed maximum number of iterations is reached (if the method diverges). Since we do not know the true error (as we do not know the true root), we can derive the following **error estimate**

$$r - x_k \approx x_{k+1} - x_k \tag{5.17}$$

for x_k that is sufficiently close to r, and terminate the iteration when the error estimate $|x_{k+1} - x_k|$ is within the error tolerance τ, i.e. $|x_{k+1} - x_k| \leq \tau$. The derivation of the error estimate is based on the mean value theorem (the MVT) as

$$f(r) - f(x_k) = f'(c_k)(r - x_k) \Rightarrow r - x_k = -\frac{f(x_k)}{f'(c_k)} \approx -\frac{f(x_k)}{f'(x_k)} = x_{k+1} - x_k$$

where c_k is between r and x_k, and the approximation above is valid if x_k (and thus c_k) is sufficiently close to r.

At the root r, we have $f(r) = 0$. If the approximation $x_k \neq r$, we have $f(x_k) \neq 0$. The difference $f(r) - f(x_k) = -f(x_k)$ is called the **residual** for the approximation x_k. When x_k converges to r, its residual converges to 0. However, we should be very cautious to stop the iteration procedure by satisfying the residual criterion $|f(x_k)| \leq \delta$, where δ is a residual tolerance, as warned by the following example.

Example 5.10 $x_k = 1 + 10^{-5}$ is a very good approximation of the root $r = 1$ of the equation $10^{10}(x - 1) = 0$, but the residual for the approximation has a very large amplitude 10^5. On the other hand, $x_k = 1 + 10^5$ is a very bad approximation of the root $r = 1$ of the equation $10^{-10}(x - 1) = 0$, but the residual for the approximation has a very small amplitude 10^{-5}. ∎

The pseudocode of Newton's method is given below

Example 5.11 Solve $x^6 - x - 1 = 0$ by conducting 5 Newton's iterations.
[**Solution:**] Let $f(x) = x^6 - x - 1$. Then $f'(x) = 6x^5 - 1$. Newton's iteration formula for this root-finding problem is

$$x_{k+1} = x_k - \frac{x_k^6 - x_k - 1}{6x_k^5 - 1}, \quad k = 0, 1, 2, \ldots$$

By the IVT, the equation has a root in $(1, 2)$ (as $f(1)f(2) < 0$). (The true root of this equation in this interval is $r = 1.1347241384 \cdots$). So we choose $x_0 = 1.5$ as the initial guess to start Newton's iteration. The results of the 5 iterations (from MATLAB) are

---

Algorithm 9 Newton's Method for solving $f(x) = 0$

$r \leftarrow x_0$
for k from 1 to n_{max} **do**
   $E_{est} \leftarrow f(r)/f'(r)$
   if $|E_{est}| \leq \tau$ **then**
      return r
   end if
   $r \leftarrow r - E_{est}$
end for

```
k   x_k                |r-x_k|        |r-x_k|/|r-x_k-1|^2
--------------------------------------------------------
0   1.50000000000000   3.652759e-01   -
1   1.30049088359046   1.657667e-01   1.2424
2   1.18148041640293   4.675628e-02   1.7016
3   1.13945559027553   4.731452e-03   2.1643
4   1.13477762523711   5.348684e-05   2.3892
5   1.13472414531622   6.914699e-09   2.4170
6   1.13472413840152   2.220446e-16   4.6440
```

■

> **Remark**   In the above example, the ratio $|r - x_k|/|r - x_{k-1}|^2$ is about 2.4 at $k = 4$ and $k = 5$, indicating quadratic convergence (see analysis below). Accordingly, the number of correct digits in x_k almost doubles at each of these two iterations $k = 4$ and $k = 5$. However, the ratio increases to about 4.6 at the iteration $k = 6$. Do you know why?

5.3.1   Convergence Analysis of Newton's Method

The convergence analysis of an iterative method establishes how the error of the next approximation x_{k+1} is related to the errors in previous approximations. In Newton's method, the approximation x_{k+1} of a root r is related with the approximation x_k by Newton's iteration formula, and so we can establish the relation between the errors $|r - x_{k+1}|$ and $|r - x_k|$.

> **Theorem 5.3 (Newton's Method Theorem)** *Let r be a simple root of $f(x) = 0$ (i.e. $f(r) = 0$ but $f'(r) \neq 0$), where $f(x) \in C^2$ near r. Then there is a positive value δ such that if an initial guess x_0 satisfies $|r - x_0| < \delta$ (x_0 is close to r within*

δ), then $|r - x_k| < \delta$ for $k = 1, 2, 3, \ldots$ *(xk stays close to r within δ) in Newton's method, and the method converges to r quadratically as*

$$\lim_{k \to \infty} \frac{|r - x_{k+1}|}{|r - x_k|^2} = \frac{|f''(r)|}{2|f'(r)|} \tag{5.18}$$

Proof *Newton's iteration formula is*

$$x_{k+1} = x_k - \frac{f(x_k)}{f'(x_k)}, \quad k = 0, 1, 2, \ldots \tag{5.19}$$

So we have

$$r - x_{k+1} = r - x_k + \frac{f(x_k)}{f'(x_k)} = \frac{f'(x_k)(r - x_k) + f(x_k)}{f'(x_k)} \tag{5.20}$$

Realizing that the numerator $f'(x_k)(r - x_k) + f(x_k)$ above is the degree-1 Taylor polynomial of $f(x)$ about x_k evaluated at $x = r$, we apply Taylor's theorem and get

$$0 = f(r) = [f(x_k) + f'(x_k)(r - x_k)] + \frac{f''(\xi_k)}{2!}(r - x_k)^2 \tag{5.21}$$

where ξ_k is between x_k and r. So we have

$$r - x_{k+1} = -\frac{f''(\xi_k)}{2f'(x_k)}(r - x_k)^2 \tag{5.22}$$

Define $C(\delta)$ as

$$C(\delta) = \frac{\max\limits_{|r-x| \leq \delta} |f''(x)|}{2 \min\limits_{|r-x| \leq \delta} |f'(x)|} \tag{5.23}$$

where δ is a positive value, and $C(\delta)$ satisfies

$$\lim_{\delta \to 0} C(\delta) = \frac{|f''(r)|}{2|f'(r)|} \tag{5.24}$$

which is finite as $f'(r) \neq 0$. If $|r - x_k| \leq \delta$, then $|r - \xi_k| < \delta$ too (as ξ_k is between x_k and r) and

$$|r - x_{k+1}| \leq C(\delta)|r - x_k|^2 \leq C(\delta)\delta|r - x_k| \tag{5.25}$$

Let $\rho = C(\delta)\delta$, then

$$|r - x_{k+1}| \leq \rho|r - x_k| \tag{5.26}$$

Note that

$$\lim_{\delta \to 0} \rho = \lim_{\delta \to 0} C(\delta)\delta = 0 \tag{5.27}$$

So we can choose δ small enough such that $\rho < 1$, then if $|r - x_0| < \delta$, we have

$$|r - x_1| \le \rho |r - x_0| < \rho \delta < \delta$$
$$|r - x_2| \le \rho |r - x_1| \le \rho^2 |r - x_0| < \rho^2 \delta < \delta$$
$$\cdots\cdots$$

$$|r - x_{k+1}| \le \rho |r - x_k| \le \rho^2 |r - x_{k-1}| \le \cdots \le \rho^{k+1} |r - x_0| < \rho^{k+1}\delta < \delta \tag{5.28}$$

and

$$0 \le \lim_{k \to \infty} |r - x_{k+1}| < \lim_{k \to \infty} \rho^{k+1}\delta = 0 \tag{5.29}$$

By the squeeze theorem, we have

$$\lim_{k \to \infty} |r - x_{k+1}| = 0 \tag{5.30}$$

so $\lim_{k \to \infty} x_k = \lim_{k \to \infty} \xi_k = r$, and we finally obtain

$$\lim_{k \to \infty} \frac{|r - x_{k+1}|}{|r - x_k|^2} = \lim_{k \to \infty} \frac{|f''(\xi_k)|}{2|f'(x_k)|} = \frac{|f''(r)|}{2|f'(r)|} \tag{5.31}$$

■

Example 5.12 Now let's look back at Example 5.11 about the root finding problem $f(x) = 0$ with $f(x) = x^6 - x - 1$. The true root is $r = 1.1347241384\cdots$, which is a simple root as $f'(r) = 6r^5 - 1 \ne 0$. We have $|f''(r)|/(2|f'(r)|) \approx 2.42$, which is about the ratios $|r - x_k|/|r - x_{k-1}|^2$ at $k = 4$ and 5 in the demo of Example 5.11. Since when x_k is close to r, we have the error relation

$$|r - x_k| \approx \frac{|f''(r)|}{2|f'(r)|}|r - x_{k-1}|^2 \approx 2.42|r - x_{k-1}|^2$$

so the relative error relation is

$$\frac{|r - x_k|}{|r|} \approx \frac{|rf''(r)|}{2|f'(r)|}\frac{|r - x_{k-1}|^2}{|r|^2} \approx 2.74\frac{|r - x_{k-1}|^2}{|r|^2}$$

Suppose x_{k-1} has about t correct decimal digits, i.e. $|r - x_{k-1}|/|r| \approx 10^{-t}$, then $|r - x_k|/|r| \approx 2.74 \times 10^{-2t} = 10^{\log_{10} 2.74 - 2t} \approx 10^{-2t}$ when t is large enough relative to $\log_{10} 2.74$. So x_k has about $2t$ correct decimal digits. This explains the approximate doubling of correct decimal digits at each late iteration in the demo. However, the ratio $|r - x_k|/|r - x_{k-1}|^2$ at $k = 6$ is increased to about 4.64 because the error at $k = 6$ is reduced to the level of roundoff error and cannot be further reduced. ■

Newton's iteration is a fixed point iteration. We will introduce fixed point iterations later, and we will look back at Newton's iteration then from the perspective of fixed point iterations.

Newton's method can also be applied to solve a system of nonlinear equations. More details will be given later in this chapter.

5.3.2   Practical Issues of Newton's Method

Practical issues of Newton's method include
- How/when to stop Newton's iteration?
- How to choose the initial guess x_0?
- What happens if the root r is multiple (not simple)?

For the first issue, we can stop Newton's iteration when the error estimate $|x_{k+1} - x_k|$ $(\approx |r - x_k|)$ is within the error tolerance τ (if the iteration converges) or when the maximum number of iterations is reached (if the iteration diverges).

As Newton's method theorem indicates, the initial guess x_0 must be within distance δ from the true root r to guarantee the convergence. The distance δ is problem dependent and may be large or small. In practice, x_0 may be chosen according to (1) the mathematical analysis of the equation, (2) the physics of the problem and (3) the graph of the residual function $f(x)$. We may combine the bisection method with Newton's method by using the bisection method first to generate a good x_0 for Newton's method.

If the root r is a multiple root with multiplicity $m \geq 2$ (i.e. $f(r) = f'(r) = \cdots f^{(m-1)}(r) = 0$ but $f^{(m)}(r) \neq 0$, or $f(x) = (x-r)^m g(x)$ with $g(r) \neq 0$), then Newton's method converges only linearly (if it converges). If we know the multiplicity m of the root in advance, we can accelerate Newton's method by the modified recurrence relation

$$x_{k+1} = x_k - m\frac{f(x_k)}{f'(x_k)}, \quad k = 0, 1, 2, \ldots \tag{5.32}$$

5.4   Secant Method

In each Newton's iteration k, there are two function evaluations $f(x_k)$ and $f'(x_k)$. In practice, $f'(x)$ may not be readily available. The secant method avoids the use of derivative $f'(x_k)$ by approximating $f(x)$ by a secant line $y = S_k(x)$ through the two points on the curve $y = f(x)$ at x_k and x_{k-1} (instead of a tangent line at x_k in Newton's method). Accordingly, the secant iteration formula can be obtained from Newton's iteration formula by replacing the slope of the tangent line, $f'(x_k)$, with the slope of the secant line, $(f(x_k) - f(x_{k-1}))/(x_k - x_{k-1})$. The secant iteration formula is

$$x_{k+1} = x_k - \frac{f(x_k)}{f(x_k) - f(x_{k-1})}(x_k - x_{k-1}), \quad k = 1, 2, 3, \ldots \tag{5.33}$$

Note that x_{k+1} is the root of the equation $S_k(x) = f(x_k) + \dfrac{f(x_k) - f(x_{k-1})}{x_k - x_{k-1}}(x - x_k) = 0$. To start the secant iteration, we need two initial guesses x_0 and x_1. An illustration of the secant method is given in Fig. 5.8. We also have the error estimate

$$|r - x_k| \approx |x_{k+1} - x_k| \tag{5.34}$$

when x_k is sufficiently close to r.

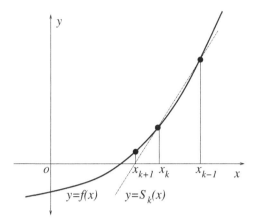

FIGURE 5.8
Illustration of the secant method.

Algorithm 10 Secant Method for solving $f(x) = 0$

$f_0 \leftarrow f(x_0)$
$f_1 \leftarrow f(x_1)$
for k from 1 to $n_{\max}$ **do**
    $E_{\text{est}} \leftarrow f_1 \cdot (x_1 - x_0)/(f_1 - f_0)$
    if $|E_{\text{est}}| \leq \tau$ **then**
        return x_1
    end if
    $x_0 \leftarrow x_1$
    $f_0 \leftarrow f_1$
    $x_1 \leftarrow x_1 - E_{\text{est}}$
    $f_1 \leftarrow f(x_1)$
end for

Here is the pseudocode for the secant method.

Example 5.13 Solve $x^6 - x - 1 = 0$ by conducting 8 secant iterations.
[Solution:] Let $f(x) = x^6 - x - 1$. The secant iteration formula for this root-finding problem is

$$x_{k+1} = x_k - \frac{x_k^6 - x_k - 1}{(x_k^6 - x_k - 1) - (x_{k-1}^6 - x_{k-1} - 1)}(x_k - x_{k-1}), \quad k = 1, 2, 3, \ldots$$

By the IVT, the equation has a root in $(1,2)$ (as $f(1)f(2) < 0$). (The true root of this equation in the interval $(1,2)$ is $r = 1.1347241384\cdots$). So we choose $x_0 = 1.5$ and $x_1 = 1.45$ as the initial two guesses to start the secant iteration. The results of the 8 iterations (from MATLAB) are

```
k   x_k                 |r-x_k|        |r-x_k|/|r-x_k-1|^1.618
----------------------------------------------------------------
0   1.50000000000000    3.652759e-01   -
1   1.45000000000000    3.152759e-01   -
2   1.28278575837153    1.480616e-01   0.9584
3   1.20499752284230    7.027338e-02   1.5453
4   1.15440124329354    1.967710e-02   1.4450
5   1.13772538414722    3.001246e-03   1.7285
6   1.13486295903299    1.388206e-04   1.6756
7   1.13472514162956    1.003228e-06   1.7495
8   1.13472413873812    3.366010e-10   1.7094
9   1.13472413840152    8.881784e-16   1.8864
```

■

Remark   In the above example, the ratio $|r - x_k|/|r - x_{k-1}|^p$, where $p = (\sqrt{5} + 1)/2 \approx 1.618$, is about 1.7 at $k = 5, 6, 7$ and 8, indicating superlinear convergence.

Theorem 5.4 (Secant Method Theorem) *Let r be a simple root of $f(x) = 0$ (i.e. $f(r) = 0$ but $f'(r) \neq 0$), where $f(x) \in C^2$ near r. Then there is a positive value δ such that if the initial guesses x_0 and x_1 satisfies $|r - x_0| < \delta$ and $|r - x_1| < \delta$, then $|r - x_k| < \delta$ for $k = 2, 3, 4, \ldots$ in the secant method, and the method converges to r superlinearly as*

$$\lim_{k \to \infty} \frac{|r - x_{k+1}|}{|r - x_k|^p} = \left| \frac{f''(r)}{2f'(r)} \right|^{p-1}, \quad p = \frac{\sqrt{5} + 1}{2} \tag{5.35}$$

Proof *The secant iteration formula is*

$$x_{k+1} = x_k - \frac{f(x_k)}{f(x_k) - f(x_{k-1})}(x_k - x_{k-1}), \quad k = 1, 2, 3, \ldots \tag{5.36}$$

We use this formula to establish the error relation among $|r - x_{k+1}|$, $|r - x_k|$ and $|r - x_{k-1}|$ as follows. We have

$$r - x_{k+1} = r - x_k + \frac{f(x_k)}{f(x_k) - f(x_{k-1})}(x_k - x_{k-1}) \tag{5.37}$$

which can be rewritten as (after some algebraic manipulation)

$$r - x_{k+1} = \frac{\left[\frac{\frac{f(x_k)}{r-x_k} - \frac{f(x_{k-1})}{r-x_{k-1}}}{x_k - x_{k-1}}\right]}{\left[\frac{f(x_k) - f(x_{k-1})}{x_k - x_{k-1}}\right]} (r - x_k)(r - x_{k-1}) = \frac{\left[\frac{g(x_k) - g(x_{k-1})}{x_k - x_{k-1}}\right]}{\left[\frac{f(x_k) - f(x_{k-1})}{x_k - x_{k-1}}\right]} (r - x_k)(r - x_{k-1})$$

(5.38)

where $g(x) = f(x)/(r-x)$. By the mean value theorem (MVT), we have

$$\frac{g(x_k) - g(x_{k-1})}{x_k - x_{k-1}} = g'(\eta_k), \quad \frac{f(x_k) - f(x_{k-1})}{x_k - x_{k-1}} = f'(\zeta_k)$$

(5.39)

where both η_k and ζ_k are between x_{k-1} and x_k, and the derivative of $g(x)$ is

$$g'(x) = \frac{f(x) + f'(x)(r-x)}{(r-x)^2}$$

(5.40)

By Taylor's theorem (treating x as the number and r as the variable), we have

$$0 = f(r) = f(x) + f'(x)(r-x) + \frac{f''(z)}{2!}(r-x)^2$$

(5.41)

where $z = z(x)$ is between x and r. So we get

$$g'(x) = -\frac{f''(z(x))}{2}, \quad g'(\eta_k) = -\frac{f''(z(\eta_k))}{2}$$

(5.42)

where $z(\eta_k)$ is between η_k (which is between x_{k-1} and x_k) and r, that is $z(\eta_k)$ is between $\min\{r, x_{k-1}, x_k\}$ and $\max\{r, x_{k-1}, x_k\}$. Finally, we obtain

$$r - x_{k+1} = -\frac{f''(z(\eta_k))}{2f'(\zeta_k)}(r - x_k)(r - x_{k-1})$$

(5.43)

For a positive value δ, we define $C(\delta)$ as

$$C(\delta) = \frac{\max\limits_{|r-x| \le \delta} |f''(x)|}{2 \min\limits_{|r-x| \le \delta} |f'(x)|}$$

(5.44)

Note that

$$\lim_{\delta \to 0} C(\delta) = \frac{|f''(r)|}{2|f'(r)|}$$

(5.45)

which is finite as $f'(r) \ne 0$. For $|r - x_{k-1}| < \delta$ and $|r - x_k| < \delta$, we then have

$$|r - x_{k+1}| \le C(\delta)|r - x_k||r - x_{k-1}|$$

(5.46)

Define

$$\rho = C(\delta)\delta \tag{5.47}$$

Since $\lim_{\delta \to 0} C(\delta) = 0$, *we can choose* δ *small enough so that*

$$\rho < 1 \tag{5.48}$$

If

$$|r - x_0| < \delta, \quad |r - x_1| < \delta \tag{5.49}$$

then

$$C(\delta)|r - x_0| < \rho = \rho^{F_0} < 1, \quad F_0 = 1$$
$$C(\delta)|r - x_1| < \rho = \rho^{F_1} < 1, \quad F_1 = 1$$
$$C(\delta)|r - x_2| < C(\delta)|r - x_1|C(\delta)|r - x_0| < \rho^{F_1+F_0} = \rho^{F_2} < 1, \quad F_2 = F_1 + F_0 = 2$$
$$C(\delta)|r - x_3| < C(\delta)|r - x_2|C(\delta)|r - x_1| < \rho^{F_2+F_1} = \rho^{F_3} < 1, \quad F_3 = F_2 + F_1 = 3$$
$$\ldots\ldots$$
$$C(\delta)|r - x_{k+1}| < C(\delta)|r - x_k|C(\delta)|r - x_{k-1}| < \rho^{F_k+F_{k-1}} = \rho^{F_{k+1}} < 1, F_{k+1} = F_k + F_{k-1}$$

$$\tag{5.50}$$

and we have

$$|r - x_k| < \rho^{F_k-1}\delta < \delta, \quad k = 2,3,4,\ldots \tag{5.51}$$

$$\lim_{k \to \infty} |r - x_k| = 0, \quad \text{i.e. } \lim_{k \to \infty} x_k = r \tag{5.52}$$

The sequence $\{F_0, F_1, F_2, F_3, \ldots\} = \{1, 1, 2, 3, \ldots\}$ *satisfying* $F_{k+1} = F_k + F_{k-1}$ $(k = 1, 2, 3, \ldots)$ *is the famous Fibonacci sequence with*

$$F_k = \frac{1}{\sqrt{5}} \left[\left(\frac{1+\sqrt{5}}{2} \right)^{k+1} - \left(\frac{1-\sqrt{5}}{2} \right)^{k+1} \right], \quad k = 0, 1, 2, 3, \ldots \tag{5.53}$$

A remark is made after the proof about the derivation of the formula.
Now we want to show that

$$\lim_{k \to \infty} \frac{|r - x_{k+1}|}{|r - x_k|^p} = \left| \frac{f''(r)}{2f'(r)} \right|^{p-1}, \quad p = \frac{\sqrt{5}+1}{2} \tag{5.54}$$

We have already shown that

$$r - x_{k+1} = -\frac{f''(z(\eta_k))}{2f'(\zeta_k)}(r - x_k)(r - x_{k-1}) \tag{5.55}$$

When k is sufficiently large, both x_{k-1} and x_k are very close to r, and we have

$$|r - x_{k+1}| \approx \frac{|f''(r)|}{2|f'(r)|}|r - x_k||r - x_{k-1}| \tag{5.56}$$

which implies a power law relation between two consecutive errors as

$$|r - x_{k+1}| \approx c|r - x_k|^p, \quad |r - x_k| \approx c|r - x_{k-1}|^p \tag{5.57}$$

where c and p are positive constants. We thus have

$$c|r - x_k|^p \approx \frac{|f''(r)|}{2|f'(r)|}|r - x_k|\left(\frac{|r - x_k|}{c}\right)^{1/p} \tag{5.58}$$

which gives

$$p = 1 + \frac{1}{p} \Rightarrow p = \frac{\sqrt{5}+1}{2} \tag{5.59}$$

$$c = \left(\frac{|f''(r)|}{2|f'(r)|}\right)^{p/(p+1)} = \left(\frac{|f''(r)|}{2|f'(r)|}\right)^{p-1} = \left(\frac{|f''(r)|}{2|f'(r)|}\right)^{(\sqrt{5}-1)/2} \tag{5.60}$$

and finally we obtain

$$\lim_{k\to\infty} \frac{|r - x_{k+1}|}{|r - x_k|^p} = \left|\frac{f''(r)}{2f'(r)}\right|^{p-1}, \quad p = \frac{\sqrt{5}+1}{2} \tag{5.61}$$

■

Remark The general term F_k ($k = 0, 1, 2, ...$) in the Fibonacci sequence $\{F_0, F_1, F_2, F_3, ...\} = \{1, 1, 2, 3, ...\}$ can be derived in different ways. One way is using linear algebra by writing the recurrence relation $F_{k+1} = F_k + F_{k-1}$ ($k = 1, 2, 3, ...$) as a discrete dynamical system

$$\begin{bmatrix} F_{k+1} \\ F_k \end{bmatrix} = \begin{bmatrix} 1 & 1 \\ 1 & 0 \end{bmatrix} \begin{bmatrix} F_k \\ F_{k-1} \end{bmatrix} \tag{5.62}$$

which gives

$$\begin{bmatrix} F_{k+1} \\ F_k \end{bmatrix} = \begin{bmatrix} 1 & 1 \\ 1 & 0 \end{bmatrix}^k \begin{bmatrix} F_1 \\ F_0 \end{bmatrix} \tag{5.63}$$

Using the diagonalization of the matrix in the discrete dynamical system, we can obtain F_k.

5.5   Fixed-Point Iteration

Newton's iteration formula for finding a root of the univariate equation $f(x) = 0$ is

$$x_{k+1} = x_k - \frac{f(x_k)}{f'(x_k)}, \quad k = 0, 1, 2, \ldots \tag{5.64}$$

The formula has the form $x_{k+1} = g(x_k)$, where

$$g(x) = x - \frac{f(x)}{f'(x)} \tag{5.65}$$

If the iteration $x_{k+1} = g(x_k)$ converges to the number x^* (i.e. $\lim_{k \to \infty} x_k = x^*$) and $f'(x^*) \neq 0$, then $x^* = g(x^*)$ and x^* is a fixed point of the equation $x = g(x)$, and x^* is also a simple root of the equation $f(x) = 0$ by

$$x^* = x^* - \frac{f(x^*)}{f'(x^*)} \Leftrightarrow f(x^*) = 0 \tag{5.66}$$

Example 5.14 Consider the equation $f(x) = 0$ with $f(x) = 2x - 2$:

$$2x - 2 = 0$$

(1) By splitting $2x$ as $4x - 2x$ and leaving the first term $4x$ at the left-hand side, the equation becomes $4x = 2x + 2$ and can be reformulated as

$$x = \frac{1}{2}x + \frac{1}{2}$$

which has the form $x = g(x)$ with $g(x) = \frac{1}{2}x + \frac{1}{2}$.
(2) By splitting $2x$ as $-2x + 4x$ and leaving the first term $-2x$ at the left-hand side, the equation becomes $-2x = -4x + 2$ and can be reformulated as

$$x = 2x - 1$$

which has the form $x = g(x)$ with $g(x) = 2x - 1$. ∎

In general, the root finding problem $f(x) = 0$ can be formulated as a fixed point finding problem $x = g(x)$ in infinitely many ways (with different $g(x)$), and the iteration starting with an initial guess x_0

$$x_{k+1} = g(x_k), \quad k = 0, 1, 2, \ldots \tag{5.67}$$

is called a **fixed-point iteration**. If the iteration converges to a fixed point x^* of $x = g(x)$, the fixed point x^* is a root of $f(x) = 0$. However, not all fixed-point iteration converges.

Example 5.15 From the previous example, the equation $2x - 2 = 0$ can be formulated as the fixed-point problems (1) $x = \frac{1}{2}x + \frac{1}{2}$ and (2) $x = 2x - 1$. Both fixed-point problems have the unique fixed point $x^* = 1$, which is the unique solution of $2x - 2 = 0$. Starting with $x_0 = 0$, the fixed point iteration $x_{k+1} = \frac{1}{2}x_k + \frac{1}{2}$ for the fixed-point problem (1) produces the sequence $\{x_k\}_{k=0}^{\infty} = \{0, 1/2, 3/4, 7/8, 15/16, \ldots\}$ with $x_k = \frac{2^k-1}{2^k}$, which converges to the fixed point $x^* = 1$ as $k \to \infty$. Starting with $x_0 = 0$, the fixed-point iteration $x_{k+1} = 2x_k - 1$ produces the sequence $\{x_k\}_{k=0}^{\infty} = \{0, -1, -3, -7, -15, \ldots\}$ with $x_k = 1 - 2^k$, which diverges. ∎

The fixed-point iteration $x_{k+1} = g(x_k)$ can be visualized as in Fig. 5.9. The fixed point iterations in the previous example are illustrated in Fig. 5.10. It seems that the convergence of a fixed point iteration in the example is related to the slope of the line $y = g(x)$.

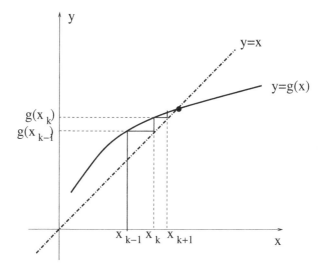

FIGURE 5.9
Illustration of a fixed point iteration.

Theorem 5.5 (Fixed-Point Iteration Theorem) *If $g(x) \in C_{[a,b]}$ and $g(x)$ is between a and b (i.e. $a \leq g(x) \leq b$) for any $x \in [a,b]$, then (1) there must exist a fixed point x^* in the interval $[a,b]$; and (2) the fixed point iteration $x_{k+1} = g(x_k)$ converges to the fixed point x^* if $x_0 \in [a,b]$ and $g'(x)$ exists with $|g'(x)| \leq \rho$, where $\rho < 1$ is a nonnegative constant.*

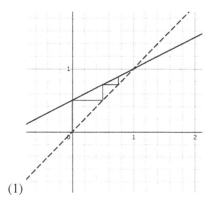 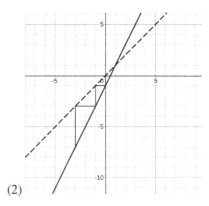

(1) (2)

FIGURE 5.10
Fixed point iteration starting with $x_0 = 0$: (1) Convergent $x_{k+1} = \frac{1}{2}x_k + \frac{1}{2}$; (2) Divergent $x_{k+1} = 2x_k - 1$.

Remark The condition $|g'(x)| \leq \rho < 1$ states that the magnitude of the slopes of the tangent lines to the curve $y = g(x)$ is less than 1, which is consistent with our illustrations in Fig. 5.10.

Remark Noticing the continuity of $g(x)$ and $a \leq g(x) \leq b$ for x in $[a, b]$, the existence of the fixed point x^* in $[a, b]$ can be shown by sketching the graphs of $y = x$ and $y = g(x)$ in the window $(x, y) \in [a, b] \times [a, b]$.

Proof Let $f(x) = x - g(x)$. Then $f(a) = a - g(a) \leq 0$ and $f(b) = b - g(b) \geq 0$. *By the intermediate value theorem. There exists x^* in $[a, b]$ such that $f(x^*) = 0$, i.e. $x^* = g(x^*)$ and x^* is a fixed point.*
We have

$$x^* = g(x^*) \tag{5.68}$$

and

$$x_{k+1} = g(x_k) \tag{5.69}$$

The subtraction of the above two equations gives

$$x^* - x_{k+1} = g(x^*) - g(x_k) \tag{5.70}$$

from which we can determine with the mean value theorem how the distance from x_{k+1} to the target x^* is related with the distance from x_k to x^* as

$$|x^* - x_{k+1}| = |g(x^*) - g(x_k)| = |g'(c_k)||x^* - x_k| \tag{5.71}$$

where c_k is a number between x_k and x^*.

If $x_0 \in [a,b]$, then $|x^* - x_1| = |g'(c_0)||x^* - x_0|$ with $c_0 \in [a,b]$ and $|g'(c_0)| \leq \rho < 1$. So we have

$$|x^* - x_1| \leq \rho|x^* - x_0| < |x^* - x_0|$$

and $x_1 \in [a,b]$. Similarly, we can show x_2, x_3, ... are in $[a,b]$ and

$$|x^* - x_2| \leq \rho|x^* - x_1| \leq \rho^2|x^* - x_0|$$

$$|x^* - x_3| \leq \rho|x^* - x_2| \leq \rho^3|x^* - x_0|$$

$$\cdots$$

$$|x^* - x_{k+1}| \leq \rho|x^* - x_k| \leq \rho^{k+1}|x^* - x_0| \tag{5.72}$$

As $0 \leq \rho < 1$, $\lim\limits_{k\to\infty} \rho^{k+1} = 0$ and

$$\lim\limits_{k\to\infty} |x^* - x_{k+1}| = \lim\limits_{k\to\infty} \rho^{k+1}|x^* - x_0| = 0 \tag{5.73}$$

So the iteration converges to the fixed point x^*. ∎

Remark Under all the conditions in the theorem, the fixed point x^* is unique because $f(x) = x - g(x)$ is increasing and intersect the x-axis only once in $[a,b]$ by $f'(x) = 1 - g'(x) > 0$ in $[a,b]$.

Remark The number ρ in the theorem is called the **contraction factor**. It is related to the speed of convergence.

Remark We can follow the similar ideas to construct iterative methods for solving a linear system and analyze the convergence of the methods (see Chapter 10).

Newton's iteration is a fixed point iteration $x_{k+1} = g(x_k)$ with $g(x) = x - f(x)/f'(x)$. It can be shown that

$$g'(x) = \frac{f(x)f''(x)}{[f'(x)]^2} \tag{5.74}$$

$$g''(x) = \frac{f''(x)}{f'(x)} - f(x)\frac{f'(x)f'''(x) - 2[f''(x)]^2}{[f'(x)]^3} \tag{5.75}$$

If the iteration converges to a simple root x^* of the equation $f(x) = 0$, then

$$g'(x^*) = 0, \quad g''(x^*) = \frac{f''(x^*)}{f'(x^*)} \tag{5.76}$$

We can reveal the quadratic convergence of the iteration as follows. We know

$$x^* - x_{k+1} = g(x^*) - g(x_k) \tag{5.77}$$

By Taylor's theorem, we have

$$g(x_k) = g(x^*) + g'(x^*)(x_k - x^*) + \frac{g''(c_k)}{2!}(x_k - x^*)^2 \tag{5.78}$$

where c_k is between x^* and x^k, and we obtain

$$x^* - x_{k+1} = g(x^*) - g(x_k) = -\frac{g''(c_k)}{2!}(x_k - x^*)^2 \tag{5.79}$$

By $\lim_{k\to\infty} x_k = x^*$ (the iteration converges to x^*), we finally have

$$\lim_{k\to\infty} \frac{|x^* - x_{k+1}|}{|x^* - x_k|^2} = \lim_{k\to\infty} \frac{|g''(c_k)|}{2!} = \frac{|f''(x^*)|}{|2f'(x^*)|} \tag{5.80}$$

as stated in Newton's method theorem.

5.6   Newton's Method for Systems of Nonlinear Equations

Consider a system of n nonlinear equations in n unknowns $x_1, x_2, ..., x_n$

$$\begin{cases} f_1(x_1, x_2, ..., x_n) & = & 0 \\ f_2(x_1, x_2, ..., x_n) & = & 0 \\ \quad\vdots \\ f_n(x_1, x_2, ..., x_n) & = & 0 \end{cases} \tag{5.81}$$

Define the vectors $\mathbf{x} = [x_1, x_2, ..., x_n]^T$ and $\mathbf{f} = [f_1, f_2, ..., f_n]^T$. We can write in the vector form for each function $f_i(x_1, x_2, ..., x_n)$ $(i = 1, 2, ..., n)$ as $f_i(\mathbf{x})$ and the system as

$$\mathbf{f}(\mathbf{x}) = \mathbf{0} \tag{5.82}$$

Let $\mathbf{r} = [r_1, r_2, ..., r_n]^T$ be a solution of the system. As in solving a scalar nonlinear equation $f(x) = 0$, Newton's method to find the solution $\mathbf{r}$ of the nonlinear system constructs a linear system to approximate the nonlinear system and solves the linear system to obtain an updated approximate solution. The construction of the linear system is based on Taylor's theorem for multivariate functions.

5.6.1   Taylor's Theorem for Multivariate Functions

Theorem 5.6 (Taylor's Theorem for Multivariate Functions) *Let $f(\mathbf{x})$ be a function that maps the point $\mathbf{x} = [x_1, x_2, \dots, x_n] \in \mathbb{U}$ in n dimensions to a scalar $f(\mathbf{x}) \in \mathbb{R}$, where $\mathbb{U}$ is a convex open set of $\mathbb{R}^n$. Suppose f has continuous partial derivatives of all orders up to $m+1$ for $\mathbf{x} \in \mathbb{U}$. Let $\mathbf{a} = [a_1, a_2, \dots, a_n]^T$ be a point in $\mathbb{U}$. Then for any $\mathbf{x} \in \mathbb{U}$,*

$$f(\mathbf{x}) = T_m(\mathbf{x}) + R_m(\mathbf{x}) \tag{5.83}$$

where the Taylor polynomial $T_m(\mathbf{x}) =$

$$\sum_{l_1 + l_2 + \cdots + l_n = l \leq m} \frac{1}{l_1! l_2! \cdots l_n!} \frac{\partial^l f(\mathbf{a})}{\partial x_1^{l_1} \partial x_2^{l_2} \cdots \partial x_n^{l_n}} (x_1 - a_1)^{l_1} (x_2 - a_2)^{l_2} \cdots (x_n - a_n)^{l_n}$$

$$\tag{5.84}$$

is Taylor's polynomial for $f(\mathbf{x})$ about $\mathbf{a}$ of degree m, and $R_m(\mathbf{x})$ is Taylor's remainder. Taylor's remainder in Lagrange form is $R_m(\mathbf{x}) =$

$$\sum_{l_1 + l_2 + \cdots + l_n = m+1} \frac{1}{l_1! l_2! \cdots l_n!} \frac{\partial^{m+1} f(\mathbf{c})}{\partial x_1^{l_1} \partial x_2^{l_2} \cdots \partial x_n^{l_n}} (x_1 - a_1)^{l_1} (x_2 - a_2)^{l_2} \cdots (x_n - a_n)^{l_n}$$

$$\tag{5.85}$$

and satisfies

$$\lim_{\mathbf{x} \to \mathbf{a}} \frac{R_m(\mathbf{x})}{\|\mathbf{x} - \mathbf{a}\|_2^m} = 0 \tag{5.86}$$

where $\mathbf{c}$ is a point between $\mathbf{a}$ and $\mathbf{x}$, i.e. $\mathbf{c} = \mathbf{a} + t(\mathbf{x} - \mathbf{a})$ with $0 < t < 1$. Let $h = \|\mathbf{x} - \mathbf{a}\|_2$, then

$$R_m(\mathbf{x}) = O(h^{m+1}) \tag{5.87}$$

Remark   A convex set in $\mathbb{R}^n$ is a region in n dimensions in which any line segment connecting two points of the region is inside the region. For example, a solid cube is convex while a solid torus is not. An open set is a region that does not contain boundary points.

Remark Note the convention that 0-th derivative corresponds to no differentiation. For example,

$$\frac{\partial^0 f(\mathbf{a})}{\partial x_{l_i}^0} = f(\mathbf{a}) \tag{5.88}$$

Remark If we introduce the notation

$$l(!) = l_1! l_2! \cdots l_n! \tag{5.89}$$

$$f^{(l)}(\mathbf{a}) = \frac{\partial^l f(\mathbf{a})}{\partial x_1^{l_1} \partial x_2^{l_2} \cdots \partial x_n^{l_n}} \tag{5.90}$$

$$(\mathbf{x} - \mathbf{a})^{(l)} = (x_1 - a_1)^{l_1} (x_2 - a_2)^{l_2} \cdots (x_n - a_n)^{l_n} \tag{5.91}$$

where $l_1, l_2, \ldots, l_n$ take all possible non-negative integers such that $l_1 + l_2 + \cdots + l_n = l$. We can write the theorem in the form

$$f(\mathbf{x}) = \sum_{l=0}^{m} \frac{f^{(l)}(\mathbf{a})}{l(!)} (\mathbf{x} - \mathbf{a})^{(l)} + \frac{f^{(m+1)}(\mathbf{c})}{(m+1)(!)} (\mathbf{x} - \mathbf{a})^{(m+1)} \tag{5.92}$$

which bears the same form as Taylor's theorem for univariate functions.

Remark Taylor's theorem for multivariate functions can be proved using Taylor's theorem for univariate functions by introducing a univariate function

$$g(t) = f(\mathbf{a} + t(\mathbf{x} - \mathbf{a})) \tag{5.93}$$

where $\mathbf{x}$ is considered fixed. Note that $g(1) = f(\mathbf{x})$ and $g(0) = f(\mathbf{x})$. So we can use Taylor's formula for $g(1)$ about 0 in one dimension to find Taylor's formula for $f(\mathbf{x})$ about $\mathbf{a}$.

Below we show a detailed proof of Taylor's theorem for three dimensions, which can be easily extended to n dimensions.

Taylor's formula for 3 dimensions ($\mathbf{x} = [x_1, x_2, x_3]^T$ and $\mathbf{a} = [a_1, a_2, a_3]^T$) is

$$
\begin{aligned}
f(\mathbf{x}) &= \sum_{l_1+l_2+l_3=l\leq m} \frac{1}{l_1!l_2!l_3!} \frac{\partial^l f(\mathbf{a})}{\partial x_1^{l_1} \partial x_2^{l_2} \partial x_3^{l_3}} (x_1-a_1)^{l_1}(x_2-a_2)^{l_2}(x_3-a_3)^{l_3} \\
&+ \sum_{l_1+l_2+l_n=m+1} \frac{1}{l_1!l_2!l_3!} \frac{\partial^{m+1} f(\mathbf{c})}{\partial x_1^{l_1} \partial x_2^{l_2} \partial x_3^{l_3}} (x_1-a_1)^{l_1}(x_2-a_2)^{l_2}(x_3-a_3)^{l_3}
\end{aligned}
$$

(5.94)

Proof *Define* $\mathbf{h} = \mathbf{x} - \mathbf{a}$ *(i.e.* $[h_1, h_2, h_3]^T = [x_1-a_1, x_2-a_2, x_3-a_3]^T$*) and* $g(t) = f(\mathbf{a}+t\mathbf{h}) = f(a_1+th_1, a_2+th_2, a_3+th_3)$*. We have* $g(1) = f(\mathbf{x})$ *and*

$$
g'(t) = \frac{\partial f(\mathbf{a}+t\mathbf{h})}{\partial x_1} h_1 + \frac{\partial f(\mathbf{a}+t\mathbf{h})}{\partial x_2} h_2 + \frac{\partial f(\mathbf{a}+t\mathbf{h})}{\partial x_3} h_3
$$

(5.95)

which can be written as

$$
g'(t) = (\mathbf{h} \cdot \nabla) f(\mathbf{a}+t\mathbf{h})
$$

(5.96)

where

$$
\mathbf{h} \cdot \nabla = h_1 \frac{\partial}{\partial x_1} + h_2 \frac{\partial}{\partial x_2} + h_3 \frac{\partial}{\partial x_3}
$$

(5.97)

We can compute $g''(t)$ *as*

$$
\begin{aligned}
g''(t) &= \left(h_1 \frac{\partial f(\mathbf{a}+t\mathbf{h})}{\partial x_1} \right)' + h_2 \left(\frac{\partial f(\mathbf{a}+t\mathbf{h})}{\partial x_2} \right)' + h_3 \left(\frac{\partial f(\mathbf{a}+t\mathbf{h})}{\partial x_3} \right)' \\
&= (\mathbf{h} \cdot \nabla) \left(h_1 \frac{\partial f(\mathbf{a}+t\mathbf{h})}{\partial x_1} \right) + (\mathbf{h} \cdot \nabla) \left(h_2 \frac{\partial f(\mathbf{a}+t\mathbf{h})}{\partial x_2} \right) + (\mathbf{h} \cdot \nabla) \left(h_3 \frac{\partial f(\mathbf{a}+t\mathbf{h})}{\partial x_3} \right) \\
&= (\mathbf{h} \cdot \nabla) \left(h_1 \frac{\partial}{\partial x_1} + h_2 \frac{\partial}{\partial x_2} + h_3 \frac{\partial}{\partial x_3} \right) f(\mathbf{a}+t\mathbf{h}) = (\mathbf{h} \cdot \nabla)^2 f(\mathbf{a}+t\mathbf{h})
\end{aligned}
$$

(5.98)

where $(\mathbf{h} \cdot \nabla)^2$ *denotes the operator obtained by applying the operator* $\mathbf{h} \cdot \nabla$ *for 2 times and can be expanded as*

$$
\left(h_1 \frac{\partial}{\partial x_1} + h_2 \frac{\partial}{\partial x_2} \right)^2 = h_1^2 \frac{\partial^2}{\partial x_1^2} + 2h_1 h_2 \frac{\partial^2}{\partial x_1 \partial x_2} + h_2^2 \frac{\partial^2}{\partial x_2^2}
$$

(5.99)

Similarly, we have

$$
g^{(l)}(t) = (\mathbf{h} \cdot \nabla)^l f(\mathbf{a}+t\mathbf{h})
$$

(5.100)

To expand the operator

$$(\mathbf{h} \cdot \nabla)^l = \left(h_1 \frac{\partial}{\partial x_1} + h_2 \frac{\partial}{\partial x_2} + h_3 \frac{\partial}{\partial x_3} \right)^l \tag{5.101}$$

we need the coefficient of the term involving

$$\left(h_1 \frac{\partial}{\partial x_1} \right)^{l_1} \left(h_2 \frac{\partial}{\partial x_2} \right)^{l_2} \left(h_3 \frac{\partial}{\partial x_3} \right)^{l_3} = h_1^{l_1} h_2^{l_2} h_3^{l_3} \frac{\partial^l}{\partial x_1^{l_1} \partial x_2^{l_2} \partial x_3^{l_3}} \tag{5.102}$$

in the expansion, where $l_1 + l_2 + l_3 = l$. Finding the coefficient can be regarded as a combination problem. We have l operators $(\mathbf{h} \cdot \nabla)$, and we choose l_1 of them to form $\left(h_1 \frac{\partial}{\partial x_1} \right)^{l_1}$, l_2 of them to form $\left(h_2 \frac{\partial}{\partial x_2} \right)^{l_2}$ and the rest l_3 of them to form $\left(h_3 \frac{\partial}{\partial x_3} \right)^{l_3}$. The total number of combinations is

$$\binom{l}{l_1} \binom{l - l_1}{l_2} \binom{l - l_1 - l_2}{l_3} = \frac{l!}{l_1! l_2! l_3!} \tag{5.103}$$

which can also be regarded as the number of distinguishable permutations of the l operators that consist of l_1 operators of the same type to form $\left(h_1 \frac{\partial}{\partial x_1} \right)^{l_1}$, l_2 of the same type to form $\left(h_2 \frac{\partial}{\partial x_2} \right)^{l_2}$ and the rest l_3 of the same type to form $\left(h_3 \frac{\partial}{\partial x_3} \right)^{l_3}$. We therefore have the expansion

$$(\mathbf{h} \cdot \nabla)^l = \left(h_1 \frac{\partial}{\partial x_1} + h_2 \frac{\partial}{\partial x_2} + h_3 \frac{\partial}{\partial x_3} \right)^l = \sum_{l_1 + l_2 + l_3 = l} \frac{l!}{l_1! l_2! l_3!} h_1^{l_1} h_2^{l_2} h_3^{l_3} \frac{\partial^l}{\partial x_1^{l_1} \partial x_2^{l_2} \partial x_3^{l_3}} \tag{5.104}$$

and

$$g^{(l)}(t) = (\mathbf{h} \cdot \nabla)^l f(\mathbf{a} + t\mathbf{h}) = \sum_{l_1 + l_2 + l_3 = l} \frac{l!}{l_1! l_2! l_3!} \frac{\partial^l f(\mathbf{a} + t\mathbf{h})}{\partial x_1^{l_1} \partial x_2^{l_2} \partial x_3^{l_3}} h_1^{l_1} h_2^{l_2} h_3^{l_3} \tag{5.105}$$

Taylor's formula for $g(1)$ about 0 is

$$g(1) = \sum_{l=0}^{m} \frac{g^{(l)}(0)}{l!} 1^l + \frac{g^{(m+1)}(c)}{(m+1)!} 1^{m+1}, \quad 0 < c < 1 \tag{5.106}$$

Substitute $g^{(l)}(0)$ into the formula, we finally obtain

$$
\begin{aligned}
f(\mathbf{x}) \quad = \quad & \sum_{l_1 + l_2 + l_3 = l \leq m} \frac{1}{l_1! l_2! l_3!} \frac{\partial^l f(\mathbf{a})}{\partial x_1^{l_1} \partial x_2^{l_2} \partial x_3^{l_3}} (x_1 - a_1)^{l_1} (x_2 - a_2)^{l_2} (x_3 - a_3)^{l_3} \\
+ \quad & \sum_{l_1 + l_2 + l_n = m+1} \frac{1}{l_1! l_2! l_3!} \frac{\partial^{m+1} f(\mathbf{c})}{\partial x_1^{l_1} \partial x_2^{l_2} \partial x_3^{l_3}} (x_1 - a_1)^{l_1} (x_2 - a_2)^{l_2} (x_3 - a_3)^{l_3}
\end{aligned}
\tag{5.107}
$$

5.6.2 Newton's Method for Nonlinear Systems

For $m = 1$ in Taylor's theorem for multivariate functions, $l_1 + l_2 + \cdots + l_n \leq 1$ is satisfied when $l_0 = l_1 = \cdots = l_n = 0$ or one of $l_1, l_2, \ldots, l_n$ is one while the others are zero. The theorem gives

$$
\begin{aligned}
f(\mathbf{x}) &= T_1(\mathbf{x}) + R_1(\mathbf{x}) \\
&= f(\mathbf{a}) + \frac{\partial f(\mathbf{a})}{\partial x_1}(x_1 - a_1) + \frac{\partial f(\mathbf{a})}{\partial x_2}(x_2 - a_2) + \cdots + \frac{\partial f(\mathbf{a})}{\partial x_n}(x_n - a_n) + O(h^2)
\end{aligned}
$$

$$(5.108)$$

where $h = \|\mathbf{x} - \mathbf{a}\|_2$.

So when $\mathbf{x}$ is close to $\mathbf{a}$, we have the linear approximation

$$
f(\mathbf{x}) \approx T_1(\mathbf{x}) = f(\mathbf{a}) + \left[\frac{\partial f(\mathbf{a})}{\partial x_1}, \frac{\partial f(\mathbf{a})}{\partial x_2}, \ldots, \frac{\partial f(\mathbf{a})}{\partial x_n} \right]
\begin{bmatrix}
x_1 - a_1 \\
x_2 - a_2 \\
\vdots \\
x_n - a_n
\end{bmatrix}
\tag{5.109}
$$

where the right-hand side is the linearization of $f(\mathbf{x})$ about $\mathbf{a}$. If we denote the row vector $\left[\frac{\partial f(\mathbf{a})}{\partial x_1}, \frac{\partial f(\mathbf{a})}{\partial x_2}, \ldots, \frac{\partial f(\mathbf{a})}{\partial x_n} \right]$ as $f'(\mathbf{a})$, which is the transpose of the gradient vector $\nabla f(\mathbf{a})$, we have

$$
f(\mathbf{x}) \approx T_1(\mathbf{x}) = f(\mathbf{a}) + f'(\mathbf{a})(\mathbf{x} - \mathbf{a}) = f(\mathbf{a}) + \nabla f(\mathbf{a})^T (\mathbf{x} - \mathbf{a}) \tag{5.110}
$$

which bears the same form as the linear approximation for a univariate function. The following n linear approximations

$$
\begin{aligned}
f_1(\mathbf{x}) &\approx f_1(\mathbf{a}) + f_1'(\mathbf{a})(\mathbf{x} - \mathbf{a}) \\
f_2(\mathbf{x}) &\approx f_2(\mathbf{a}) + f_2'(\mathbf{a})(\mathbf{x} - \mathbf{a}) \\
&\vdots \\
f_n(\mathbf{x}) &\approx f_n(\mathbf{a}) + f_n'(\mathbf{a})(\mathbf{x} - \mathbf{a})
\end{aligned}
$$

can be written as

$$
\mathbf{f}(\mathbf{x}) \approx \mathbf{f}(\mathbf{a}) + \mathbf{f}'(\mathbf{a})(\mathbf{x} - \mathbf{a}) \tag{5.111}
$$

where

$$
\mathbf{f}'(\mathbf{a}) =
\begin{bmatrix}
\frac{\partial f_1(\mathbf{a})}{\partial x_1} & \frac{\partial f_1(\mathbf{a})}{\partial x_2} & \cdots & \frac{\partial f_1(\mathbf{a})}{\partial x_n} \\
\frac{\partial f_2(\mathbf{a})}{\partial x_1} & \frac{\partial f_2(\mathbf{a})}{\partial x_2} & \cdots & \frac{\partial f_2(\mathbf{a})}{\partial x_n} \\
\vdots & \vdots & \ddots & \vdots \\
\frac{\partial f_n(\mathbf{a})}{\partial x_1} & \frac{\partial f_n(\mathbf{a})}{\partial x_2} & \cdots & \frac{\partial f(\mathbf{a})}{\partial x_n}
\end{bmatrix}
\tag{5.112}
$$

is call the **Jacobian matrix** of **f** at **a**.

Newton's method is based on the linear approximation

$$\mathbf{f}(\mathbf{x}) \approx \mathbf{f}(\mathbf{a}) + \mathbf{f}'(\mathbf{a})(\mathbf{x} - \mathbf{a}) \tag{5.113}$$

Let $\mathbf{x}_k$ be a current approximation of the solution **r** to the nonlinear system $\mathbf{f}(\mathbf{x}) = \mathbf{0}$. With $\mathbf{a} = \mathbf{x}_k$, we have

$$\mathbf{f}(\mathbf{x}) \approx \mathbf{f}(\mathbf{x}_k) + \mathbf{f}'(\mathbf{x}_k)(\mathbf{x} - \mathbf{x}_k) \tag{5.114}$$

We then set the linear approximation of $\mathbf{f}(\mathbf{x})$ be zero

$$\mathbf{f}(\mathbf{x}_k) + \mathbf{f}'(\mathbf{x}_k)(\mathbf{x} - \mathbf{x}_k) = 0 \tag{5.115}$$

which is a linear system for **x** whose solution is taken as the next approximation $\mathbf{x}_{k+1}$ of the solution **r** to the nonlinear system $\mathbf{f}(\mathbf{x}) = \mathbf{0}$. Let $\mathbf{h}_k = \mathbf{x}_{k+1} - \mathbf{x}_k$, we solve for $\mathbf{h}_k$ from

$$\mathbf{f}'(\mathbf{x}_k)\mathbf{h}_k = -\mathbf{f}(\mathbf{x}_k) \tag{5.116}$$

which is a linear system with the Jacobian matrix $\mathbf{f}'(\mathbf{x_k})$ as the coefficient matrix. After $\mathbf{h}_k$ is solved, we then get

$$\mathbf{x}_{k+1} = \mathbf{x}_k + \mathbf{h}_k \tag{5.117}$$

To summarize, Newton's iteration for nonlinear system $\mathbf{f}(\mathbf{x}) = \mathbf{0}$ is

$$\mathbf{f}'(\mathbf{x}_k)\mathbf{h}_k = -\mathbf{f}(\mathbf{x}_k), \quad \mathbf{x}_{k+1} = \mathbf{x}_k + \mathbf{h}_k \quad (k = 0, 1, 2, \ldots) \tag{5.118}$$

Example 5.16 Conduct 3 steps of Newton's iteration to solve the nonlinear system

$$\begin{cases} x_1^2 + x_2^2 - 2 &= 0 \\ x_1^2 - x_2 &= 0 \end{cases}$$

starting with the initial guess $[0.5, 0.5]^T$.

[Solution:] As shown in Fig. 5.11, the nonlinear system has two solutions $[-1, -1]^T$ and $[1, 1]^T$. Denote the nonlinear system as $\mathbf{f}(\mathbf{x}) = \mathbf{0}$, where

$$\mathbf{f}(\mathbf{x}) = \begin{bmatrix} x_1^2 + x_2^2 - 2 \\ x_1^2 - x_2 \end{bmatrix}$$

The Jacobian matrix at **x** is

$$\mathbf{f}'(\mathbf{x}) = \begin{bmatrix} \frac{\partial f_1(\mathbf{x})}{\partial x_1} & \frac{\partial f_1(\mathbf{x})}{\partial x_2} \\ \frac{\partial f_2(\mathbf{x})}{\partial x_1} & \frac{\partial f_2(\mathbf{x})}{\partial x_2} \end{bmatrix} = \begin{bmatrix} 2x_1 & 2x_2 \\ 2x_1 & -1 \end{bmatrix}$$

Newton's iteration for this system goes as follows.

Initial: $\mathbf{x}_0 = [0.5, 0.5]^T$.

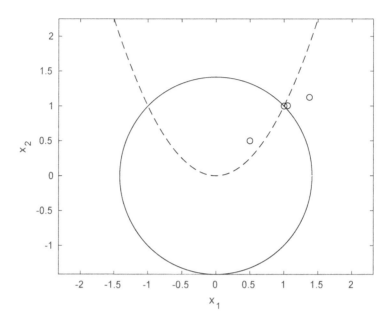

FIGURE 5.11
Illustration of Newton's method to solve a nonlinear system. The curves are the graphs of the nonlinear equation, and the open circles are approximate solutions from Newton's iteration.

Step 1:

$$\mathbf{f}'(\mathbf{x}_0) = \begin{bmatrix} 1 & 1 \\ 1 & -1 \end{bmatrix}, \quad \mathbf{f}(\mathbf{x}_0) = \begin{bmatrix} -1.5 \\ -0.25 \end{bmatrix}$$

$$\mathbf{f}'(\mathbf{x}_0)\mathbf{h}_0 = -\mathbf{f}(\mathbf{x}_0) \Rightarrow \mathbf{h}_0 \approx \begin{bmatrix} 0.8750 \\ 0.6250 \end{bmatrix}, \quad \mathbf{x}_1 = \mathbf{x}_0 + \mathbf{h}_0 \approx \begin{bmatrix} 1.3750 \\ 1.1250 \end{bmatrix}$$

Step 2:

$$\mathbf{f}'(\mathbf{x}_1) \approx \begin{bmatrix} 2.7500 & 2.2500 \\ 2.7500 & -1.0000 \end{bmatrix}, \quad \mathbf{f}(\mathbf{x}_1) \approx \begin{bmatrix} 1.1562 \\ 0.7656 \end{bmatrix}$$

$$\mathbf{f}'(\mathbf{x}_1)\mathbf{h}_1 = -\mathbf{f}(\mathbf{x}_1) \Rightarrow \mathbf{h}_1 \approx \begin{bmatrix} -0.3221 \\ -0.1202 \end{bmatrix}, \quad \mathbf{x}_2 = \mathbf{x}_1 + \mathbf{h}_1 \approx \begin{bmatrix} 1.0529 \\ 1.0048 \end{bmatrix}$$

Step 3:

$$\mathbf{f}'(\mathbf{x}_1) \approx \begin{bmatrix} 2.1058 & 2.0096 \\ 2.1058 & -1.0000 \end{bmatrix}, \quad \mathbf{f}(\mathbf{x}_2) \approx \begin{bmatrix} 0.1182 \\ 0.1038 \end{bmatrix}$$

$$\mathbf{f}'(\mathbf{x}_2)\mathbf{h}_2 = -\mathbf{f}(\mathbf{x}_2) \Rightarrow \mathbf{h}_2 \approx \begin{bmatrix} -0.0516 \\ -0.0048 \end{bmatrix}, \quad \mathbf{x}_3 = \mathbf{x}_2 + \mathbf{h}_2 \approx \begin{bmatrix} 1.0013 \\ 1.0000 \end{bmatrix}$$

∎

We may stop Newton's iteration when

$$\|\mathbf{x}_{k+1} - \mathbf{x}_k\|_2 \leq \tau \tag{5.119}$$

where τ is the error tolerance because, just as in the univariate case, we can estimate the error vector as

$$\mathbf{r} - \mathbf{x}_k \approx \mathbf{x}_{k+1} - \mathbf{x}_k \tag{5.120}$$

when $\mathbf{x}_k$ is sufficiently close to the exact solution $\mathbf{r}$.

Now we can wrap up Newton's method for nonlinear systems with the pseudocode for the method.

Algorithm 11 Newton's Method for solving $\mathbf{f}(\mathbf{x}) = \mathbf{0}$

$\mathbf{r} \leftarrow \mathbf{x}_0$
for k from 1 to $n_{\max}$ **do**
   $\mathbf{b} \leftarrow -\mathbf{f}(\mathbf{r})$
   $A \leftarrow \mathbf{f}'(\mathbf{r})$
   solve $A\mathbf{h} = \mathbf{b}$ for $\mathbf{h}$
   $E_{\text{est}} \leftarrow \|\mathbf{h}\|_2$
   if $E_{\text{est}} \leq \tau$ **then**
      return $\mathbf{r}$
   end if
   $\mathbf{r} \leftarrow \mathbf{r} + \mathbf{h}$
end for

5.7 Unconstrained Optimization

This section touches some ideas about optimization. In calculus, we distinguish a local minimum and a global minimum, as exemplified in Fig. 5.12. In practice, we often need to minimize functions, which may be the cost of a product or the error of an approximation (see the least squares problems in Chapter 11). The minimization of a function subject to no constraints is an unconstrained optimization problem.

By Taylor's theorem for multivariate functions with $m = 2$ in Eq. (5.83), we have

$$f(\mathbf{x}) = T_2(\mathbf{x}) + O(h^3) \tag{5.121}$$

where $h = \|\mathbf{x} - \mathbf{a}\|_2$. We know

$$T_2(\mathbf{x}) = T_1(\mathbf{x}) + \sum_{l_1 + l_2 + \cdots + l_n = 2} \frac{1}{l_1! l_2! \cdots l_n!} \frac{\partial^2 f(\mathbf{a})}{\partial x_1^{l_1} \partial x_2^{l_2} \cdots \partial x_n^{l_n}} (x_1 - a_1)^{l_1} (x_2 - a_2)^{l_2} \cdots (x_n - a_n)^{l_n}$$

$$= T_1(\mathbf{x}) + \sum_{i=1}^{n} \frac{1}{2} \frac{\partial^2 f(\mathbf{a})}{\partial x_i^2} (x_i - a_i)^2 + \sum_{i \neq j; i,j=1}^{n} \frac{\partial^2 f(\mathbf{a})}{\partial x_i \partial x_j} (x_i - a_i)(x_j - a_j)$$

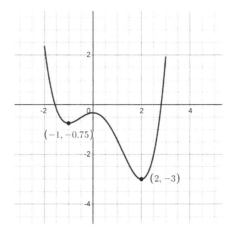

FIGURE 5.12
Examples of local minimums -0.75 and -3 at $x = -1$ and $x = 2$, respectively, and a global minimum -3 at $x = 2$ (the global minimum is also a local minimum in this case).

$$= T_1(\mathbf{x}) + \frac{1}{2}\sum_{i=1}^{n}\sum_{j=1}^{n}\frac{\partial^2 f(\mathbf{a})}{\partial x_i \partial x_j}(x_i - a_i)(x_j - a_j) \tag{5.122}$$

where

$$T_1(\mathbf{x}) = f(\mathbf{a}) + \nabla f(\mathbf{a})^T(\mathbf{x} - \mathbf{a}) \tag{5.123}$$

and the term of the double summations is the quadratic form $(\mathbf{x} - \mathbf{a})^T H(\mathbf{x} - \mathbf{a})$ with the matrix H given by

$$H = \begin{bmatrix} \frac{\partial^2 f(\mathbf{a})}{\partial x_1^2} & \frac{\partial^2 f(\mathbf{a})}{\partial x_1 \partial x_2} & \cdots & \frac{\partial^2 f(\mathbf{a})}{\partial x_1 \partial x_n} \\ \frac{\partial^2 f(\mathbf{a})}{\partial x_2 \partial x_1} & \frac{\partial^2 f(\mathbf{a})}{\partial x_2^2} & \cdots & \frac{\partial^2 f(\mathbf{a})}{\partial x_2 \partial x_n} \\ \vdots & \vdots & \ddots & \vdots \\ \frac{\partial^2 f(\mathbf{a})}{\partial x_n \partial x_1} & \frac{\partial^2 f(\mathbf{a})}{\partial x_n \partial x_2} & \cdots & \frac{\partial^2 f(\mathbf{a})}{\partial x_n^2} \end{bmatrix} \tag{5.124}$$

which is called the **Hessian matrix** of f at $\mathbf{a}$. The Hessian matrix is symmetric by Clairaut's theorem (if the second-order partial derivatives are continuous near $\mathbf{a}$). So we obtain

$$f(\mathbf{x}) = f(\mathbf{a}) + \nabla f(\mathbf{a})^T(\mathbf{x} - \mathbf{a}) + \frac{1}{2}(\mathbf{x} - \mathbf{a})^T H(\mathbf{x} - \mathbf{a}) + O(h^3) \tag{5.125}$$

or with $\mathbf{h} = \mathbf{x} - \mathbf{a}$

$$f(\mathbf{a} + \mathbf{h}) = f(\mathbf{a}) + \nabla f(\mathbf{a})^T \mathbf{h} + \frac{1}{2}\mathbf{h}^T H \mathbf{h} + O(h^3) \tag{5.126}$$

from which, we immediately have the following theorem

Theorem 5.7 *If* $f(\mathbf{x})$ *has bounded continuous partial derivatives up to third order, then* $f(\mathbf{x})$ *has a local minimum at* $\hat{\mathbf{x}}$ *if*

$$\nabla f(\hat{\mathbf{x}}) = \mathbf{0} \tag{5.127}$$

(i.e. $\hat{\mathbf{x}}$ *is a critical point), and*

$$\mathbf{h}^T H \mathbf{h} > 0 \tag{5.128}$$

for any nonzero vector $\mathbf{h}$, *i.e. the Hessian matrix* H *at* $\hat{\mathbf{x}}$ *is SPD – symmetric positive definite.*

Remark This theorem bears the similarity with the second derivative test for the univariate case in which $f(x)$ has a local minimum at $\hat{x}$ if $f'(\hat{x}) = 0$ and $f''(\hat{x}) > 0$ and can be understood similarly as in the univariate case.

We use an example in one dimension to introduce some ideas for finding a local minimum. We then extend these ideas to multiple dimensions.

Example 5.17 Find the value $\hat{x}$ such that $f(x) = x^2 - 2x + 2$ has a local minimum at $x = \hat{x}$.
[Solution 1:] From calculus, we know $f(\hat{x})$ is a local extremum if $f'(\hat{x}) = 0$, i.e. $\hat{x}$ is a critical point of $f(x)$, and $f(x)$ has a local minimum (instead of maximum) at $\hat{x}$ if $f''(\hat{x}) > 0$, i.e. $f(x)$ is concave up at $\hat{x}$. So we can find $\hat{x}$ by solving the root finding problem $f'(x) = 2x - 2 = 0$, which leads to $\hat{x} = 1$. Since $f''(1) = 2 > 0$, $f(x)$ has a local minimum $f(1) = 1$ (also global minimum in this case) at $\hat{x} = 1$, as shown in Fig. 5.13. Note that a general root finding problem $f'(x) = 0$ can be solved by Newton's method.
[Solution 2:] Suppose we have a current guess x_k of $\hat{x}$. We can find the next better guess x_{k+1} by properly moving away from x_k. The direction of the move can be seen from the linear approximation

$$f(x_{k+1}) \approx f(x_k) + f'(x_k)(x_{k+1} - x_k)$$

for x_{k+1} near x_k. Let $h_k = x_{k+1} - x_k$. To make $f(x_{k+1}) < f(x_k)$, we want

$$f'(x_k)h_k < 0$$

So h_k must have the opposite sign with $f'(x_k)$, i.e. $h_k = -\alpha_k f'(x_k)$ with some positive constant $\alpha_k > 0$.

If we $x_k = 3$ as in the case shown in Fig. 5.13, then $f'(x_k) = f'(3) = 4 > 0$ implies $f(x)$ is increasing locally near $x_k = 3$, and $h_k < 0$ makes the next guess $x_{k+1} < x_k$ to reduce $f(x)$. We now know which direction to move, but we also need to determine α_k to know what distance to move. With $x_k = 3$ in Fig. 5.13, if we make move from $x_k = 3$ such that $x_{k+1} = 2$, we make the progress to reduce $f(x)$ such that $f(x_{k+1}) < f(x_k)$, but if we make a too large move such that $x_{k+1} = -2$, then $f(x_{k+1}) > f(x_k)$. Note that the linear approximation

$$f(x_{k+1}) \approx f(x_k) + f'(x_k)(x_{k+1} - x_k)$$

holds only if x_{k+1} is near x_k. A very small move can guarantee the reduction of $f(x)$, but the progress to reach the minimum can be very slow. In practice, we can do the co-called **line search** (along the x-axis in this univariate optimization problem) to find a suitable step size. ■

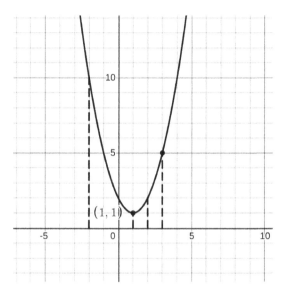

FIGURE 5.13
Searching for the global minimum of a function.

Similar as the Solution 1 in the previous example, finding a critical point $\hat{\mathbf{x}}$ of $f(\mathbf{x})$ to minimize $f(\mathbf{x})$ leads to a root finding problem

$$\nabla f(\mathbf{x}) = 0 \tag{5.129}$$

which is in general a system of nonlinear equations, and we can apply Newton's method to solve the system. Note that the Jacobian matrix in this case is the Hessian matrix of $f(\mathbf{x})$. Here is how Newton's method works for unconstrained minimization: starting from $\mathbf{x}_0$, for $k = 1, 2, \ldots$ until convergence, solve for $\mathbf{h}_k$ from

$$H(\mathbf{x}_k)\mathbf{h}_k = -\nabla f(\mathbf{x}_k) \tag{5.130}$$

to update the approximation as

$$\mathbf{x}_{k+1} = \mathbf{x}_k + \mathbf{h}_k \tag{5.131}$$

where $H(\mathbf{x}_k)$ is the Hessian at $\mathbf{x}_k$.

Similarly as the Solution 2 in the previous example, we can choose a direction and step size to move at the current guess $\mathbf{x}_k$ to reduce $f(\mathbf{x})$ toward the minimum. Recall from calculus that the negative gradient vector $-\nabla f(\mathbf{x}_k)$ gives the steepest descent direction for $f(\mathbf{x})$ at $\mathbf{x}_k$. So we can choose

$$\mathbf{h}_k = -\alpha_k \nabla f(\mathbf{x}_k) \tag{5.132}$$

with some positive constant $\alpha_k > 0$ and compute the next guess as

$$\mathbf{x}_{k+1} = \mathbf{x}_k + \mathbf{h}_k \tag{5.133}$$

This algorithm is called the **gradient descent (GD)** algorithm. Many sophisticated optimization algorithms are based on it. Again, we can use line search along the parametric line $\mathbf{x} = \mathbf{x}_k + t(-\nabla f(\mathbf{x}_k))$, where t is the parameter, and $-\nabla f(\mathbf{x}_k)$ is the direction vector of the line, to determine α_k such that at $t = \alpha_k$, $f(\mathbf{x})$ is minimum along the line. The line search problem is a univariate minimization problem.

Remark In other related methods, the moving step is set as

$$\mathbf{h}_k = -\alpha_k B_k^{-1} \nabla f(\mathbf{x}_k) \tag{5.134}$$

where B_k^{-1} is a SPD (symmetric positive definite) matrix so that

$$-\nabla f(\mathbf{x}_k)^T \mathbf{h}_k = \alpha_k \nabla f(\mathbf{x}_k)^T B_k^{-1} \nabla f(\mathbf{x}_k) > 0 \tag{5.135}$$

Note that the positive inner product (or the dot product) of $-\nabla f(\mathbf{x})_k$ and $\mathbf{h}_k$ implies the direction of $\mathbf{h}_k$ is near the steepest descent direction $-\nabla f(\mathbf{x}_k)$ as the angle between $\mathbf{h}_k$ and the steepest descent direction $-\nabla f(\mathbf{x}_k)$ is acute. Choosing B_k as the identity matrix I results in the gradient descent method. Choosing B_k as the Hessian matrix $H(\mathbf{x}_k)$ with $\alpha_k = 1$ results in Newton's method.

5.8 Software

MATLAB has the built-in functions `fzero` to find a root of a univariate equation and `roots` to find the roots of a polynomial. In Python, the class `scipy.optimize` provides functions for minimizing objective functions, possibly subject to constraints, and includes solvers for root finding.

PETSc (`https://www.mcs.anl.gov/petsc/`) is a suite of data structures and routines for the solution of scientific applications modeled by partial differential equations. It also includes various nonlinear solvers and contains the Tao (called PETSc/Tao) optimization software library.

5.9   Exercises

Exercise 5.1 (1) Use the intermediate value theorem to prove that the equation $e^x = -x$ has at least one real root. (2) Use Rolle's theorem to show that this equation has only one real root.

Exercise 5.2 The graph of a continuous function $f(x)$ is shown in Fig. 5.14. Conduct 4 iterations of the bisection method in Table 5.2 to find an approximation of the root of $f(x) = 0$.

k	a	sign(f(a))	b	sign(f(b))	c	sign(f(c))	b-c
1							
2							
3							
4							

TABLE 5.2
Bisection iteration.

Exercise 5.3 (1) Show that equation $e^x = 3x$ has a unique root between 0 and 1. (2) Conduct iterations of the bisection method in a table as below (Table 5.3) until you find an approximation of this root with the error less than 0.04. Write down your final approximation and its error upper bound.

k	a	sign(f(a))	b	sign(f(b))	c	sign(f(c))	b-c
1							
2							
3							
4							

TABLE 5.3
Bisection iterations.

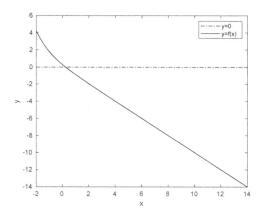

FIGURE 5.14
Graph of $y = f(x)$ for the bisection iteration.

Exercise 5.4 Consider the bisection method for finding the root of the equation $f(x) = 0$ with the initial interval $[a, b]$. If $a = -2.7$, $b = 5.3$, and a, b and f satisfy the conditions for the method, how many bisection iterations are needed to guarantee the final approximation of the root has error less than the error tolerance 2^{-20}?

Exercise 5.5 Consider the nonlinear equation $e^{-x} = x - 2$.
(1) Sketch the curves $y = e^{-x}$ and $y = x - 2$ in the same figure to show that the equation has a unique solution.
(2) The equation $e^{-x} = x - 2$ can be rewritten as a root finding problem $e^{-x} - x + 2 = 0$. Use the IVT (the Intermediate Value Theorem) to show that the equation $e^{-x} - x + 2 = 0$ has a root.
(3) Starting with the interval $[0, 4]$, conduct the bisection method to find an approximation of the root such that the error of the approximation is less than 0.2. Organize your bisection steps in a table as Table 5.4.
(4) If we want the error of the final approximation is less than 2^{-30}, how many middle points need to be computed?

k	a	sign(f(a))	b	sign(f(b))	c	sign(f(c))	b-c
1							
2							
3							
4							

TABLE 5.4
Bisection steps.

Exercise 5.6 The bisection method is started with the initial interval $[a, b]$ with $b < 0$. Devise a stopping criterion for the method such that the root is determined with *relative* error less than ε.

Exercise 5.7 Determine if each sequence below converges or not, and the order of convergence if it converges. (1) $x_k = 1/3^k$, $k = 0, 1, 2, \ldots$ (2) $x_k = 2^{-2^k}$, $k = 0, 1, 2, \ldots$ (3) $x_k = 3^{-3^k}$, $k = 0, 1, 2, \ldots$ (4) $x_k = \cos(k\pi)$, $k = 0, 1, 2, \ldots$

Exercise 5.8 To find the square root $\sqrt{R}$, where the constant $R > 0$, we can solve the nonlinear equation $x^2 - R = 0$ using Newton's method. (1) Write down Newton's iteration formula for solving this equation. What arithmetic is required in your formula? (It should not involve the calculation of square roots.) (2) Using your iteration formula to find an approximation of $\sqrt{2}$ (i.e. $R = 2$) by conducting 4 iterations with an initial guess $x_0 = 1$. How does your final approximation compare with the value of $\sqrt{2}$ from your calculator? (3) In a similar way, write down the Newton's iteration formula to find the cubic root $\sqrt[3]{R}$.

Exercise 5.9 We approximate $\sqrt[m]{a}$ by solving the equation $x^m - a = 0$ using Newton's method, where integer $m \geq 2$ and real $a > 0$ are constants.
(1) Show that $\sqrt[m]{a}$ is a simple root of the equation $x^m - a = 0$.
(2) Write down Newton's iteration for finding the root $\sqrt[m]{a}$. What is the convergence order of your Newton's iteration if it converges?
(3) Plot the graph of $f(x) = x^m - a$ for $m = 3$ and $a = 2$. Conduct one Newton iteration on your graph to locate x_{k+1} from $x_k = 1$.
(4) Starting from $x_0 = 2/3$, conduct two Newton's iterations for finding $\sqrt[3]{2}$ and write your x_1 and x_2 as fractions.
(5) In your Newton's iteration for (4), if the absolution error in x_k is $|\sqrt[3]{2} - x_k| = 10^{-4}$, which is your estimate of the absolute error in x_{k+1}: 10^{-2}, 10^{-4}, 10^{-8} or 10^{-16}?

Exercise 5.10 The solution $1/B$ of the equation $1 - \dfrac{1}{Bx} = 0$, where B is a nonzero constant, is the reciprocal of B. (1) Write down Newton's iteration formula for solving the equation to find the reciprocal. Algebraically simplify your iteration to show that divisions can be avoided in the iteration to find $1/B$. (2) Approximate the reciprocal of 5 (i.e. $B = 5$) by conducting 4 iterations with $x_0 = 0.1$ and $x_0 = 1$ using your simplified iteration formula, respectively. (3) State how you can approximate A/B (i.e. division of A by B) without any divisions?

Exercise 5.11 To find the square root $\sqrt{R}$, where the constant $R > 0$, we can solve the nonlinear equation $x^2 - R = 0$ using Newton's method. (1) Use Newton's iteration for solving this equation to show that

$$|x_{k+1}^2 - R| = \frac{1}{4x_k^2}|x_k^2 - R|^2$$

(2) What is order of convergence of Newton's method in this case?

Exercise 5.12 The exact root of the equation $(x-r)^m = 0$, where r is a real constant and $m \geq 2$ is a positive integer, is r. (1) What is the multiplicity of the root r? (2) Write down Newton's iteration for solving this equation. (3) Use Newton's iteration to show that

$$|x_{k+1} - r| = \frac{m-1}{m}|x_k - r|$$

(4) What is the order of convergence of Newton's method in this case? (5) Conduct 10 Newton's iterations for the equation $(x-1)^2 = 0$ with $x_0 = 2$ and observe how the error reduces after each iteration.

Exercise 5.13 Consider the equation $(x-2)^5 e^x = 0$.
(1) What is the root of this equation? What is the multiplicity of the root?
(2) Write down Newton's iteration for finding the root.
(3) What is the convergence order of your Newton's iteration if the iteration converges?
(4) Use the iteration formula in (2) to show that

$$|x_{k+1} - 2| = \left|\frac{x_k + 2}{x_k + 3}\right| |x_k - 2|$$

(5) If the absolute error of the current approximation is 10^{-4}, use the formula in (4) to estimate the absolute error in the next approximation in your Newton's iteration. Is your estimate consistent with your answer in (3)?

Exercise 5.14 Consider the equation $5 - \frac{1}{x} = 0$. (1) Conduct two Newton's iterations with $x_0 = 0.1$ and $x_0 = 1$, respectively. (2) Use the graph of $f(x) = 5 - \frac{1}{x}$ to explain the former converges while the latter diverges.

Exercise 5.15 Suppose Newton's iteration converges to a simple root r. If in the iteration $|r - x_4| = 2 \times 10^{-3}$ and $|r - x_5| \approx 10^{-6}$, what is $|r - x_6|$ approximately?

Exercise 5.16 The solution $1/B$ of the equation $B - \frac{1}{x} = 0$, where B is a nonzero constant, is the reciprocal of B. (1) Write down the iteration formula of the secant method for solving the equation to find the reciprocal. (2) Algebraically simplify your formula to show that no divisions are needed in your simplified iteration formula. (3) Approximate the reciprocal of 5 (with $B = 5$) by conducting 4 iterations with $x_0 = 0.1$ and $x_1 = 0.12$. (4) State how you can approximate A/B without using any divisions?

Exercise 5.17 Describe how you can use the MATLAB function `fzero` to find the root of the equation $f(x) = 0$.

Exercise 5.18 Consider the equation $x = \frac{\pi}{2}\sin x$. (1) Sketch the graphs of $y = x$ and $y = \frac{\pi}{2}\sin x$ to find the true solutions. (2) Conduct the fixed-point iteration $x_{k+1} = \frac{\pi}{2}\sin x_k$ starting from $x_0 = 1.5$ until the approximation has three correct decimal digits. (3) Why do you expect that the fixed-point iteration $x_{k+1} = \frac{\pi}{2}\sin x_k$ converges to a fixed point x^* from a starting point x_0 close enough to x^*?

Exercise 5.19 Consider the fixed-point iteration $x_{k+1} = x_k^2 - \frac{3}{16}$. (1) What are the fixed points? (2) Conduct 10 steps if the fixed-point iteration starting from $x_0 = 0.2$ and 0.8, respectively. (3) For which x_0 is the iteration convergent? Why? (4) For which x_0 is the iteration divergent? Why?

Exercise 5.20 Suppose that $f(x,y) = T_2(x,y) + R_2(x,y)$ by Taylor's theorem, where $T_2(x,y)$ is Taylor's polynomial of degree 2 for $f(x,y)$ about (a,b), and $R_2(x,y)$ is Taylor's remainder. Write down the expressions for $T_2(x,y)$ and $R_2(x,y)$.

Exercise 5.21 Suppose that $f(x,y) = T_2(x,y) + R_2(x,y)$ by Taylor's theorem, where $T_2(x,y)$ is Taylor's polynomial of degree 2 for $f(x,y)$ about (a,b), and $R_2(x,y)$ is Taylor's remainder. Write down the expressions for $T_2(x,y)$ and $R_2(x,y)$ when $f(x,y) = e^x \sin y + y \ln(x+1)$ and $(a,b) = (0,0)$.

Exercise 5.22 Suppose Newton's method is applied to a nonsingular square linear system $Ax = b$. (1) Write down Newton's iteration. (2) How many iterations does Newton's method need to converge to the true solution $A^{-1}b$ starting from an initial guess $x_0 \neq A^{-1}b$?

Exercise 5.23 Consider the system

$$\begin{cases} x_1^3 + x_2^3 + x_3^3 &= 0 \\ x_1^2 + x_2^2 + x_3^2 &= 2 \\ x_1 x_2 + x_1 x_3 + x_2 x_3 &= -1 \end{cases}$$

Find the Jacobian matrix of the system at $x = [1, 2, 3]^T$.

Exercise 5.24 Consider the system

$$\begin{cases} x_1 x_2 - 1 &= 0 \\ x_2 - 1 &= 0 \end{cases}$$

(1) What is the true solution of the system? (2) What happens if we apply Newton's method to this system with the initial guess $x_0 = [0,0]^T$? (3) Conduct 2 steps of Newton's method with the initial guess $x_0 = [0,1]^T$.

Exercise 5.25 Consider the system

$$\begin{cases} x_1^2 + x_2 - 1 &= 0 \\ x_1 - x_2^2 &= 0 \end{cases}$$

(1) Write down the Jacobian matrix for the system. (2) Conduct 3 steps of Newton's method with the initial guess $x_0 = [0,0]^T$ to solve the system.

Exercise 5.26 Suppose the gradient descent is used to find the minimum value of $f(x,y)$. At the point $x_0 = (0,0)$, we know the gradient and Hessian of f as

$$\nabla f(x_0) = \begin{bmatrix} -5 \\ 1 \end{bmatrix}, \quad H(x_0) = \begin{bmatrix} 6 & -1 \\ -1 & 2 \end{bmatrix}$$

What is the next point x_1 to approximate the location of the minimum value?

Exercise 5.27 Find the minimum value of $f(x_1, x_2) = 3x_1^2 - 6x_1 + x_2^2 - 4x_2 + 17$ algebraically. Then find it using the linear system involving the gradient and Hessian of f.

Exercise 5.28 Consider the minimization of $f(x, y) = (x-1)^2 + 2(y-2)^2 + 3$. (1) Sketch level curves $f(x, y) = c$ for $c = 3$, 5 and 7, respectively. (2) What is the minimum value of f? (3) What is the Hessian matrix of f? (4) Suppose the gradient decent is applied to find the minimum, find the direction of the steepest descent at the point $\mathbf{x}_k = (2, 1)$. (5) Denote the direction in (4) as the vector $\mathbf{d}_k$. If a line search is conducted along $\mathbf{d}_k$ to find the next point $\mathbf{x}_{k+1} = \mathbf{x}_k + \alpha_k \mathbf{d}_k$. Find the value of α_k such that f is the smallest along $\mathbf{d}_k$.

5.10   Programming Problems

Problem 5.1 Write a MATLAB m-function `bisection.m` for approximating the root of a nonlinear function using the bisection method over a given interval. Add brief comments at suitable places in `bisection.m` to explain your code. Then write a test m-script `test_bisect.m` to test your bisection solver on the equation $10 + x^3 - 12\cos(x) = 0$, to find a root within an error tolerance of $\varepsilon = 10^{-4}$ using a starting interval of $[-5, 5]$. Why is the starting interval $[-5, 5]$ a valid interval for the method? Theoretically predict the number of bisection iterations. Is the result from your test consistent with your prediction?

Problem 5.2 Create a MATLAB m-function named `newton.m` that performs Newton's method to approximate the root of a provided equation to a specified tolerance. The m-function should have the calling syntax

```
function [r] = newton(f, df, x0, nmax, etol, ishow)
```

where the output `r` is the final approximation of the root you have found, `f` is the name of a provided nonlinear residual function in the root-finding problem, `df` is the name of the derivative of that function, `x0` is the initial guess, `nmax` is the number of allowed iterations, `etol` is the error tolerance, and `ishow` is a flag to enable/disable printing of iteration information (such as the iteration step and the approximation at the step) to the screen during the Newton solve. The m-function terminates when the max number of iterations `nmax` is reached or the absolute value of the error estimate (difference between two consecutive approximations) is less than `etol`.

Create a MATLAB m-script named `test_newton.m` to verify that your `newton.m` function works as desired on the root-finding problem $x^2(x-3)(x+1) = 0$. In your test, start with initial guesses of $x_0 = \{-3, 1, 2, 4\}$, and solve the problem to error tolerances of $\{10^{-6}, 10^{-10}\}$ (i.e. $4 \times 2 = 8$ tests in total) with `ishow` turned on. Allow a maximum of 100 iterations. (1) How does the performance of Newton's method change as you vary the initial guess? (2) How does Newton's method change as you vary the tolerance? (3) Explain your results.

Problem 5.3 Create a MATLAB m-function named `secant.m` that performs the secant method to approximate the root of a provided equation to a specified tolerance. The m-function should have the calling syntax

```
function [r] = secant(f, x0, x1, nmax, etol, ishow)
```

where the output `r` is the final approximation of the root you have found, `f` is the name of a provided nonlinear residual function in the root-finding problem, `x0` and `x1` are the two initial guesses, `nmax` is the number of allowed iterations, `etol` is the error tolerance and `ishow` is a flag to enable/disable printing of iteration information (such as the iteration step and the approximation at the step) to the screen during the solve. The m-function terminates when the max number of iterations `nmax` is reached or the absolute value of the error estimate (difference between two consecutive approximations) is less than `etol`.

Create a MATLAB m-script named `test_secant.m` to verify that your `secant.m` function works as desired on the root-finding problem $(x-3)^2(x+2) = 0$. In your test, start with initial guesses of $(x_0, x_1) = (-3, -2.5)$ and $(4, 3.5)$ and solve the problem to error tolerances of $\{10^{-6}, 10^{-10}\}$ (i.e. $2 \times 2 = 4$ tests in total) with `ishow` turned on. Allow a maximum of 100 iterations. (1) How does the performance of the secant method change as you vary the initial guesses? (2) How does the secant method change as you vary the tolerance? (3) Explain your results.

Problem 5.4 In celestial mechanics, Kepler's equation may be used to determine the position of an object in an elliptical orbit. This nonlinear equation is

$$\varepsilon \sin(\omega) - \omega = t,$$

where

- $\varepsilon = \sqrt{1 - \frac{b^2}{a^2}}$ is the object's orbital eccentricity,
- t is proportional to time and
- ω is the sweeping angle of the object around its elliptical orbit.

In this problem, we will use the parameters $a = 2.0$ and $b = 1.25$ to define the orbit for our object. We wish to solve this equation to find the angle ω at various time t to obtain $\omega(t)$, ω as a function of t.

Create a MATLAB m-script named `kepler.m`, in which you perform the following tasks. For each time value of t in $\{0, 0.001, 0.002, \ldots, 10\}$:

(1) Use Newton's method to solve Kepler's equation for ω (You need to figure out a way to pass the value of t to your Newton's method). Each solve should use a tolerance of 10^{-5}, a maximum of 10 Newton iterations, and should disable output of iteration information to the screen (as you're solving $10/0.001 + 1 = 10001$ nonlinear problems). The initial guess for each solve should be the solution from the previous solve for the previous value of t (why this is a good choice of the initial guess?). Start the first solve with an initial guess of $\omega = 0$.

(2) Using the formula for the radial position of an object at angle ω in it's elliptical orbit,

$$r(\omega) = \frac{ab}{\sqrt{(b\cos\omega)^2 + (a\sin\omega)^2}}$$

compute the Cartesian coordinates of the object, $x(t) = r\cos(\omega)$ and $y(t) = r\sin(\omega)$. Store the values of t, $x(t)$ and $y(t)$ so that you can use them to create three plots: $x(t)$ vs t, $y(t)$ vs t and $y(t)$ vs $x(t)$.

Problem 5.5 Write a program to apply Newton's method to solve the following system

$$\begin{cases} x_1^2 + x_2^2 - 10x_1 + 8 & = & 0 \\ x_1 x_2^2 + x_1 - 10x_2 + 8 & = & 0 \end{cases}$$

starting with $\mathbf{x}_0 = [0.7, 0.5]^T$. Conduct the iteration until $\|\mathbf{x}_{k+1} - \mathbf{x}_k\|_2 \le 0.5 \times 10^{-5}$.

Problem 5.6 To find the points of intersection of three spheres, we need to solve the following nonlinear system

$$\begin{cases} (x-a_1)^2 + (y-b_1)^2 + (z-c_1)^2 & = & r_1^2 \\ (x-a_2)^2 + (y-b_2)^2 + (z-c_2)^2 & = & r_2^2 \\ (x-a_3)^2 + (y-b_3)^2 + (z-c_3)^2 & = & r_3^2 \end{cases}$$

Such a system is relevant for GPS (Global Positioning System) to find the location (x, y, z) of a receiver that receives signal from satellites located at (a_i, b_i, c_i) ($i = 1, 2, 3$) with the distance r_i away. Write a program to solve such a system using Newton's method for given (a_i, b_i, c_i) and r_i, $i = 1, 2, 3$. Test your program with (a_i, b_i, c_i) and r_i, $i = 1, 2, 3$ chosen such that the solution is known and can be compared with your numerical result.

Problem 5.7 Suppose that three objects are fixed at $(-\sqrt{3}/2, 0, 0)$, $(\sqrt{3}/2, 0, 0)$ and $(0, 1, 0)$. The location (x, y, z) of a fourth object is determined by minimizing the following potential energy of the four-object configuration

$$U(x, y, z) = \sum_{i=1}^{3} \left(\frac{1}{r_{i4}^{12}} - \frac{2}{r_{i4}^6} \right)$$

where r_{i4} is the distance between the i-th object and the fourth object. Find the location of the fourth object to minimize the potential energy U.

6

Interpolation

This chapter describes some basic polynomial interpolation methods, including monomial, Lagrange and Newton's interpolation. We present monomial interpolation toward the polynomial interpolant existence and uniqueness theorem and Newton's interpolation toward the derivation of interpolation error. We discuss the behavior of interpolation error to motivate spline interpolation. Finally, we introduce spline interpolation, trigonometric interpolation and discrete Fourier transform.

A mom measured her daughter's heights at new year's days and obtained the following data (Table 6.1)

Year	2015	2016	2018	2019
Height [cm]	153	154	161	165

TABLE 6.1
Heights at new year's days.

She forgot to do it in year 2017. Can you estimate her daughter's height at the new year's day in 2017? Since the mom measured the height at a particular day (the new year's day) of a year, we may regard the daughter's height y ([cm]) as the function of the year x and denote the function as $y = f(x)$. Note that we are given in the table only the data (x_i, y_i) ($y_i = f(x_i)$, $i = 0, 1, 2, 3$) generated by the function $f(x)$ instead of the function $f(x)$ itself. However, we can seek a function $y = p(x)$ that reproduces the data and then use $y = p(x)$ to estimate the height for a year missing in the data (not listed in the table).

For example, it can be easily verified that the following polynomial $p(x)$ can reproduce the data in the table

$$
\begin{aligned}
p(x) \ = \ & 153\frac{(x-2016)(x-2018)(x-2019)}{(2015-2016)(2015-2018)(2015-2019)} \\
+ \ & 154\frac{(x-2015)(x-2018)(x-2019)}{(2016-2015)(2016-2018)(2016-2019)} \\
+ \ & 161\frac{(x-2015)(x-2016)(x-2019)}{(2018-2015)(2018-2016)(2018-2019)} \\
+ \ & 165\frac{(x-2015)(x-2016)(x-2018)}{(2019-2015)(2019-2016)(2019-2018)}
\end{aligned}
\tag{6.1}
$$

DOI: 10.1201/9781003201694-6

We will soon learn how such $p(x)$ is obtained (by verifying that $p(x)$ reproduces the data, you may already get an idea for constructing $p(x)$). So we can use $p(x)$ to estimate the height for year 2017 as $p(2017) = 157$.

The process of finding the function $p(x)$ to reproduce the given data is called **interpolation**. Evaluation of $p(x)$ at a value x gives an **estimation/approximation** of the true value $f(x)$. The error $f(x) - p(x)$ is called **interpolation error**.

6.1   Terminology of Interpolation

In interpolation, we are given **data**, and we seek a function to reproduce the data. Here we consider only data as given in Table 6.2, say the measurements of a quantity y as we alter a parameter x or the data generated by the function $y = f(x)$ (which may be unknown or hard to be operated on): $y_i = f(x_i)$, $i = 0, 1, \ldots, n$.

x	x_0	x_1	$\cdots$	x_n
y	y_0	y_1	$\cdots$	y_n

TABLE 6.2
$(n+1)$ data (data points).

We may also call the data as **data points** and list the $(n+1)$ data points as

$$\{(x_i, y_i)\}_{i=0}^n = \{(x_0, y_0), (x_1, y_1), \ldots, (x_n, y_n)\} \qquad (6.2)$$

The values $x_0, x_1, \ldots, x_n$ are called **nodes**. We assume the nodes are distinct. Without loss of generality, from now on we may also assume

$$x_0 < x_1 < \cdots < x_n$$

Interpolation is a process to find a function $y = p(x)$ that reproduces the given data (i.e. the graph of the function $y = p(x)$ passes through the given data points). We say the function $y = p(x)$ interpolates the data and is called an **interpolant**. To be an interpolant, $y = p(x)$ must satisfy the $(n+1)$ **interpolating conditions**

$$y_i = p(x_i), \quad i = 0, 1, \ldots, n \qquad (6.3)$$

If we restrict interpolation of the $(n+1)$ data points to be a polynomial of degree $\leq n$ (i.e. at most n), then the interpolation is called **polynomial interpolation**.

Example 6.1 Find a polynomial of degree ≤ 1 to interpolate the two distinct data point (x_0, y_0) and (x_1, y_1), where the nodes $x_0 < x_1$.
[Solution:] Let the interpolating polynomial be $y = p_1(x)$. Since $p_1(x)$ is a polynomial

of degree ≤ 1, it is a linear polynomial of the general form $p_1(x) = c_0 + c_1 x$ that satisfies the 2 interpolating conditions

$$y_0 = c_0 + c_1 x_0$$

$$y_1 = c_0 + c_1 x_1$$

which can be solved for the slope c_1 and the y-intercept c_0:

$$c_1 = \frac{y_1 - y_0}{x_1 - x_0}, \quad c_0 = y_0 - \frac{y_1 - y_0}{x_1 - x_0} x_0$$

So we have

$$p_1(x) = y_0 + \frac{y_1 - y_0}{x_1 - x_0}(x - x_0) \tag{6.4}$$

which is simply the line through the two given points (x_0, y_0) and (x_1, y_1). ∎

Remark In the above example, the interpolation is called **linear interpolation**; and $p_1(x)$ can be rewritten as

$$p_1(x) = y_0 \frac{x - x_1}{x_0 - x_1} + y_1 \frac{x - x_0}{x_1 - x_0} \tag{6.5}$$

which is a linear combination of two so-called Lagrange basis functions

$$L_0(x) = \frac{x - x_1}{x_0 - x_1}, \quad L_1(x) = \frac{x - x_0}{x_1 - x_0} \tag{6.6}$$

and if $y_0 = y_1$, the slope $c_1 = 0$ and $p_1(x) = y_0$ (assuming $y_0 \neq 0$) is a constant polynomial of the true degree 0 (< 1).

Interpolation is a method commonly used for **curve fitting**. Another method commonly used for curve fitting is **least squares fitting**, which is described in Chapter 11. The difference in the goals of these two methods is illustrated in Fig. 6.1. In interpolation, we seek a function to pass through all the data points, while in least squares fitting, we seek a function to represent the overall feature/trend of the data points.

6.2 Polynomial Space

To do polynomial interpolation, we first look at how we can build a polynomial. We consider polynomials with real coefficients. The set of all polynomials of degree $\leq n$

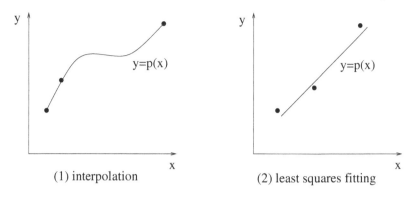

FIGURE 6.1
Curving fitting.

is called a function space (set of functions), or specifically a polynomial space denoted as $\mathbb{P}_n$.

A polynomial of degree $\leq n$ can be written in the general form as

$$p_n(x) = c_0 + c_1 x + c_2 x^2 + \cdots + c_n x^n. \tag{6.7}$$

The coefficients of $p_n(x)$ can form a vector $[c_0, c_1, \ldots, c_n]^T$ in $\mathbb{R}^{n+1}$. Each polynomial in $\mathbb{P}_n$ therefore corresponds to a vector in $\mathbb{R}^{n+1}$ and vice versa. The polynomial space $\mathbb{P}_n$ is **isomorphic** to $\mathbb{R}^{n+1}$ because there is a bijection (one-to-one correspondence) between the two spaces and the bijection preserves addition and scalar multiplication.

We can build a polynomial in the polynomial space $\mathbb{P}_n$ by forming a linear combination of the **basis** functions of the space $\mathbb{P}_n$. The general form $p_n(x) = c_0 + c_1 x + c_2 x^2 + \cdots + c_n x^n$ can be regarded as a linear combination (using the weights c_0, $c_1, \ldots, c_n$) of the basis functions in the set $\{1, x, \ldots, x^n\}$, which is called the **monomial basis** of $\mathbb{P}_n$.

Soon we will introduce other bases of $\mathbb{P}_n$ for building a polynomial of degree $\leq n$. Below we introduce two **orthogonal bases** for $\mathbb{P}_n$ when $x \in [-1, 1]$. To do so, we introduce inner products over $[-1, 1]$ in $\mathbb{P}_n$ to define orthogonality.

Definition 6.1 *Let $p_n(x)$ and $q_n(x)$ be members of $\mathbb{P}_n$ (i.e. $p_n(x) \in \mathbb{P}_n$ and $q_n(x) \in \mathbb{P}_n$), the inner product of $p_n(x)$ and $q_n(x)$ over $[-1, 1]$ with respect to the weight function $w(x) \geq 0$ is defined as*

$$\langle p_n, q_n \rangle_w = \int_{-1}^{1} p_n(x) q_n(x) w(x) \, dx$$

We say $p_n(x)$ and $q_n(x)$ are orthogonal if $\langle p_n, q_n \rangle_w = 0$.

Remark Note that the inner product of two vectors $\mathbf{u} = [u_1, u_2, \ldots, u_n]^T$ and $\mathbf{v} = [v_1, v_2, \ldots, v_n]^T$ in $\mathbb{R}^n$, denoted as $\mathbf{u}^T \mathbf{v}$, $\mathbf{u} \cdot \mathbf{v}$ (the dot product) or $\langle \mathbf{u}, \mathbf{v} \rangle$, is defined as a sum

$$\langle \mathbf{u}, \mathbf{v} \rangle = \sum_{k=1}^{n} u_k v_k$$

We know the definite integral is the limit of a Riemann sum. So the above definition of the inner product in the polynomial space (or a function space in general) is an extension of the inner product in the vector space.

6.2.1 Chebyshev Basis

The Chebyshev basis for $\mathbb{P}_n$ over $[-1, 1]$ is the set $\{Y_0(x), Y_1(x), \ldots, Y_n(x)\}$, where the Chebyshev polynomials

$$Y_0(x) = 1$$

$$Y_1(x) = x$$

$$Y_2(x) = 2x^2 - 1$$

$$Y_3(x) = 4x^3 - 3x$$

$$\ldots$$

$$Y_n(x) = \cos(n \cos^{-1} x) \tag{6.8}$$

and they satisfy the triple recurrence relation

$$Y_k(x) = 2xY_{k-1}(x) - Y_{k-2}(x), \quad k = 2, 3, \ldots \tag{6.9}$$

The graphs of the first few Chebyshev polynomials are shown in Fig. 6.2.

Remark We can prove the recurrence relation $Y_k(x) = 2xY_{k-1}(x) - Y_{k-2}(x)$ using the definition $Y_n(x) = \cos(n \cos^{-1} x)$ and trigonometric sum and difference formulas for cosines (give it a try). We can use the recurrence relation $Y_k(x) = 2xY_{k-1}(x) - Y_{k-2}(x)$ with $Y_0(x) = 1$ and $Y_1(x) = x$ to prove that the definition $Y_n(x) = \cos(n \cos^{-1} x)$ does define a polynomial of degree n (show it).

The Chebyshev basis is an orthogonal basis because

$$\langle Y_i, Y_j \rangle_w = \begin{cases} 0, & i \neq j \\ \pi, & i = j = 0 \\ \pi/2, & i = j \neq 0 \end{cases} \tag{6.10}$$

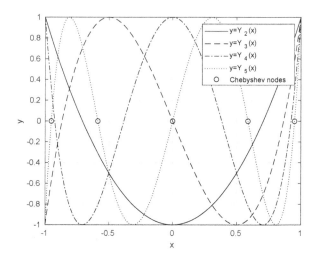

FIGURE 6.2
Graphs of Chebyshev polynomials.

where the weight $w(x) = 1/\sqrt{1-x^2}$.

The Chebyshev polynomial $Y_{n+1}(x)$ of degree $(n+1)$ has $(n+1)$ distinct real roots $x_0, x_1, \ldots, x_n$ in $(-1, 1)$, which are derived as follows

$$Y_{n+1}(x) = \cos[(n+1)\cos^{-1}x] = 0 \Rightarrow (n+1)\cos^{-1}x = \frac{2i+1}{2}\pi, \quad i \in \mathbb{Z}$$

$$\Rightarrow x_i = \cos\left(\frac{2i+1}{2n+2}\pi\right), \quad i = 0, 1, \ldots, n \quad (6.11)$$

Note that $i = 0, 1, \ldots, n$ correspond to the distinct values of $\cos\left(\frac{2i+1}{2n+2}\pi\right)$. When these $(n+1)$ distinct roots are used as the nodes in polynomial interpolation, they are called **Chebyshev nodes**. Shown in Fig. 6.3 are the $n+1 = 5$ Chebyshev nodes, which are also marked as the x-intercepts of $Y_5(x)$ in Fig. 6.2.

It can be shown that these Chebyshev nodes solve the following so-called minimax problem

$$\min_{\{a_0, a_1, \cdots, a_n\}} \max_{-1 \leq x \leq 1} |(x - a_0)(x - a_1) \cdots (x - a_n)| \quad (6.12)$$

i.e. the minimax problem is solved when $a_0, a_1, \ldots, a_n$ are the above Chebyshev nodes $x_0, x_1, \ldots, x_n$. Later, we will use this fact to help reduce interpolation error by Chebyshev nodes.

Remark In this remark, we show that the minimax problem is solved by Chebyshev nodes.

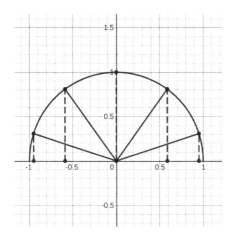

FIGURE 6.3

The $n+1 = 5$ Chebyshev nodes locate at the five solid dots on the horizontal axis. They are the cosine values of five angles in the unit circle.

By the definition $Y_n(x) = \cos(n\cos^{-1}x)$ $(-1 \le x \le 1)$, we have $|Y_n(x)| \le 1$. By the recurrence relation $Y_k(x) = 2xY_{k-1}(x) - Y_{k-2}(x)$ with $Y_0(x) = 1$ and $Y_1(x) = x$, we know the leading term in $Y_n(x)$ is $2^{n-1}x^n$ for $n \ge 1$. By the fundamental theorem of algebra, we have

$$\frac{1}{2^{n-1}}|Y_n(x)| = |(x-x_0)(x-x_1)\cdots(x-x_n)| \tag{6.13}$$

where $x_0, x_1, \dots, x_n$ are the n zeros of $Y_n(x)$ (Chebyshev nodes). So we have

$$|(x-x_0)(x-x_1)\cdots(x-x_n)| = \frac{1}{2^{n-1}}|Y_n(x)| \le \frac{1}{2^{n-1}}, \quad -1 \le x \le 1 \tag{6.14}$$

Note that the upper bound $1/2^{n-1}$ is sharp and can be reached when $|Y_n(x)| = |\cos(n\cos^{-1}x)| = 1$, i.e.

$$n\cos^{-1}x = k\pi, \quad k = 0,1,2,\dots,(n-1), n \Rightarrow$$

$$x = \cos 0 = 1, \cos\frac{\pi}{n}, \cos\frac{2\pi}{n}, \cdots \frac{(n-1)\pi}{n}, \cos\pi = -1 \tag{6.15}$$

Now consider the polynomial

$$Q_n(x) = (x-a_1)(x-a_2)\cdots(x-a_n) \tag{6.16}$$

We can show that the minimax problem

$$\min_{\{a_0,a_1,\cdots,a_n\}} \max_{-1 \le x \le 1} |Q_n(x)| \tag{6.17}$$

is solved by the Chebyshev nodes $x_0, x_1, \ldots, x_n$ (i.e. when $Q_n(x) = \frac{1}{2^{n-1}}Y_n(x) = (x-x_0)(x-x_1)\cdots(x-x_n)$, the maximum value of $Q_n(x)$ for $-1 \le x \le 1$ is the minimum). We know $\frac{1}{2^{n-1}}|Y_n(x)| \le \frac{1}{2^{n-1}}$, $-1 \le x \le 1$. If we can show that $|Q_n(x)| \ge \frac{1}{2^{n-1}}$ for $-1 \le x \le 1$ when $Q_n(x) \ne \frac{1}{2^{n-1}}Y_n(x)$, then we know $\frac{1}{2^{n-1}}Y_n(x)$ gives the solution of the minimax problem.

We can use proof by contradiction. Assume $|Q_n(x)| < \frac{1}{2^{n-1}}$ for $-1 \le x \le 1$ and $Q_n(x) \ne \frac{1}{2^{n-1}}Y_n(x)$. Let

$$F(x) = Q_n(x) - \frac{1}{2^{n-1}}Y_n(x) \tag{6.18}$$

Then $F(x)$ is a nonzero polynomial of degree $< n$ as both $Q_n(x)$ and $\frac{1}{2^{n-1}}Y_n(x)$ have the same leading term x^n. So $F(x)$ has at most $n-1$ zeros by the fundamental theorem of algebra. Consider the values of $F(x)$ at the $(n+1)$ points $x = \cos 0 = 1, \cos\frac{\pi}{n}, \cos\frac{2\pi}{n}, \cdots \frac{(n-1)\pi}{n}, \cos\pi = -1$ where $|Y_n(x)| = |\cos(n\cos^{-1}x)| = 1$. Since $|Q_n(x)| < \frac{1}{2^{n-1}}$ by the assumption, we have

$$F(1) = Q_n(1) - \frac{1}{2^{n-1}}Y_n(1) = Q_n(1) - \frac{1}{2^{n-1}} < 0$$

$$F\left(\cos\frac{\pi}{n}\right) = Q_n\left(\cos\frac{\pi}{n}\right) - \frac{1}{2^{n-1}}Y_n\left(\cos\frac{\pi}{n}\right) = Q_n\left(\cos\frac{\pi}{n}\right) + \frac{1}{2^{n-1}} > 0$$

$$F\left(\cos\frac{2\pi}{n}\right) = Q_n\left(\cos\frac{2\pi}{n}\right) - \frac{1}{2^{n-1}}Y_n\left(\cos\frac{2\pi}{n}\right) = Q_n\left(\cos\frac{2\pi}{n}\right) - \frac{1}{2^{n-1}} < 0$$

$$\vdots$$

The signs of the above values alternate, leading to n sign changes of $F(x)$. So $F(x)$ must have n zeros by the intermediate value theorem, which is in contradiction with the fact that $F(x)$ is a nonzero polynomial of degree $< n$ and has at most $(n-1)$ zeros. Finally we conclude that no such $|Q_n(x)| < \frac{1}{2^{n-1}}$, and $\frac{1}{2^{n-1}}Y_n(x)$ gives the solution of the minimax problem.

6.2.2 Legendre Basis

Legendre basis for $\mathbb{P}_n$ over $[-1,1]$ is the set $\{P_0(x), P_1(x), \ldots, P_n(x)\}$, where the Legendre polynomials

$$P_0(x) = 1$$

$$P_1(x) = x$$

$$P_2(x) = \frac{1}{2}(3x^2 - 1)$$

$$P_3(x) = \frac{1}{2}(5x^3 - 3x)$$

$$\cdots$$

$$P_k(x) = \frac{1}{2^k k!} \frac{d^k}{dx^k} [(x^2 - 1)^k]$$ (6.19)

where the last formula is called **Rodrigues' formula**. Legendre polynomials satisfy the triple recurrence relation

$$(k+1)P_{k+1}(x) = (2k+1)P_k(x) - kP_{k-1}(x)$$ (6.20)

The graphs of the first few Legendre polynomials are shown in Fig. 6.4.

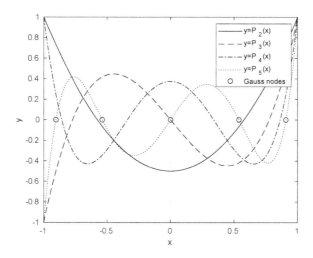

FIGURE 6.4
Graphs of Legendre polynomials.

Remark The Legendre polynomial $P_k(x)$ is defined as a polynomial solution of the Legendre equation

$$[(1 - x^2)y']' + k(k+1)y = 0, \quad -1 \leq x \leq 1$$ (6.21)

with $P_k(1) = 1$. The Legendre equation

$$[(1 - x^2)y']' + \lambda y = 0, \quad -1 \leq x \leq 1$$ (6.22)

with a bounded solution $y = y(x)$ on $(-1, 1)$ is called a singular Sturm-Liouville problem. The Legendre polynomial $P_k(x)$ is an eigenfunction corresponding to the eigenvalue $\lambda = k(k+1)$ of the problem.

It can be verified by induction that the Legendre polynomial $P_k(x)$ given by Rodrigues' formula satisfies the Legendre equation. The triple recurrence relation $(k+1)P_{k+1}(x) = (2k+1)P_k(x) - kP_{k-1}(x)$ can be proved using Rodrigues' formula.

Legendre basis is an orthogonal basis because

$$\langle P_i, P_j \rangle = \int_{-1}^{1} P_i(x)P_j(x)\mathrm{d}x = \left\{ \begin{array}{ll} 0, & i \neq j \\ 2/(2i+1), & i = j \end{array} \right. \tag{6.23}$$

The Legendre polynomial $P_{n+1}(x)$ of degree $(n+1)$ has $(n+1)$ distinct real roots x_0, $x_1, \ldots, x_n$ (see the x-intercepts in Fig. 6.4) over $(-1,1)$. These $(n+1)$ distinct roots are called **Gauss nodes** in Chapter 7 for numerical integration. The $n+1 = 5$ Gauss nodes are marked in Fig. 6.4.

Remark The orthogonality of Legendre polynomials can be shown using the fact that they are eigenfunctions of Sturm-Liouville problem

$$[(1-x^2)y']' + \lambda y = 0, \quad -1 \leq x \leq 1 \tag{6.24}$$

Consider two different $P_i(x)$ and $P_j(x)$. They satisfy

$$[(1-x^2)P_i(x)']' + \lambda_i P_i(x) = 0 \tag{6.25}$$
$$[(1-x^2)P_j(x)']' + \lambda_j P_j(x) = 0 \tag{6.26}$$

where $\lambda_i = i(i+1) \neq \lambda_j = j(j+1)$. It can be shown that

$$(\lambda_i - \lambda_j)P_i(x)P_j(x) = [(x^2-1)(P_i'(x)P_j(x) - P_j'(xP_i(x)))]' \tag{6.27}$$

which can be integrated from -1 to 1 to reveal the orthogonality.

Remark To prove that the Legendre polynomial $P_{n+1}(x)$ of degree $(n+1)$ has $(n+1)$ distinct real roots $x_0, x_1, \ldots, x_n$ over $(-1,1)$, we can use Rodrigues's formula with Rolle's theorem. By Rodrigues' formula,

$$P_{n+1}(x) = \frac{1}{2^{n+1}(n+1)!}\frac{\mathrm{d}^{n+1}}{\mathrm{d}x^{n+1}}[(x^2-1)^{n+1}] \tag{6.28}$$

Note that $x^2 - 1 = 0$ at $x = -1$ and 1, and $\frac{\mathrm{d}^m}{\mathrm{d}x^m}[(x^2-1)^{n+1}]$ has a factor (x^2-1) whenever for $m < n+1$. Using Rolle's theorem, we know
- $\frac{\mathrm{d}}{\mathrm{d}x}[(x^2-1)^{n+1}]$ has a zero $r_{11} \in (-1,1)$ by $x^2 - 1 = 0$ at $x = -1$ and 1.

- $\frac{d^2}{dx^2}[(x^2-1)^{n+1}]$ has two simple zeros r_{21} and r_{22} over $(-1,1)$ by $\frac{d}{dx}[(x^2-1)^{n+1}] = 0$ at -1, r_{11} and 1. Recall that $\frac{d}{dx}[(x^2-1)^{n+1}]$ equals 0 at -1 and 1 because it has the factor (x^2-1).
- $\frac{d^3}{dx^3}[(x^2-1)^{n+1}]$ has three simple zeros r_{31}, r_{32} and r_{33} over $(-1,1)$ by $\frac{d^2}{dx^2}[(x^2-1)^{n+1}] = 0$ at -1, r_{21}, r_{22} and 1.
- $\vdots$
- $\frac{d^{n+1}}{dx^{n+1}}[(x^2-1)^{n+1}]$ has $(n+1)$ simple zeros $x_0, x_1, \ldots, x_n$ over $(-1,1)$.

6.3 Monomial Interpolation

Monomial interpolation seeks a polynomial $p_n(x)$ of degree $\leq n$ in the general form as a linear combination of monomials $1, x, \ldots, x^n$, i.e.

$$p_n(x) = c_0 + c_1 x + c_2 x^2 + \cdots + c_n x^n \tag{6.29}$$

to interpolate the $(n+1)$ distinct data points $\{(x_i, y_i)\}_{i=0}^{n} = \{(x_0, y_0), (x_1, y_1), \ldots, (x_n, y_n)\}$ with distinct nodes $x_0, x_1, \ldots, x_n$.

To determine $p_n(x)$, we need to find the $(n+1)$ coefficients $c_0, c_1, \ldots, c_n$ by satisfying the $(n+1)$ interpolating conditions

$$p_n(x_i) = y_i, \quad i = 0, 1, \ldots, n \tag{6.30}$$

Example 6.2 Interpolate the 3 data points $(-1, 1)$, $(0, 1)$ and $(1, 3)$ by a polynomial of degree ≤ 2.

[Solution:] Let the interpolating polynomial be

$$p_2(x) = c_0 + c_1 x + c_2 x^2$$

It must satisfy the 3 interpolating conditions

$$p_2(-1) = 1: \quad c_0 + c_1(-1) + c_2(-1)^2 = 1$$
$$p_2(0) = 1: \quad c_0 + c_1(0) + c_2(0)^2 = 1$$
$$p_2(1) = 3: \quad c_0 + c_1(1) + c_2(1)^2 = 1$$

So the three coefficients c_0, c_1 and c_2 in $p_2(x)$ satisfy the linear system

$$\begin{bmatrix} 1 & (-1) & (-1)^2 \\ 1 & (0) & (0)^2 \\ 1 & (1) & (1)^2 \end{bmatrix} \begin{bmatrix} c_0 \\ c_1 \\ c_2 \end{bmatrix} = \begin{bmatrix} 1 \\ 1 \\ 3 \end{bmatrix}$$

which can be solved to give $c_0 = 1$, $c_1 = 1$ and $c_2 = 1$. So $p_2(x) = 1 + x + x^2$. ∎

Remark If the y-values of the data in the above example are changed to the same value 1, then $c_0 = 1$, $c_1 = 0$, $c_2 = 0$ and $p_2(x) = 1$. The true degree of $p_2(x)$ now becomes 0 (≤ 2), and the curve $y = p_2(x)$ is just a horizontal line $y = 1$ through the 3 data points.

The monomial interpolation problem of finding $p_n(x)$ to interpolate the $(n+1)$ distinct data points $\{(x_i, y_i)\}_{i=0}^{n}$ can be solved similarly as in the above example. The $(n+1)$ interpolating conditions are

$$p_n(x_0) = y_0 : \quad c_0 + c_1(x_0) + c_2(x_0)^2 + \cdots + c_n(x_0)^n = y_0$$

$$p_n(x_1) = y_0 : \quad c_0 + c_1(x_1) + c_2(x_1)^2 + \cdots + c_n(x_1)^n = y_1$$

$$p_n(x_2) = y_0 : \quad c_0 + c_1(x_2) + c_2(x_2)^2 + \cdots + c_n(x_2)^n = y_2$$

$$\cdots\cdots$$

$$p_n(x_n) = y_0 : \quad c_0 + c_1(x_n) + c_2(x_n)^2 + \cdots + c_n(x_n)^n = y_n$$

So the $(n+1)$ coefficients $c_0, c_1, c_2, ..., c_n$ in $p_n(x)$ satisfy a linear system of order $(n+1)$

$$
\begin{bmatrix}
1 & (x_0) & (x_0)^2 & \cdots & (x_0)^n \\
1 & (x_1) & (x_1)^2 & \cdots & (x_1)^n \\
1 & (x_2) & (x_2)^2 & \cdots & (x_2)^n \\
\vdots & \vdots & \vdots & \vdots & \vdots \\
1 & (x_n) & (x_n)^2 & \cdots & (x_n)^n
\end{bmatrix}
\begin{bmatrix}
c_0 \\ c_1 \\ c_2 \\ \vdots \\ c_n
\end{bmatrix}
=
\begin{bmatrix}
y_0 \\ y_1 \\ y_2 \\ \vdots \\ y_n
\end{bmatrix}
\tag{6.31}
$$

Let's denote the coefficient matrix of the linear system as V. The matrix V is called a **Vandermonde matrix**, and its columns can be generated by the vector $[x_0, x_1, x_2, \ldots, x_n]^T$ by applying entry-wise powers from 0 to n.

It is known that the determinant of the Vandermonde matrix V is

$$
\begin{aligned}
\det V &= [(x_1 - x_0)(x_2 - x_0) \cdots (x_n - x_0)][(x_2 - x_1)(x_3 - x_1) \cdots (x_n - x_1)] \cdots \\
&\quad [(x_{n-1} - x_{n-2})(x_n - x_{n-2})][(x_n - x_{n-1})] \\
&= \prod_{i=0}^{n} \prod_{j=i+1}^{n} (x_j - x_i)
\end{aligned}
\tag{6.32}
$$

where the product symbol $\prod$ is similar to the summation symbol $\sum$ but with product operation as $\prod_{i=0}^{n} a_i = a_0 a_1 \cdots a_n$. Since the nodes $x_0, x_1, ..., x_n$ are distinct, $\det V \neq 0$ and V is invertible/nonsingular. The linear system therefore has a unique solution. So $c_0, c_1, ..., c_n$ are determined uniquely. We then have the following polynomial interpolant existence and uniqueness theorem.

Theorem 6.1 (Polynomial Interpolant Existence and Uniqueness) *For any real data $\{(x_i, y_i)\}_{i=0}^{n}$ with distinct nodes x_0, x_1, ... , x_n, there exists a unique polynomial $p_n(x)$ of degree $\leq n$ that interpolates the data.*

Remark Note that the uniqueness requires the degree is at most n for $(n+1)$ data. For example, that $p_1(x) = 1 + x$, $p_2(x) = 1 + x^2$ and $p_3(x) = 1 + x^3$ all interpolate the 2 data $(0, 1)$ and $(1, 2)$ does not violate the theorem. For the 2 data, only polynomial of degree ≤ 1 is unique (meaning there are not two different polynomials of degree ≤ 1 that interpolate the same 2 data), an obvious fact in geometry that there is only one line through two distinct points.

6.4 Lagrange Interpolation

We can construct an interpolating polynomial without solving coupled linear equations (as in monomial interpolation). We do so by using a linear combination of a particular basis, the Lagrange basis, to build the interpolating polynomial.

Example 6.3 The polynomial $p_2(x)$ of degree ≤ 2 that interpolates the 3 data points (x_0, y_0), (x_1, y_1) and (x_2, y_2) can be constructed as

$$p_2(x) = y_0 L_0(x) + y_1 L_1(x) + y_2 L_2(x)$$

where $L_0(x)$, $L_1(x)$ and $L_2(x)$ form a Lagrange basis and are called Lagrange basis functions or cardinal polynomials of Lagrange form. They are polynomials of degree 2 given by

$$L_0(x) = \frac{(x - x_1)(x - x_2)}{(x_0 - x_1)(x_0 - x_2)} \tag{6.33}$$

$$L_1(x) = \frac{(x - x_0)(x - x_2)}{(x_1 - x_0)(x_1 - x_2)} \tag{6.34}$$

$$L_2(x) = \frac{(x - x_0)(x - x_1)}{(x_2 - x_0)(x_2 - x_1)} \tag{6.35}$$

The linear combination of the three degree-2 cardinal polynomials of Lagrange form as in $p_2(x)$ is certainly a polynomial of degree ≤ 2. Take a close look at a Lagrange basis function, say $L_1(x)$. We find out the following features.

- The numerator of $L_1(x)$ has all the binomial factors $(x - x_0)$ and $(x - x_2)$ except $(x - x_1)$. So $L_1(x) = 0$ at all the nodes x_0 and x_2 except x_1.

- The denominator of $L_1(x)$ is just its numerator evaluated at x_1. So $L_1(x) = 1$ at the node x_1.

The other Lagrange basis functions have the similar features. So a Lagrange basis function satisfy a very important property

$$L_i(x_j) = \delta_{ij} = \begin{cases} 0, & \text{if } j \neq i \\ 1, & \text{if } j = i \end{cases} \tag{6.36}$$

where δ_{ij} is the entry of the identity matrix at row i and column j, and is called the **Kronecker delta**. So $p_2(x) = y_0 L_0(x) + y_1 L_1(x) + y_2 L_2(x)$ constructed in this way satisfies

$$p_2(x_i) = y_0 L_0(x_i) + y_1 L_1(x_i) + y_2 L_2(x_i) = y_i, \quad i = 0, 1, 2$$

and $p_2(x)$ is polynomial of degree ≤ 2 that interpolates the given 3 data. ∎

We can easily extend the construction process in the above example to the general case. To interpolate $(n+1)$ data points $\{(x_i, y_i)\}_{i=0}^n = \{(x_0, y_0), (x_1, y_1), \ldots, (x_n, y_n)\}$ with distinct nodes $x_0, x_1, \ldots, x_n$, we can construct the interpolating polynomial $p_n(x)$ of degree $\leq n$ in Lagrange form as

$$p_n(x) = y_0 L_0(x) + y_1 L_1(x) + \cdots + y_n L_n(x) = \sum_{i=0}^n y_i L_i(x) \tag{6.37}$$

where $L_i(x)$ $(i = 0, 1, \ldots, n)$ is a Lagrange basis function that is constructed as

$$L_i(x) = \frac{(x - x_0)(x - x_1) \cdots (x - x_{i-1})(x - x_{i+1}) \cdots (x - x_n)}{(x_i - x_0)(x_i - x_1) \cdots (x_i - x_{i-1})(x_i - x_{i+1}) \cdots (x_i - x_n)}$$
$$= \frac{\prod_{k=0, k \neq i}^n (x - x_k)}{\prod_{k=0, k \neq i}^n (x_i - x_k)} \tag{6.38}$$

The numerator of $L_i(x)$ is the product of all the binomial factors $(x - x_0)$, $(x - x_1)$, ..., $(x - x_n)$ except $(x - x_i)$, and the denominator of $L_i(x)$ is its numerator evaluated at x_i. Clearly $L_i(x)$ is a polynomial of degree n, and it has the following property

$$L_i(x_j) = \delta_{ij} = \begin{cases} 0, & \text{if } j \neq i \\ 1, & \text{if } j = i \end{cases}; \quad \text{where } i = 0, 1, \ldots, n; j = 0, 1, \ldots, n \tag{6.39}$$

So $p_n(x)$, as a linear combinations of the Lagrange basis functions of degree n, is a polynomial of degree $\leq n$, and it satisfies

$$p_n(x_j) = y_j, \quad j = 0, 1, \ldots, n \tag{6.40}$$

i.e. it interpolates the given $(n+1)$ distinct data.

Example 6.4 Find a polynomial of degree ≤ 2 in Lagrange form to interpolate the data $(-1, 1)$, $(0, 1)$ and $(1, 3)$, then simplify it into the standard form to compare with

the result in Example 6.2. Why must the current result be same as that in Example 6.2?
[**Solution:**] The interpolating polynomial in Lagrange form is

$$p_2(x) = 1L_0(x) + 1L_1(x) + 3L_2(x)$$

where

$$L_0(x) = \frac{(x-0)(x-1)}{(-1-0)(-1-1)} = \frac{x^2 - x}{2}$$

$$L_1(x) = \frac{(x+1)(x-1)}{(0+1)(0-1)} = \frac{x^2 - 1}{-1}$$

$$L_0(x) = \frac{(x+1)(x-0)}{(1+1)(1-0)} = \frac{x^2 + x}{2}$$

After simplification, $p_2(x) = 1 + x + x^2$. It must be the same as the result in Example 6.2 because of the Polynomial Interpolant Existence and Uniqueness Theorem. Shown in Fig. 6.5 is the graph of three Lagrange basis functions, which demonstrate the property given in Eq. (6.39).

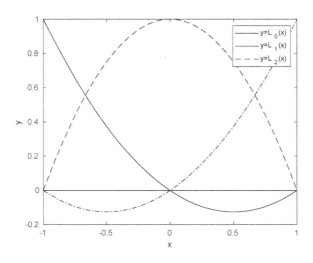

FIGURE 6.5
Three Lagrange basis functions for the nodes $x_0 = -1$, $x_1 = 0$ and $x_2 = 1$.

■

6.5 Newton's Interpolation

In Newton's interpolation, an interpolating polynomial is built by forming a linear combination of Newton polynomial basis functions.

Example 6.5 Interpolate the following 3 data points by a polynomial of degree ≤ 2.

x	-1	0	1
y	1	1	3

[**Solution:**] This time, we find the interpolating polynomial $p_2(x)$ of degree ≤ 2 in Newton form as a linear combination of three Newton basis functions as

$$p_2(x) = d_0 N_0(x) + d_1 N_1(x) + d_2 N_2(x)$$

where the Newton basis functions are constructed using the given nodes as follows

$$N_0(x) = 1$$

$$N_1(x) = (x - x_0) = (x - (-1))$$
$$N_2(x) = (x - x_0)(x - x_1) = (x - (-1))(x - 0)$$

and the constant weights d_0, d_1 and d_2 can be determined in turn from the interpolating conditions as

$$p_2(x_0) = y_0 : d_0 + 0 + 0 = y_0 \Rightarrow d_0 = 1$$

$$p_2(x_1) = y_1 : d_0 + d_1(x_1 - x_0) + 0 = y_1 \Rightarrow d_1 = \frac{y_1 - d_0}{x_1 - x_0} = 0$$

$$p_2(x_2) = y_2 : d_0 + d_1(x_2 - x_0) + d_2(x_2 - x_0)(x_2 - x_1) = y_2$$

$$\Rightarrow d_2 = \frac{y_2 - d_0 - d_1(x_2 - x_0)}{(x_2 - x_0)(x_2 - x_1)} = 1$$

Again, $p_2(x)$ can be simplified to $p_2(x) = 1 + x + x^2$, which is the same as in Examples 6.2 and 6.4 because of the Polynomial Interpolant Existence and Uniqueness Theorem.

Note that

- The Newton basis function $N_i(x) = (x - x_0) \cdots (x - x_{i-1})$ (except $N_0(x) = 1$) is the product of the binomials and is a polynomial of degree i with zeros $x_0, \ldots, x_{i-1}$ (i.e. $N_i(x) = 0$ at all the nodes from x_0 to x_{i-1}).
- The weight d_i in the linear combination of the Newton basis functions is determined using forward substitution after $d_0, \ldots, d_{i-1}$ are determined. If we write the interpolating conditions as a linear system in the unknowns weights, the system is lower triangular as

$$\begin{bmatrix} 1 & 0 & 0 \\ 1 & (x_1 - x_0) & 0 \\ 1 & (x_2 - x_0) & (x_2 - x_0)(x_2 - x_1) \end{bmatrix} \begin{bmatrix} d_0 \\ d_1 \\ d_2 \end{bmatrix} = \begin{bmatrix} y_0 \\ y_1 \\ y_2 \end{bmatrix} \qquad (6.41)$$

∎

x	x_0	x_1	$\cdots$	x_n
y	$f(x_0)$	$f(x_1)$	$\cdots$	$f(x_n)$

TABLE 6.3
Data (data points) generated by the function $y = f(x)$.

Now let's interpolate the $(n+1)$ distinct data in Table 6.3 using a polynomial of degree $\leq n$ in Newton form. The data are generated by the function $f(x)$ at the distinct nodes $x_0, x_1, \ldots, x_n$.

Similar as the previous example, $p_n(x)$ is

$$p_n(x) = d_0 N_0(x) + d_1 N_1(x) + \cdots + d_n N_n(x) \tag{6.42}$$

where the Newton basis functions are

$$N_0(x) = 1$$

$$N_1(x) = (x - x_0)$$

$$N_2(x) = (x - x_0)(x - x_1)$$

$$\cdots$$

$$N_n(x) = (x - x_0)(x - x_1)\cdots(x - x_{n-1}) \tag{6.43}$$

We have

$$p_n(x) = d_0 + d_1(x - x_0) + \cdots + d_n(x - x_0)(x - x_1)\cdots(x - x_{n-1}) \tag{6.44}$$

Note that $N_i(x) = (x - x_0)\cdots(x - x_{i-1})$ $(i = 1,\ldots,n)$ is a polynomial of degree i satisfying

$$N_i(x_j) = 0, \quad j = 0, 1, \ldots, i - 1 \tag{6.45}$$

and can be written in a compact way as

$$N_i(x) = \prod_{j=0}^{i-1}(x - x_j) \tag{6.46}$$

The constant weights $d_0, d_1, \ldots, d_n$ are determined in turn from the interpolating conditions $p_n(x_0) = f(x_0)$, $p_n(x_1) = f(x_1)$, $\ldots$, $p_n(x_n) = f(x_n)$. The linear system for the weights is the following lower triangular system by the property in Eq. (6.45).

$$
\begin{bmatrix}
1 & & & & \\
1 & N_1(x_1) & & & \\
1 & N_1(x_2) & N_2(x_2) & & \\
\vdots & \vdots & \vdots & \ddots & \\
1 & N_1(x_n) & N_2(x_n) & \cdots & N_n(x_n)
\end{bmatrix}
\begin{bmatrix}
d_0 \\
d_1 \\
d_2 \\
\vdots \\
d_n
\end{bmatrix}
=
\begin{bmatrix}
f(x_0) \\
f(x_1) \\
f(x_2) \\
\vdots \\
f(x_n)
\end{bmatrix}
\tag{6.47}
$$

The weights d_0, d_1, ... , d_n are determined in turn using forward substitution as follows. Define the polynomial consisting of the first $(i+1)$ terms in $p_n(x)$ as $p_i(x)$ $(i = 0,1,\ldots,n)$, where

$$p_i(x) = d_0N_0(x) + d_1N_1(x) + \cdots + d_iN_i(x)$$
$$= d_0 + d_1(x - x_0) + \cdots + d_i(x - x_0)(x - x_1)\cdots(x - x_{i-1}) \tag{6.48}$$

We have

$$p_n(x_0) = f(x_0) : d_0 = f(x_0)$$

$$p_n(x_1) = f(x_1) : d_0 + d_1N_1(x_1) = f(x_1) \Rightarrow d_1 = \frac{f(x_1) - d_0}{N_1(x_1)} = \frac{f(x_1) - p_0(x_1)}{N_1(x_1)}$$

$$\cdots\cdots$$

After d_0, d_1, ... , d_{i-1} are determined, d_i $(i = 1,2,\ldots,n)$ is determined from the interpolating condition $p_n(x_i) = f(x_i)$ as

$$p_n(x_i) = f(x_i) : d_0 + d_1N_1(x_i) + \cdots + d_iN_i(x_i) = f(x_i) \Rightarrow \tag{6.49}$$

$$d_i = \frac{f(x_i) - (d_0 + d_1N_1(x_i) + \cdots + d_{i-1}N_{i-1}(x_i))}{N_i(x_i)} = \frac{f(x_i) - p_{i-1}(x_i)}{N_i(x_i)} \tag{6.50}$$

Note that the expression of d_i only involves the nodes x_0, x_1, ..., x_i and the function values $f(x_0)$, $f(x_1)$, ..., $f(x_i)$. It is therefore also denoted as

$$d_i = f[x_0, x_1, \ldots, x_i] \tag{6.51}$$

and is called the **divided difference** of order i for f. In addition, the polynomial $p_i(x)$ interpolates the first $(i+1)$ data $\{(x_k, f(x_k))\}_{k=0}^i$.

The polynomial $p_n(x)$ determined above interpolates the given $(n+1)$ data. If we add one more data point $(x_{n+1}, f(x_{n+1}))$ to the initial data set so that the new data set has $(n+2)$ data as in Table 6.4. Then the polynomial $p_{n+1}(x)$ of degree $\leq (n+1)$

					added data
x	x_0	x_1	$\cdots$	x_n	x_{n+1}
y	$f(x_0)$	$f(x_1)$	$\cdots$	$f(x_n)$	$f(x_{n+1})$

TABLE 6.4
Data (data points) generated by the function $y = f(x)$.

that interpolates the expanded data set is simply

$$p_{n+1}(x) = p_n(x) + d_{n+1}N_{n+1}(x) = p_n(x) + d_{n+1}(x - x_0)(x - x_1)\cdots(x - x_n) \tag{6.52}$$

where d_{n+1} is determined from the additional interpolating condition

$$p_{n+1}(x_{n+1}) = f(x_{n+1}) \Rightarrow d_{n+1} = \frac{f(x_{n+1}) - p_n(x_{n+1})}{N_{n+1}(x_{n+1})} = f[x_0, x_1, \ldots, x_{n+1}] \tag{6.53}$$

It can be easily verified that $p_{n+1}(x_i) = p_n(x_i) = f(x_i)$ for $i = 0, 1, \ldots, n$, so $p_{n+1}(x)$ satisfies all the interpolating conditions.

So after $p_n(x)$ is known, we can use it to obtain $p_{n+1}(x)$ as above to interpolate one more data point. Newton's interpolation provides an **adaptive** way to expand an existing polynomial with one more term to interpolate one more data point. By the polynomial interpolant existence and uniqueness theorem, it does not matter what form $p_n(x)$ (e.g. Lagrange form or Newton form) takes as long as it is an interpolating polynomial of degree $\leq n$.

Example 6.6 $p_2(x) = 1 + x + x^2$ interpolates the first three data points in the following table. Find the polynomial $p_3(x)$ of degree ≤ 3 that interpolate all the four data.

x	-1	0	1	2
y	1	1	3	19

[Solution:] We can expand $p_2(x) = 1 + x + x^2$ to form $p_3(x)$ as

$$p_3(x) = p_2(x) + d_3(x - x_0)(x - x_1)(x - x_2) = p_2(x) + d_3(x + 1)x(x - 1)$$

Clearly $p_3(x)$ also interpolates the first three data points because $p_3(x_i) = p_2(x_i) = y_i$ for $i = 0, 1, 2$. We determine d_3 so that $p_3(x)$ also interpolate the extra data point $(x_3, y_3) = (2, 19)$ as the following

$$p_3(x_3) = y_3 : \ p_2(x_3) + d_3(x_3 - x_0)(x_3 - x_1)(x_3 - x_2) = y_3 \Rightarrow d_3 = \frac{19 - p_2(2)}{(2+1)2(2-1)} = 2$$

Finally, we have

$$p_3(x) = p_2(x) + d_3(x + 1)x(x - 1) = 1 + x + x^2 + 2(x + 1)x(x - 1)$$

The interpolating polynomial $p_n(x)$ in Newton's form can be written in nested form to be evaluated by Horner's method. The nested form is

$$p_n(x) = d_0 + (x - x_0)(d_1 + (x - x_1)(d_2 + \cdots + (x - x_{n-2})(d_{n-1} + d_n(x - x_{n-1})) \cdots))$$

$$(6.54)$$

6.6 Interpolation Error

We can approximate the function $y = f(x)$ by a polynomial $y = p_n(x)$ of degree $\leq n$ that interpolates the $(n + 1)$ data generated by $y = f(x)$ at the distinct nodes $x_0, x_1, \ldots, x_n$. The error $f(x) - p_n(x)$ in the approximation is called polynomial interpolation error.

Remark In Chapter 3, we use a Taylor polynomial to approximate a function and analyze the error (Taylor's remainder) in the approximation using Taylor's theorem. Later, we will compare Taylor polynomial approximation with interpolating polynomial approximation.

Before we give the formula for interpolation error, we list a few reasons why we want to approximate $f(x)$ by the interpolating polynomial $p_n(x)$.

- If $f(x)$ is given as tabulated data, we can estimate $f(x)$ at a non-node x-value using $p_n(x)$, and approximate the derivatives of $f(x)$ by the derivatives of $p_n(x)$.
- We can derive a numerical differentiation rule $f'(x) \approx p'_n(x)$ that involves only the function values of $f(x)$ at nodes (also called **stencil points** in numerical differentiation).
- We can derive a numerical integration rule $\int_a^b p_n(x)dx$ that involves only the function values of $f(x)$ at nodes to approximate the integral $\int_a^b f(x)dx$. This is very useful when an antiderivative $f(x)$ does not have a closed form.

6.6.1 Error in Polynomial Interpolation

Theorem 6.2 (Interpolation Error for General Nodes) *Let $p_n(x)$ be the polynomial of degree $\leq n$ that interpolates the $(n+1)$ data generated by $f(x)$ at the distinct nodes $x_0, x_1, \ldots, x_n$. Assume, without loss of generality, that $x_0 < x_1 < \cdots < x_n$. Then*

$$f(x) - p_n(x) = \frac{f^{(n+1)}(\xi)}{(n+1)!}(x-x_0)(x-x_1)\cdots(x-x_n) \tag{6.55}$$

where $\xi \in [\min\{x_0,x\}, \max\{x_n,x\}]$.

Remark

$$f(x) - p_n(x) = \frac{f^{(n+1)}(\xi)}{(n+1)!}N_{n+1}(x) \tag{6.56}$$

where $N_{n+1}(x) = (x-x_0)(x-x_1)\cdots(x-x_n)$ has zeros at $x_0, x_1, \ldots, x_n$. The interpolation error equals 0 at the nodes $x_0, x_1, \ldots, x_n$, which must be so due to the interpolating conditions.

Proof *The polynomial $p_n(x)$ interpolates the data in Table 6.5.*

x	x_0	x_1	$\cdots$	x_n
y	$f(x_0)$	$f(x_1)$	$\cdots$	$f(x_n)$

TABLE 6.5
Data (data points) generated by the function $y = f(x)$.

If we add one more data $(x_{n+1}, f(x_{n+1}))$, where x_{n+1} can be anywhere, we can expand $p_n(x)$ to $p_{n+1}(x)$ by Newton's interpolation to interpolate the expanded data set as

$$p_{n+1}(x) = p_n(x) + d_{n+1}N_{n+1}(x) \tag{6.57}$$

where d_{n+1} is determined to satisfy the extra interpolating condition

$$p_{n+1}(x_{n+1}) = f(x_{n+1}) \tag{6.58}$$

and

$$d_{n+1} = f[x_0, x_1, \ldots, x_n] \tag{6.59}$$

We thus have

$$p_n(x_{n+1}) + d_{n+1}N_{n+1}(x_{n+1}) = f(x_{n+1}) \Rightarrow$$

$$f(x_{n+1}) - p_n(x_{n+1}) = d_{n+1}N_{n+1}(x_{n+1}) \tag{6.60}$$

Define the function $\phi(x)$

$$\phi(x) = f(x) - p_{n+1}(x) = f(x) - p_n(x) - d_{n+1}N_{n+1}(x) \tag{6.61}$$

Clearly $\phi(x) = 0$ at the $(n+2)$ nodes $x_0, x_1, \ldots, x_n$, and x_{n+1} because of the interpolating conditions. We can order these $(n+2)$ nodes from the smallest to the largest, depending on the location of x_{n+1}, and then apply Rolles's theorem between any two consecutive nodes. So $\phi'(x) = 0$ has $(n+1)$ zeros between $\min\{x_0, x_{n+1}\}$ and $\max\{x_n, x_{n+1}\}$. Applying Rolles's theorem between any two consecutive zeros of $\phi'(x)$, we then know $\phi''(x)$ has n zeros. Doing so repeatedly, we know at last $\phi^{(n+1)}(x) = 0$ has one zero ξ between $\min\{x_0, x_{n+1}\}$ and $\max\{x_n, x_{n+1}\}$.

Note that $p_n(x)$ is a polynomial of degree $\leq n$ and $N_{n+1}(x) = (x - x_0)(x - x_1)\cdots(x - x_n)$ is a polynomial of degree $(n+1)$ with the leading term x^{n+1}, so $p_n^{(n+1)}(x) = 0$, $N_{n+1}^{(n+1)}(x) = (n+1)!$, and $\phi^{(n+1)}(x) = f^{(n+1)}(x) - d_{n+1}n!$. So we have

$$\phi^{(n+1)}(\xi) = f^{(n+1)}(\xi) - d_{n+1}(n+1)! = 0 \Rightarrow d_{n+1} = \frac{f^{(n+1)}(\xi)}{(n+1)!} \tag{6.62}$$

Finally we have

$$d_{n+1} = f[x_0, x_1, \ldots, x_n] = \frac{f^{(n+1)}(\xi)}{(n+1)!} \tag{6.63}$$

and

$$f(x_{n+1}) - p_n(x_{n+1}) = d_{n+1}N_{n+1}(x_{n+1}) = \frac{f^{(n+1)}(\xi)}{(n+1)!}N_{n+1}(x_{n+1}) \tag{6.64}$$

Since x_{n+1} can be anywhere and can be replaced by x, we then obtain

$$f(x) - p_n(x) = \frac{f^{(n+1)}(\xi)}{(n+1)!}N_{n+1}(x) \tag{6.65}$$

where $\xi \in [\min\{x_0,x\}, \max\{x_n,x\}]$.

Example 6.7 Let $p_1(x)$ be a linear polynomial that interpolates $f(x) = e^x$ at -1 and 1. Find an upper bound for the absolute interpolation error $|f(x) - p_1(x)|$ when $x \in [-1,1]$.

[Solution:] By the interpolation error theorem, we have

$$|f(x) - p_1(x)| = \frac{|f''(\xi)|}{2!}|(x - (-1))(x - 1)|$$

where $\xi \in [-1,1]$, $f(x) = e^x$, and $x \in [-1,1]$.

For $\xi \in [-1,1]$, $|f''(\xi)| = e^\xi \le e$.

For $x \in [-1,1]$, $|(x - (-1))(x - 1)| = -(x+1)(x-1) = 1 - x^2 \le 1$.

So we have

$$|f(x) - p_1(x)| \le \frac{e}{2!} \cdot 1 = \frac{e}{2}$$

∎

Theorem 6.3 (Interpolation Error for Equally-Spaced Nodes) *Let $p_n(x)$ be the polynomial of degree $\le n$ that interpolates the $(n+1)$ data generated by $f(x)$ at the equally-spaced nodes $x_0, x_1, \ldots, x_n$, where $x_i - x_{i-1} = h = (x_n - x_0)/n$, $i = 1, 2, \ldots, n$. Then for $x \in [x_0, x_n]$,*

$$|f(x) - p_n(x)| \le \frac{M}{4(n+1)}h^{n+1} \tag{6.66}$$

where $M = \max_{x_0 \le x \le x_n} |f^{(n+1)}(x)|$.

Remark By the previous theorem for general nodes, we prove the current theorem for equally-spaced nodes by proving that if $x \in [x_0, x_n]$

$$|(x - x_0)(x - x_1) \cdots (x - x_n)| \le \frac{n!}{4}h^{n+1} \tag{6.67}$$

This is left as an exercise, Exercise 6.12. (Hint: Suppose x is between x_j and x_{j+1}, then look at the upper bounds of the distances $|x - x_i|$ for $i \leq j - 1$ and $i \geq j + 2$ and the product $|(x - x_j)(x - x_{j+1})|$.)

Example 6.8 Find the error upper bound if $f(x) = \sin x$ is approximated by the polynomial $p_{10}(x)$ of degree ≤ 10 that interpolates $f(x) = \sin x$ at 11 equally-spaced nodes $x_0 = 0$, $x_1 = 0.1$, $x_2 = 0.2$, ..., $x_{10} = 1$.
[**Solution:**] By the interpolation error theorem for equally-spaced nodes, we have

$$|f(x) - p_{10}(x)| \leq \frac{M}{4(10 + 1)} h^{10+1}$$

where $h = (x_n - x_0)/10 = (1 - 0)/10$, and $M = \max_{0 \leq x \leq 1} |f^{(11)}(x)| = \max_{0 \leq x \leq 1} |-\cos x| = 1$. So we have

$$|f(x) - p_{10}(x)| \leq \frac{1}{44} 10^{-11}$$

∎

6.6.2   Behavior of Interpolation Error

Let $p_n(x)$ be the polynomial of degree $\leq n$ that interpolates the $(n + 1)$ data generated by $f(x)$ at the distinct nodes x_0, x_1, ... , x_n. Assume, without loss of generality, that $x_0 < x_1 < \cdots < x_n$. By the interpolation error theorem for general nodes, the interpolation error is

$$f(x) - p_n(x) = \frac{f^{(n+1)}(\xi)}{(n+1)!} N_{n+1}(x) \tag{6.68}$$

where $\xi \in [\min\{x_0, x\}, \max\{x_n, x\}]$ and $N_{n+1}(x) = (x - x_0)(x - x_1) \cdots (x - x_n)$. So the behavior of interpolation error is determined by both factors $\frac{f^{(n+1)}(\xi)}{(n+1)!}$ and $N_{n+1}(x)$.

6.6.2.1   Equally-Spaced Nodes

Shown in Fig. 6.6 is the graph of $N_7(x)$ with 7 equally-spaced nodes $x_0 = -1, x_1, \dots ,$ $x_6 = 1$, which is oscillatory with peaks and valleys within $[x_0, x_6] = [-1, 1]$. The peak and valley near the ending nodes $x_0 = -1$ and $x_6 = 1$ have larger amplitudes relative to the others. For x outside $[x_0, x_6] = [-1, 1]$, $N_7(x)$ ascends or descends sharply as x moves away from the ending nodes.

In general, the behavior of $N_{n+1}(x)$ with equally-spaced nodes x_0, x_1, ... , x_n is similar to this particular example $N_7(x)$. The Newton basis function $N_{n+1}(x)$ intersects the x axis for $(n + 1)$ times at the nodes x_0, x_1, ... , x_n and is oscillatory. Accordingly, the interpolation error $f(x) - p_n(x)$, as a function of x, equals to 0 at the nodes x_0, x_1, ... , x_n and is oscillatory too.

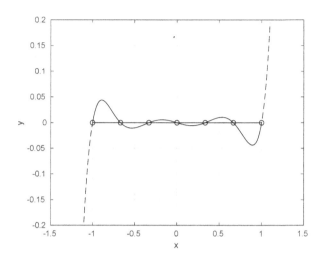

FIGURE 6.6
The graph of $N_7(x)$ with 7 equally-spaced nodes.

If the nodes $x_0, x_1, \dots, x_n$ are equally spaced, $|N_{n+1}(x)|$ has peaks within $[x_0, x_n]$, and the two peaks near the ending nodes x_0 and x_n have relatively large amplitudes. For x within $[x_0, x_n]$ and near x_0 and x_n, the absolute interpolation error $|f(x) - p_n(x)|$ may have two large peaks too. Fig. 6.7 indicates that if we increase the number of nodes in the same interval, the two large peaks in $N_{n+1}(x)$ look more pronounced (but with smaller amplitudes in this case).

> **Remark**   The maximum amplitude of $|N_{n+1}(x)| = |(x - x_0)(x - x_1) \cdots (x - x_n)|$ with equally spaced nodes $x_0 = a, x_1, \dots, x_n = b$ may increase or decrease with the number of the nodes, $(n + 1)$. We have $|N_{n+1}(x)| \leq \frac{n!}{4} h^{n+1}$, where $h = \frac{b-a}{n}$.

For x outside $[x_0, x_n]$, $N_{n+1}(x)$ ascends or descends rapidly as x is out of the interval and moves away from the ending nodes x_0 and x_n. Approximation of $f(x)$ by the interpolating polynomial $p_n(x)$ at a value of x outside $[x_0, x_n]$ is called **extrapolation**, and we say the value of $f(x)$ is extrapolated. Accordingly the absolute error in extrapolation, $|f(x) - p_n(x)|$, where x is outside $[x_0, x_n]$, grows rapidly as x moves away from the ending nodes x_0 and x_n. So extrapolation may not be reliable, and we should avoid it if possible.

Now let's look at the effect of the factor $\frac{f^{(n+1)}(\xi)}{(n+1)!}$ on the interpolation error, where $\xi = \xi(x)$ depends on x. By increasing $(n + 1)$, the number of nodes (or n, the degree of an interpolating polynomial), this factor generally decreases because of $(n + 1)!$ at its denominator. However, there are exceptions. One famous exception is the **Runge's**

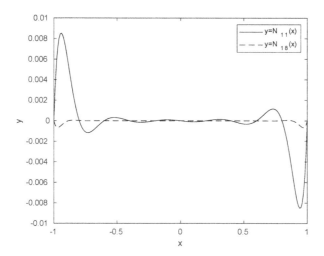

FIGURE 6.7
The graphs of $N_{11}(x)$ and $N_{18(x)}$ with equally-spaced nodes.

phenomenon, which occurs when the **Runge function**

$$f(x) = \frac{1}{1 + 25x^2} \tag{6.69}$$

is interpolated using more and more equally-spaced nodes between -1 and 1.

As seen from Fig. 6.8, as the degree of the interpolating polynomial increases from $n = 10$ to $n = 17$ (the number of equally spaced nodes increases from $(n+1) = 11$ to 18), the maximum absolute interpolation error increases instead. From Fig. 6.7, we know the maximum amplitude of the other factor $|N_{n+1}(x)|$ in the interpolation error decreases as $(n+1)$ increases from 11 to 18 in this case, so the increase of the interpolation error is caused by the factor $\frac{f^{(n+1)}(\xi)}{(n+1)!}$. Shown in Fig. 6.9 are the graphs of $\frac{|f^{(n+1)}(x)|}{(n+1)!}$ for $x \in [-1, 1]$. Note that the upper bound of the factor, $\max_{-1 \le x \le 1} \frac{|f^{(n+1)}(x)|}{(n+1)!}$, grows with $(n+1)$ (even though the increase of the upper bound does not necessarily increases the factor $\frac{f^{(n+1)}(\xi)}{(n+1)!}$ at a fixed value of x as $\xi = \xi(x)$ locates at different location when $(n+1)$ increases).

Remark The interpolation error is mathematical approximation error due to the approximation of $f(x)$ by $p_n(x)$ that interpolates $f(x)$. Because of the polynomial interpolant existence and uniqueness theorem, this error does not depend on what method is used to find and evaluate $p_n(x)$.

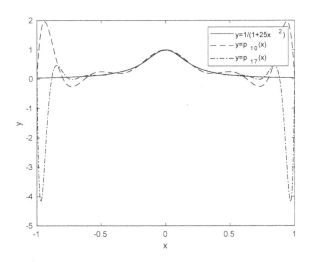

FIGURE 6.8
Runge's phenomenon.

Remark The effect of roundoff error on polynomial interpolation is method dependent. For example, to find $p_n(x)$ in monomial interpolation, we need to solve a linear system with a Vandermonde matrix as the coefficient matrix. The linear system becomes more and more ill-conditioned with the increases of its order (see Chapter 4), so the roundoff error introduced in the computation of $p_n(x)$ of a high degree n can get amplified dramatically. The lower triangular linear system in Newton's iteration is also ill-conditioned and is worse than the Vandermonde system in monomial interpolation when $(n+1)$ is large. Lagrange interpolation does not need to solve a linear system to find $p_n(x)$, but in the evaluation of $p_n(x)$ in Lagrange form, roundoff error can also get amplified. Shown in Fig. 6.10 demonstrates the effect different interpolation methods on errors. Clearly, we should avoid interpolation with equally spaced nodes by polynomials of very high degrees.

6.6.2.2 Chebyshev Nodes

We cannot do much about $f^{(n+1)}(\xi)$ to reduce the interpolation error

$$f(x) - p_n(x) = \frac{f^{(n+1)}(\xi)}{(n+1)!}(x - x_0)(x - x_1)\cdots(x - x_n) \qquad (6.70)$$

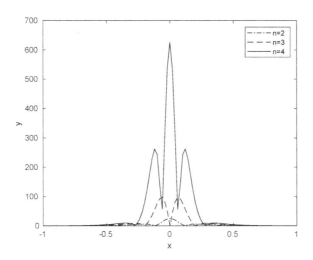

FIGURE 6.9

$\frac{|f^{(n+1)}(x)|}{(n+1)!}$ for the Runge function $f(x)$.

as $f(x)$ is the function (generally unknown) that generates the data and ξ is unknown. So we want to choose the nodes to minimize the maximum of

$$N_{n+1}(x) = (x - x_0)(x - x_1) \cdots (x - x_n) \tag{6.71}$$

which is a monic polynomial whose leading coefficient of the highest order term is 1.

As mentioned before, the Chebyshev nodes

$$t_i = \cos\left(\frac{2i+1}{2n+2}\pi\right), \quad i = 0, 1, \ldots, n \tag{6.72}$$

solve the **minmax problem**

$$\min_{\{t_0, t_1, \cdots, t_n\}} \max_{-1 \le t \le 1} |(t - t_0)(t - t_1) \ldots (t - t_n)| \tag{6.73}$$

So to interpolate $F(t)$ on the interval $[-1, 1]$, it is preferable to generate data $\{(t_i, y_i)\}_{i=0}^{n}$ by $F(t)$ at the Chebyshev nodes $\{t_i\}_{i=0}^{n}$. Shown in Fig. 6.3 are the $n + 1 = 5$ Chebyshev nodes.

If we want to interpolate the function $f(x)$ on a general interval $[a, b]$, we can apply the linear mapping

$$x = g(t) = \frac{b-a}{2}t + \frac{b+a}{2}, \quad -1 \le t \le 1, \quad a \le x \le b \tag{6.74}$$

to map between the intervals $[a, b]$ and $[-1, 1]$ (the same mapping is used in Chapter 7 for change of intervals of definite integrals). We let $f(x)$ generate data at x_i given by

$$x_i = \frac{b-a}{2}t_i + \frac{b+a}{2}, \quad i = 0, 1, \ldots, n \tag{6.75}$$

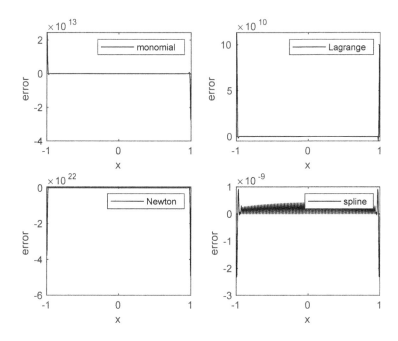

FIGURE 6.10
Error in the interpolation of $f(x) = \cos x$ by different methods using $(n+1) = 101$ equally spaced points in $[-1,1]$. The last plot is for cubic spline interpolation, which will be introduced soon.

where t_i are the Chebyshev nodes on $[-1,1]$

$$t_i = \cos\left(\frac{2i+1}{2n+2}\pi\right), \quad i = 0,1,\ldots,n \tag{6.76}$$

Then the interpolation of $f(x)$ for $x \in [a,b]$ is equivalent to the interpolation of $F(t) = f(g(t))$ for $t \in [-1,1]$.

Now let's compare Chebyshev nodes and equally-spaced nodes in the interpolation of the Runge function

$$f(x) = \frac{1}{1+25x^2}$$

over the interval $[-1,1]$. As seen from Fig. 6.11, the maximum interpolation error with Chebyshev nodes has much smaller amplitude than the same number of equally spaced nodes.

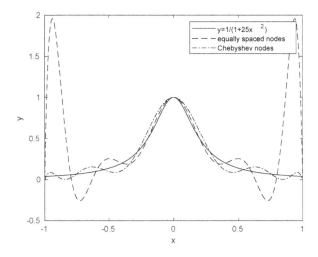

FIGURE 6.11
Interpolation of the Runge function by $n + 1 = 11$ Chebyshev nodes and 11 equally spaced nodes.

6.7   Spline Interpolation

High-order interpolating polynomials are oscillatory and can cause large interpolation error. In addition, changing one or few data points of a data set can dramatically alter the entire interpolating polynomial. Instead of interpolating a given data set by one high-degree polynomial, we may connect many low-degree polynomials that locally interpolate a few data of the set to achieve a global interpolation of the whole data set, leading to a piecewise polynomial interpolation.

6.7.1   Piecewise Linear Interpolation

A very common example of piecewise polynomial interpolation is piecewise linear interpolation. The piecewise linear interpolation is used by graphing devices to sketch the graph of the function $y = f(x)$. The graph of $y = f(x)$ is plotted by connecting consecutive data points generated by $y = f(x)$ with line segments, as demonstrated in Fig. 6.12. So the plotted graph is actually the graph of the piecewise linear interpolant, which interpolates and approximates $y = f(x)$.

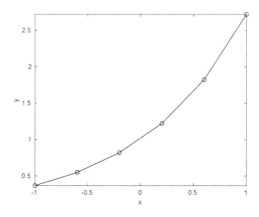

FIGURE 6.12
The graph of $y = e^x$ is plotted by connecting consecutive data points (open circles)
generated by $y = e^x$ with line segments.

The piecewise linear interpolant that interpolates the data $\{(x_i, f(x_i))\}_{i=0}^n = \{(x_0, f(x_0)), (x_1, f(x_1)), \ldots, (x_n, f(x_n))\}$, where $x_0 < x_1 < \cdots < x_n$, is

$$
S_{1,n}(x) = \begin{cases}
f(x_0)\frac{x-x_1}{x_0-x_1} + f(x_1)\frac{x-x_0}{x_1-x_0}, & x \in [x_0, x_1] \\
f(x_1)\frac{x-x_2}{x_1-x_2} + f(x_2)\frac{x-x_1}{x_2-x_1}, & x \in [x_1, x_2] \\
\quad\vdots \\
f(x_{n-1})\frac{x-x_n}{x_{n-1}-x_n} + f(x_n)\frac{x-x_{n-1}}{x_n-x_{n-1}}, & x \in [x_{n-1}, x_n]
\end{cases}
\tag{6.77}
$$

where the i-th piece ($i = 1, 2, \ldots, n$) on the i-th interval $[x_{i-1}, x_i]$ is simply a linear
polynomial that interpolates $(x_{i-1}, f(x_{i-1}))$ and $(x_i, f(x_i))$ in Lagrange form

$$
f(x_{i-1})\frac{x-x_i}{x_{i-1}-x_i} + f(x_i)\frac{x-x_{i-1}}{x_i-x_{i-1}}, \quad x \in [x_{i-1}, x_i]
\tag{6.78}
$$

To evaluate $S_{1,n}(x)$ at a given x ($x_0 \le x \le_n$), we need to determine in which subinterval
$[x_{i-1}, x_i]$ the value of x falls, and then use the piece over this subinterval for evaluation.

Remark To determine x ($x \in [x_0, x_n]$) falls in which of the n subintervals
formed by $x_0 < x_1 < \cdots < x_n$, we can use linear search or binary search. In the
linear search, x is compared in turn with $x_1, x_2, \ldots$ till $x \le x_i$ for some i, and the
subinterval to be found is $[x_{i-1}, x_i]$. In the worst case, the linear search requires
$O(n)$ comparisons.
    The idea of the binary search is similar to that of the bisection method for root
finding. In each step of the binary search, x is compared with the element in the
middle of an ordered list of elements to determine if x falls to the left or right of
this middle element to obtain a new list which is left or right half of the previous

list. Starting with the list $x_0, x_1, ..., x_n$, repeatedly apply the binary search steps till there is only one element in a final list. The subinterval to be found is between this last element and the previous middle element. The binary search reduces the number of comparisons in the worst case to $O(\log_2 n)$ (see the similar result in the Bisection Method Theorem, Theorem 5.2).

Theorem 6.4 (Accuracy of Piecewise Linear Interpolant) *Let $S_{1,n}(x)$ be the piecewise linear polynomial that interpolates the function $f(x)$ at the distinct nodes $x_0 < x_1 < \cdots < x_n$. Let x be a point in the subinterval $[x_{i-1}, x_i]$ (where $i = 1$, $2, ...,$ or n) and $h = x_i - x_{i-1}$. Assume $f(x) \in C^2_{[x_{i-1},x_i]}$ (i.e. $f''(x)$ is continuous on $[x_{i-1}, x_i]$). Then*

$$|f(x) - S_{1,n}(x)| \leq \frac{M}{8}h^2 \tag{6.79}$$

where $M = \max_{x_{i-1} \leq x \leq x_i} |f''(x)|$.

Remark This theorem is just the special case of Theorem 6.3 with $n + 1 = 2$.

6.7.2 Cubic Spline

Historically, a spline was an elastic metal or wood strip that was bent to pass through predefined points to make a curve for shipbuilding. Mathematically, a cubic spline is a function that is constructed of piecewise cubic polynomials. Cubic spline interpolation is interpolation by a cubic spline.

Given an interval $[a,b]$, we partition it into n subintervals by $n + 1$ nodes

$$a = x_0 < x_1 < \cdots < x_{n-1} < x_n = b \tag{6.80}$$

A function $y = S_{3,n}(x)$ is a cubic spline on the partitioned interval $[a,b]$ if it is a polynomial of degree ≤ 3 (cubic) on each of the n subinterval $[x_{i-1}, x_i]$ ($i = 1, 2, \ldots, n$) and the function $S_{3,n}(x)$ and its 1st and 2nd derivatives $S'_{3,n}(x)$ and $S''_{3,n}(x)$ are continuous at the $n - 1$ interior nodes $x_1, x_2, \ldots, x_{n-1}$ (so $S_{3,n}(x)$, $S'_{3,n}(x)$ and $S''_{3,n}(x)$ are continuous on $[a,b]$). We may write $S_{3,n}(x_i)$ in a piecewise manner as

$$S_{3,n}(x) = \begin{cases} s_1(x), & x \in [x_0, x_1] \\ s_2(x), & x \in [x_1, x_2] \\ \vdots \\ s_n(x), & x \in [x_{n-1}, x_n] \end{cases} \tag{6.81}$$

where $s_i(x) = c_{i0} + c_{i1}x + c_{i2}x^2 + c_{i3}x^3$ $(i = 1, 2, \ldots, n)$ is a polynomial of degree ≤ 3 (cubic) with 4 coefficients. Obviously we need to determine $4n$ coefficients to determine $S_{3,n}(x)$. The continuity requirements at the interior partition points can be written as

$$\begin{aligned}
s_i(x_i) &= s_{i+1}(x_i), & i = 1, 2, \ldots, n-1 \\
s_i'(x_i) &= s_{i+1}'(x_i), & i = 1, 2, \ldots, n-1 \\
s_i''(x_i) &= s_{i+1}''(x_i), & i = 1, 2, \ldots, n-1
\end{aligned} \tag{6.82}$$

which gives $3(n-1)$ linear equations for the $4n$ coefficients in $S_{3,n}(x)$. In the next subsection, we describe how to determine an interpolating $S_{3,n}(x)$.

Example 6.9 Let $S(x)$ be a piecewise function on $[0,4]$ defined as

$$S(x) = \begin{cases} s_1(x) = 2x^3, & x \in [0,1] \\ s_2(x) = x^3 + 3x^2 - 3x + 1 & x \in [1,2] \\ s_3(x) = 9x^2 - 15x + 9, & x \in [2,4] \end{cases}$$

Is $S(x)$ a cubic spline?
[Solution:] We first check the degree requirement. $S(x)$ is a polynomial of degree ≤ 3 on each subinterval. We then check the continuity requirement.
  (1) We check the continuity of $S(x)$ at the interior nodes 1 and 2 as follows

$$s_1(1) = 2, \quad s_2(1) = 1 + 3 - 3 + 1 = 2$$

$$s_2(2) = 8 + 12 - 6 + 1 = 15, \quad s_3(2) = 36 - 30 + 9 = 15$$

So $S(x)$ is continuous on $[0,4]$.
  (2) We have

$$S'(x) = \begin{cases} s_1'(x) = 6x^2, & x \in [0,1] \\ s_2'(x) = 3x^2 + 6x - 3 & x \in [1,2] \\ s_3'(x) = 18x - 15, & x \in [2,4] \end{cases}$$

We check the continuity of $S'(x)$ as follows

$$s_1'(1) = 6, \quad s_2'(1) = 3 + 6 - 3 = 6$$

$$s_2'(2) = 12 + 12 - 3 = 21, \quad s_3'(2) = 36 - 15 = 21$$

So $S'(x)$ is continuous on $[0,4]$.
  (3) We have

$$S''(x) = \begin{cases} s_1''(x) = 12x, & x \in [0,1] \\ s_2''(x) = 6x + 6 & x \in [1,2] \\ s_3''(x) = 18, & x \in [2,4] \end{cases}$$

We check the continuity of $S''(x)$ as follows

$$s_1''(1) = 12, \quad s_2''(1) = 6 + 6 = 12$$

$$s_2''(2) = 12 + 6 = 18, \quad s_3''(2) = 18$$

So $S''(x)$ is continuous on $[0,4]$.
  Finally we can conclude that $S(x)$ is cubic spline on $[0,4]$. ∎

6.7.3 Cubic Spline Interpolation

After we know what a cubic spline is, we can use it to interpolate data. We use $S_{3,n}(x)$ to interpolate the data $\{(x_i, y_i)\}_{i=0}^n$, where $S_{3,n}(x)$ is a cubic spline on the interval $[x_0, x_n]$ partitioned by the $n+1$ interpolating nodes $x_0, x_1, \ldots, x_n$. Before we describe how to determine $S_{3,n}(x)$, we show the need of so-called **boundary conditions** and describe some commonly used boundary conditions.

The $n+1$ interpolating conditions $S_{3,n}(x_i) = y_i$ $(i = 0, 1, \ldots, n)$ give $n+1$ linear equations for the $4n$ coefficients in $S_{3,n}(x)$. Together with $3(n-1)$ linear equations from the continuity requirements for $S_{3,n}(x)$, $S'_{3,n}(x)$ and $S''_{3,n}(x)$ at the $(n-1)$ interior nodes, there are $(n+1) + 3(n-1) = 4n - 2$ linear equations for the $4n$ coefficients. To uniquely determine $S_{3,n}(x)$, we need two more linear equations. So we apply two boundary conditions for $S_{3,n}(x)$ to enforce the behavior of $S_{3,n}(x)$ at the two ending nodes (boundaries) x_0 and x_n. Below is a list of options.

- **Free boundaries**: We enforce the boundary conditions

$$S''_{3,n}(x_0) = 0, \quad S''_{3,n}(x_n) = 0 \tag{6.83}$$

The interpolating cubic spline $S_{3,n}(x)$ with free boundaries is called a **natural cubic spline** (even though the boundary conditions are not natural if the data are generated by $f(x)$ with $f''(x_0) \neq 0$ or $f''(x_n) \neq 0$).

- **Clamped boundaries**: If the slopes at the boundaries are available as m_0 and m_n, we can use the following boundary conditions

$$S'_{3,n}(x_0) = m_0, \quad S'_{3,n}(x_n) = m_n \tag{6.84}$$

- **Not-a-knot boundaries**: We enforce the continuity of the 3rd derivative of $S_{3,n}(x)$ at the first and last interior nodes x_1 and x_{n-1} and have

$$s_1'''(x_1) = s_2'''(x_1), \quad s_{n-1}'''(x_{n-1}) = s_n'''(x_{n-1}) \tag{6.85}$$

Not-a-knot boundary conditions do not force $S_{3,n}(x)$ to satisfy some conditions that the data generating function $f(x)$ does not satisfy.

- **Periodic boundaries**: If the data generating function $f(x)$ is periodic (the data are periodic), we may apply periodic boundary conditions

$$S'_{3,n}(x_0) = S'_{3,n}(x_n), \quad S''_{3,n}(x_0) = S''_{3,n}(x_n) \tag{6.86}$$

We now describe how to determine a natural cubic spline $S_{3,n}(x)$ (with the free boundaries) to interpolate the data $\{(x_i, y_i)\}_{i=0}^n$. For other boundary conditions, the process is similar. Instead of directly deriving equations for $4n$ coefficients in $S_{3,n}(x)$, we can simplify the process by starting with the $n+1$ unknowns

$$z_i = S''_{3,n}(x_i), \quad i = 0, 1, \ldots, n$$

So $S''_{3,n}(x)$ is a piecewise linear polynomial that interpolates the data $\{(x_i, z_i)\}_{i=0}^n$. On the subinterval $[x_{i-1}, x_i]$ $(i = 1, 2, \ldots, n)$, $S''_{3,n}(x) = s_i''(x)$ is a linear polynomial that interpolates (x_{i-1}, z_{i-1}) and (x_i, z_i) and has the Lagrange form

$$S''_{3,n}(x) = s_i''(x) = z_{i-1}\frac{x - x_i}{x_{i-1} - x_i} + z_i\frac{x - x_{i-1}}{x_i - x_{i-1}} = -z_{i-1}\frac{x - x_i}{h_i} + z_i\frac{x - x_{i-1}}{h_i} \tag{6.87}$$

where h_i is the length of the subinterval:

$$h_i = x_i - x_{i-1}$$

We integrate $s_i''(x)$ twice to obtain $s_i(x)$ and write $s_i(x)$ in the following form for easier algebraic manipulation

$$s_i(x) = -z_{i-1}\frac{(x-x_i)^3}{6h_i} + z_i\frac{(x-x_{i-1})^3}{6h_i} - C_i(x-x_i) + D_i(x-x_{i-1}) \tag{6.88}$$

where C_i and D_i are integration constants. As shown below, we can use the interpolating conditions $s_i(x_{i-1}) = y_{i-1}$ and $s_i(x_i) = y_i$ to determine C_i and D_i, and finally use the continuity of $S_{3,n}'(x)$ to derive linear equations for the unknowns z_i ($i = 0, 1, \ldots, n$).

After the expressions for C_i and D_i obtained from $s_i(x_{i-1}) = y_{i-1}$ and $s_i(x_i) = y_i$ are substituted in $s_i(x)$, we have

$$\begin{aligned}
s_i(x) &= -z_{i-1}\frac{(x-x_i)^3}{6h_i} + z_i\frac{(x-x_{i-1})^3}{6h_i} \\
&\quad - \left(\frac{y_{i-1}}{h_i} - z_{i-1}\frac{h_i}{6}\right)(x-x_i) + \left(\frac{y_i}{h_i} - z_i\frac{h_i}{6}\right)(x-x_{i-1}) \tag{6.89}
\end{aligned}$$

which can be differentiated to give

$$s_i'(x) = -z_{i-1}\frac{(x-x_i)^2}{2h_i} + z_i\frac{(x-x_{i-1})^2}{2h_i} - \left(\frac{y_{i-1}}{h_i} - z_{i-1}\frac{h_i}{6}\right) + \left(\frac{y_i}{h_i} - z_i\frac{h_i}{6}\right) \tag{6.90}$$

$$\begin{aligned}
s_{i+1}'(x) &= -z_i\frac{(x-x_{i+1})^2}{2h_{i+1}} + z_{i+1}\frac{(x-x_i)^2}{2h_{i+1}} \\
&\quad - \left(\frac{y_i}{h_{i+1}} - z_i\frac{h_{i+1}}{6}\right) + \left(\frac{y_{i+1}}{h_{i+1}} - z_{i+1}\frac{h_{i+1}}{6}\right) \tag{6.91}
\end{aligned}$$

Then the continuity condition $s_i'(x_i) = s_{i+1}'(x_i)$ for $i = 1, 2, \ldots, n-1$ requires

$$\begin{aligned}
&z_i\frac{h_i}{2} - \left(\frac{y_{i-1}}{h_i} - z_{i-1}\frac{h_i}{6}\right) + \left(\frac{y_i}{h_i} - z_i\frac{h_i}{6}\right) \\
&= -z_i\frac{h_{i+1}}{2} - \left(\frac{y_i}{h_{i+1}} - z_i\frac{h_{i+1}}{6}\right) + \left(\frac{y_{i+1}}{h_{i+1}} - z_{i+1}\frac{h_{i+1}}{6}\right) \tag{6.92}
\end{aligned}$$

which can be arranged as

$$z_i\left(\frac{h_i}{2} - \frac{h_i}{6} + \frac{h_{i+1}}{2} - \frac{h_{i+1}}{6}\right) + z_{i-1}\frac{h_i}{6} + z_{i+1}\frac{h_{i+1}}{6} = \frac{y_{i-1}-y_i}{h_i} + \frac{-y_i+y_{i+1}}{h_{i+1}} \tag{6.93}$$

Let

$$b_i = 6\left(\frac{y_{i+1}-y_i}{h_{i+1}} - \frac{y_i-y_{i-1}}{h_i}\right)$$

We finally obtain $(n-1)$ linear equations for the $n+1$ unknowns z_i $(i = 0, 1, \ldots, n)$

$$h_i z_{i-1} + 2(h_i + h_{i+1})z_i + h_{i+1}z_{i+1} = b_i, \quad i = 1, 2, \ldots, n-1 \qquad (6.94)$$

We then add the two boundary conditions to form a square linear system for all the $(n+1)$ unknowns z_i $(i = 0, 1, \ldots, n)$.

With the natural boundary conditions, we have two more equations $z_0 = 0$ and $z_n = 0$ and the following square linear system

$$
\begin{bmatrix}
1 & 0 & & & & \\
h_1 & 2(h_1 + h_2) & h_2 & & & \\
& h_2 & 2(h_2 + h_3) & h_3 & & \\
& & \ddots & \ddots & \ddots & \\
& & & h_{n-1} & 2(h_{n-1} + h_n) & h_n \\
& & & & 0 & 1
\end{bmatrix}
\begin{bmatrix}
z_0 \\ z_1 \\ z_2 \\ \vdots \\ z_{n-1} \\ z_n
\end{bmatrix}
=
\begin{bmatrix}
0 \\ b_1 \\ b_2 \\ \vdots \\ b_{n-1} \\ 0
\end{bmatrix}
$$

$$(6.95)$$

which leads to a symmetric tridiagonal linear system for the $n-1$ unknowns z_i $(i = 1, \ldots, n-1)$

$$
\begin{bmatrix}
2(h_1 + h_2) & h_2 & & & \\
h_2 & 2(h_2 + h_3) & h_3 & & \\
& \ddots & \ddots & \ddots & \\
& & h_{n-2} & 2(h_{n-2} + h_{n-1}) & h_{n-1} \\
& & & h_{n-1} & 2(h_{n-1} + h_n)
\end{bmatrix}
\begin{bmatrix}
z_1 \\ z_2 \\ \vdots \\ z_{n-2} \\ z_{n-1}
\end{bmatrix}
$$
$$
=
\begin{bmatrix}
b_1 \\ b_2 \\ \vdots \\ b_{n-2} \\ b_{n-1}
\end{bmatrix}
$$

$$(6.96)$$

We solve this tridiagonal linear system to find z_i $(i = 1, \ldots, n-1)$. Together with $z_0 = z_n = 0$, we can write down the interpolating cubic spline as $S_{3,n}(x) = s_i(x)$ for $x \in [x_{i-1}, x_i]$ with $i = 1, 2, \ldots, n$, where

$$
\begin{aligned}
s_i(x) &= -z_{i-1}\frac{(x - x_i)^3}{6h_i} + z_i\frac{(x - x_{i-1})^3}{6h_i} \\
&\quad - \left(\frac{y_{i-1}}{h_i} - z_{i-1}\frac{h_i}{6}\right)(x - x_i) + \left(\frac{y_i}{h_i} - z_i\frac{h_i}{6}\right)(x - x_{i-1})
\end{aligned}
\qquad (6.97)
$$

Remark If the interpolating nodes are equally spaced with $h = x_i - x_{i-1}$ $(i = 1, 2, \ldots, n)$, the linear system becomes

$$
\begin{bmatrix}
4 & 1 & & & & \\
1 & 4 & 1 & & & \\
& \ddots & \ddots & \ddots & & \\
& & 1 & 4 & 1 \\
& & & 1 & 4
\end{bmatrix}
\begin{bmatrix}
z_1 \\
z_2 \\
\vdots \\
z_{n-2} \\
z_{n-1}
\end{bmatrix}
=
\begin{bmatrix}
d_1 \\
d_2 \\
\vdots \\
d_{n-2} \\
d_{n-1}
\end{bmatrix}
\tag{6.98}
$$

where

$$
d_i = 6\frac{y_{i+1} - 2y_i + y_{i-1}}{h^2}, \quad i = 1, 2, \ldots, n-1
$$

Example 6.10 Find a natural cubic spline $S_{3,n}(x)$ to interpolate the data $(x_0, y_0) = (-1, 1)$, $(x_1, y_1) = (0, 0)$ and $(x_2, y_2) = (1, 3)$. Then evaluate the spline at $x = 0.5$.
[Solution:] In this case, the interpolating nodes are equally spaced with $h = 1$, we know $z_0 = z_2 = 0$ and the only unknown z_1 satisfies

$$
4z_1 = d_1
$$

where

$$
d_1 = 6\frac{y_2 - 2y_1 + y_0}{h^2} = 6\frac{3 - 2 \cdot 0 + 1}{1^2} = 24
$$

So $z_1 = 6$. We have $S_{3,n}(x) = s_1(x)$ for $x \in [x_0, x_1] = [-1, 0]$ and $S_{3,n}(x) = s_2(x)$ for $x \in [x_1, x_2] = [0, 1]$, where

$$
s_1(x) = -z_0\frac{(x - x_1)^3}{6h} + z_1\frac{(x - x_0)^3}{6h} - \left(\frac{y_0}{h} - z_0\frac{h}{6}\right)(x - x_1) + \left(\frac{y_1}{h} - z_1\frac{h}{6}\right)(x - x_0)
$$

$$
= (x + 1)^3 - (x - 0) - (x + 1) = (x + 1)^3 - 2x - 1
$$

$$
s_2(x) = -z_1\frac{(x - x_2)^3}{6h} + z_2\frac{(x - x_1)^3}{6h} - \left(\frac{y_1}{h} - z_1\frac{h}{6}\right)(x - x_2) + \left(\frac{y_2}{h} - z_2\frac{h}{6}\right)(x - x_1)
$$

$$
= -(x - 1)^3 + (x - 1) + 3(x - 0) = -(x - 1)^3 + 4x - 1
$$

Since $0.5 \in [0, 1]$, $S_{3,n}(0.5) = s_2(0.5) = 1.125$. ■

Regarding the interpolation error of cubic spline interpolation, we have the following theorem.

Theorem 6.5 (Accuracy of Cubic Spline Interpolation) *Let $S_{3,n}(x)$ be a cubic spline that interpolate $f(x)$ at $(n+1)$ equally spaced nodes $a = x_0 < x_1 <$ $\cdots < x_n = b$ with $x_i - x_{i-1} = h = \dfrac{b-a}{n}$ ($i = 1,2,\ldots,n$). Then for $x \in [a,b]$,*

$$|f(x) - S_{3,n}(x)| \leq Ch^2 \tag{6.99}$$

where C is a positive constant that depends on $f''(a)$, $f''(b)$ and $\max_{a \leq x \leq b} |f^{(4)}(x)|$. If the boundary conditions of the cubic spline are

$$S_{3,n}''(a) = f''(a), \quad S_{3,n}''(b) = f''(b)$$

(for example, free boundary conditions for $f(x)$ that satisfies $f''(a) = f''(b) = 0$), or

$$S_{3,n}'(a) = f'(a), \quad S_{3,n}'(b) = f'(b)$$

(i.e. clamped boundary conditions matching with $f'(a)$ and $f'(b)$), or

$$S_{3,n}(x_0^*) = f(x_0^*), \quad S_{3,n}(x_{n-1}^*) = f(x_{n-1}^*)$$

where $x_0^* \in (x_0, x_1)$ and $x_{n-1}^* \in (x_{n-1}, x_n)$ (i.e. two extra interpolating end conditions are added), then

$$|f(x) - S_{3,n}(x)| \leq Ch^4 \tag{6.100}$$

Remark The proof of this theorem is beyond the scope of this textbook.

6.8   Discrete Fourier Transform (DFT)

Suppose the data to be interpolated are generated by a periodic function $y = f(x)$. It is a natural choice to construct an interpolant using periodic basis functions. Let's consider a periodic function $y = F(t)$ over $[0, 2\pi]$ with the period 2π and the $(n+1)$ data points $\{(t_i, y_i)\}_{i=0}^n$ generated by F at the following equally spaced nodes

$$t_i = ih, \quad h = \frac{2\pi}{n+1}, \quad i = 0, 1, \ldots, n \tag{6.101}$$

as illustrated in Fig. 6.13. Note that the interval $[0, 2\pi]$ is divided into $(n+1)$ subintervals and $t_n = 2\pi - h \neq 2\pi$. The information at $t = 2\pi$ is the same as at $t = 0$ due to the periodicity. So it is redundant and excluded from the interpolation. We assume $n + 1 = 2m$, i.e. the number of data points is even.

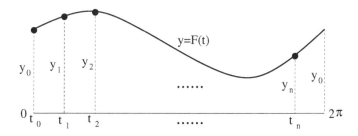

FIGURE 6.13
Periodic data at equally spaced interpolation nodes.

To begin, we need a periodic basis over $[0, 2\pi]$. We first introduce the following trigonometric basis functions

$$\phi_0(t) = 1, \quad \phi_{2k-1}(t) = \sin kt, \quad \phi_{2k}(t) = \cos kt, \quad k = 1, 2, \dots \qquad (6.102)$$

These basis functions are obviously periodic on $[0, 2\pi]$. In addition, they satisfy

$$\int_0^{2\pi} \phi_0(t)\phi_0(t)\mathrm{d}t \;=\; 2\pi \qquad (6.103)$$

$$\int_0^{2\pi} \phi_j(t)\phi_j(t)\mathrm{d}t \;=\; \pi, \quad j \neq 0 \qquad (6.104)$$

$$\int_0^{2\pi} \phi_j(t)\phi_k(t)\mathrm{d}t \;=\; 0, \quad j \neq k \qquad (6.105)$$

and are pairwise orthogonal on $[0, 2\pi]$ by the last identity above.

Remark

$$\int_0^{2\pi} \phi_j(t)\phi_k(t)\mathrm{d}t = 0, \quad j \neq k \qquad (6.106)$$

can be proved using the following product-to-sum formulas

$$2\cos jt \cos kt \;=\; \cos(j-k)t + \cos(j+k)t \qquad (6.107)$$
$$2\sin jt \sin kt \;=\; \cos(j-k)t - \cos(j+k)t \qquad (6.108)$$
$$2\cos jt \sin kt \;=\; \sin(j+k)t - \sin(j-k)t \qquad (6.109)$$

To interpolate the $n+1 = 2m$ periodic data $\{(t_i, y_i)\}_{i=0}^{n}$, we construct the following trigonometric interpolant (also called a trigonometric polynomial)

$$
\begin{aligned}
G_n(t) &= \frac{1}{2}[a_0\phi_0(t) + a_m\phi_{2m}(t)] + \sum_{k=1}^{m-1}[b_k\phi_{2k-1}(t) + a_k\phi_{2k}(t)] \\
&= \frac{1}{2}(a_0 + a_m\cos mt) + \sum_{k=1}^{m-1}(b_k\sin kt + a_k\cos kt)
\end{aligned}
\tag{6.110}
$$

Remark In $p_n(t)$, we include the basis function $\phi_{2m}(t) = \cos mt$ (the term $a_m\cos mt$) instead of $\phi_{2m-1}(t) = \sin mt$ (the term $b_m\sin mt$) because at all the nodes $t_i = ih = i\frac{2\pi}{n+1} = i\frac{\pi}{m}$, $\phi_{2m-1}(t_i) = \sin mt_i = \sin i\pi = 0$ $(i = 0, 1, \ldots, n)$.

Remark The terms $a_0\phi_0(t) = a_0$ and $a_m\phi_{2m}(t) = a_m\cos mt$ are multiplied by $1/2$ so that we can have one general formulas for all the coefficients/weights a_k $(k = 0, 1, 2, \ldots, m)$, as derived below.

Remark $G_n(t)$ interpolates $F(t)$ and satisfies the interpolating conditions $G_n(t_i) = y_i$ $(i = 0, 1, 2, \ldots, n)$. We have

$$
G_n(t_i) = \frac{1}{2}(a_0 + a_m\cos mt_i) + \sum_{k=1}^{m-1}(b_k\sin kt_i + a_k\cos kt_i)
\tag{6.111}
$$

Since $t_i = i2\pi/(n+1)$, we have $[k + l(n+1)]t_i = kt_i + li2\pi$, where l is an integer. If we denote

$$
\bar{k} = k + l(n+1), \quad k = 0, 1, \ldots, m
\tag{6.112}
$$

By the periodicity of sine and cosine, we then know the function

$$
\bar{G}_n(t) = \frac{1}{2}(a_0\cos\bar{0}t + a_m\cos\bar{m}t) + \sum_{k=1}^{m-1}(b_k\sin\bar{k}t + a_k\cos\bar{k}t)
\tag{6.113}
$$

also satisfies $\bar{G}_n(t_i) = G_n(t_i) = y_i$ $(i = 0, 1, 2, \ldots, n)$ and interpolates $F(t)$, where $\bar{0} = 0 + l(n+1) = l(n+1)$ and $\bar{m} = m + l(n+1)$. Note that even though $\bar{G}_n(t)$ has the same values as $G_n(t)$ at the interpolating nodes t_i, $\bar{G}_n(t)$ with $l \neq 0$ oscillates more than $G_n(t)$ between any two consecutive interpolating nodes, leading to the so called **aliasing phenomenon**.

To determine the $n + 1 = 2m$ coefficients/weights a_k and b_k, we use the $n + 1$ interpolating conditions $G_n(t_i) = y_i$ ($i = 0, 1, 2, \ldots, n$) as usual, and we end up with the following linear system

$$
\begin{bmatrix}
1 & \sin t_0 & \cos t_0 & \cdots & \sin(m-1)t_0 & \cos(m-1)t_0 & \cos m t_0 \\
1 & \sin t_1 & \cos t_1 & \cdots & \sin(m-1)t_1 & \cos(m-1)t_1 & \cos m t_1 \\
\vdots & \vdots & \vdots & & \vdots & \vdots & \vdots \\
1 & \sin t_n & \cos t_n & \cdots & \sin(m-1)t_n & \cos(m-1)t_n & \cos m t_n
\end{bmatrix}
\begin{bmatrix}
a_0/2 \\ b_1 \\ a_1 \\ \vdots \\ b_{m-1} \\ a_{m-1} \\ a_m/2
\end{bmatrix}
=
\begin{bmatrix}
y_0 \\ y_1 \\ \vdots \\ y_n
\end{bmatrix}
$$

$$(6.114)$$

which can be written as a vector equation

$$
\frac{1}{2} a_0 \mathbf{u}_0 + b_1 \mathbf{u}_1 + a_1 \mathbf{u}_2 + \cdots + b_{m-1} \mathbf{u}_{2(m-1)-1} + a_{m-1} \mathbf{u}_{2(m-1)} + \frac{1}{2} a_m \mathbf{u}_{2m} = \mathbf{y}
$$

$$(6.115)$$

where, with $k = 1, 2, \ldots, m - 1$

$$
\mathbf{u}_0 = \begin{bmatrix} 1 \\ 1 \\ \vdots \\ 1 \end{bmatrix}, \quad
\mathbf{u}_{2k-1} = \begin{bmatrix} \sin k t_0 \\ \sin k t_1 \\ \vdots \\ \sin k t_n \end{bmatrix}, \quad
\mathbf{u}_{2k} = \begin{bmatrix} \cos k t_0 \\ \cos k t_1 \\ \vdots \\ \cos k t_n \end{bmatrix}, \quad
\mathbf{u}_{2m} = \begin{bmatrix} \cos m t_0 \\ \cos m t_1 \\ \vdots \\ \cos m t_n \end{bmatrix}, \quad
\mathbf{y} = \begin{bmatrix} y_0 \\ y_1 \\ \vdots \\ y_n \end{bmatrix}
$$

Clearly each vector $\mathbf{u}_j$ above is a vector formed by evaluating the basis function $\phi_j(t)$ at the interpolating nodes t_0, t_1, ..., t_n. Because of the pairwise orthogonality of the basis functions $\phi_j(t)$ and $\phi_k(t)$ over the interval $[0, 2\pi]$, where $j \neq k$, there is the corresponding pairwise orthogonality of the vectors $\mathbf{u}_j$ and $\mathbf{u}_k$ in the vector space $\mathbb{R}^{n+1}$, that is $\mathbf{u}_j^T \mathbf{u}_k = 0$ (or $\mathbf{u}_j \cdot \mathbf{u}_k = 0$), which can be written as

$$
\sum_{i=0}^{n} \cos j t_i \cos k t_i = 0, \quad j \neq k \; (j = 0, 1, \ldots, m; \; k = 0, 1, \ldots, m) \qquad (6.116)
$$

$$
\sum_{i=0}^{n} \sin j t_i \sin k t_i = 0, \quad j \neq k \; (j = 1, 2, \ldots, m-1; \; k = 1, 2, \ldots, m-1)
$$

$$(6.117)$$

$$
\sum_{i=0}^{n} \cos j t_i \sin k t_i = 0, \quad j \neq k \; (j = 0, 1, \ldots, m; \; k = 1, 2, \ldots, m-1) \qquad (6.118)
$$

Remark Again, the above results can be shown using the product-to-sum formulas. For example,

$$
\sum_{i=0}^{n} \cos j t_i \cos k t_i = \sum_{i=0}^{n} \frac{1}{2} [\cos(j+k)t_i + \cos(j-k)t_i] \qquad (6.119)
$$

We below show that $\sum_{i=0}^{n} \cos(j+k)t_i = 0$.

Let $l = j+k$ and $S = \sum_{i=0}^{n} \cos lt_i$. Note that $0 < l < 2m = n+1$ (as $j \neq k$) and $\sin(lh/2) \neq 0$ (as $0 < lh/2 = l\pi/(n+1) < \pi$). We have

$$2\sin\left(\frac{lh}{2}\right)S = \sum_{i=0}^{n} 2\sin\left(\frac{lh}{2}\right)\cos lt_i = \sum_{i=0}^{n}[\sin l(t_i+h/2) - \sin l(t_i-h/2)]$$

(6.120)

Since $t_i + h/2 = t_{i+1} - h/2$, the right-hand side summation above is a finite telescoping series and can be condensed as

$$[2\sin(lh/2)]S = \sin l(t_n+h/2) - \sin l(t_0 - h/2)$$ (6.121)

which is zero due to $(t_n+h/2) - (t_0 - h/2) = t_n - t_0 + h = 2\pi$ and the periodicity of $\sin lt$. So we can conclude $S = 0$, i.e. $\sum_{i=0}^{n} \cos(j+k)t_i = 0$. Similarly, we can show $\sum_{i=0}^{n} \cos(j-k)t_i = 0$. We thus have proved the orthogonality

$$\sum_{i=0}^{n} \cos jt_i \cos kt_i = 0, \quad j \neq k \ (j = 0,1,\ldots,m; \ k = 0,1,\ldots,m)$$ (6.122)

Because of the pairwise orthogonality, the $2m \ (= n+1)$ vectors $\mathbf{u}_0, \mathbf{u}_1, \ldots, \mathbf{u}_{2m-2}$ and $\mathbf{u}_{2m}$ (excluding $\mathbf{u}_{2m-1}$) form an orthogonal basis for $\mathbb{R}^{n+1}$. The $2m$ weights a_k ($k = 0,1,\ldots,m$) and b_k ($k = 1,2,\ldots,m-1$) in writing $\mathbf{y}$ as a linear combination of these basis vectors as

$$\frac{1}{2}a_0\mathbf{u}_0 + b_1\mathbf{u}_1 + a_1\mathbf{u}_2 + \cdots + b_{m-1}\mathbf{u}_{2(m-1)-1} + a_{m-1}\mathbf{u}_{2(m-1)} + \frac{1}{2}a_m\mathbf{u}_{2m} = \mathbf{y}$$

(6.123)

can be easily as

$$a_0 = \frac{2\mathbf{u}_0^T\mathbf{y}}{\mathbf{u}_0^T\mathbf{u}_{2k}} = \frac{2\sum_{i=0}^{n} y_i}{\sum_{i=0}^{n} 1}, \quad a_m = \frac{2\mathbf{u}_{2m}^T\mathbf{y}}{\mathbf{u}_{2m}^T\mathbf{u}_{2m}} = \frac{2\sum_{i=0}^{n} y_i \cos mt_i}{\sum_{i=0}^{n} \cos^2 mt_i}$$ (6.124)

$$a_k = \frac{\mathbf{u}_{2k}^T\mathbf{y}}{\mathbf{u}_{2k}^T\mathbf{u}_{2k}} = \frac{\sum_{i=0}^{n} y_i \cos kt_i}{\sum_{i=0}^{n} \cos^2 kt_i}, \quad b_k = \frac{\mathbf{u}_{2k-1}^T\mathbf{y}}{\mathbf{u}_{2k-1}^T\mathbf{u}_{2k-1}} = \frac{\sum_{i=0}^{n} y_i \sin kt_i}{\sum_{i=0}^{n} \sin^2 kt_i},$$

$$k = 1,2,\ldots,m-1$$ (6.125)

Note that

$$\sum_{i=0}^{n} 1 = n+1 = 2m \tag{6.126}$$

$$\sum_{i=0}^{n} \cos^2 mt_i = \sum_{i=0}^{n} \cos^2 i\pi = \sum_{i=0}^{n} 1 = n+1 = 2m \tag{6.127}$$

$$\sum_{i=0}^{n} \cos^2 kt_i = \sum_{i=0}^{n} \frac{1}{2}(1 + \cos 2kt_i) = \frac{1}{2}(n+1) = m, \quad k = 1, 2, \ldots, m-1 \tag{6.128}$$

$$\sum_{i=0}^{n} \sin^2 kt_i = \sum_{i=0}^{n} \frac{1}{2}(1 - \cos 2kt_i) = \frac{1}{2}(n+1) = m, \quad k = 1, 2, \ldots, m-1 \tag{6.129}$$

So we have the general formulas

$$a_k = \frac{1}{m} \sum_{i=0}^{n} y_i \cos kt_i, \quad k = 0, 1, \ldots, m \tag{6.130}$$

$$b_k = \frac{1}{m} \sum_{i=0}^{n} y_i \sin kt_i, \quad k = 1, 2, \ldots, m-1 \tag{6.131}$$

where again $2m = n+1$.

We can now summarize the above results as follows. Given the data $\{(t_i, y_i)\}_{i=0}^{n}$ generated by $F(t)$ at the following equally spaced nodes

$$t_i = ih, \quad h = \frac{2\pi}{n+1}, \quad i = 0, 1, \ldots, n \tag{6.132}$$

where $n+1 = 2m$ is even, the discrete Fourier transform (DFT) of the data gives the $2m$ Fourier coefficients

$$a_k = \frac{1}{m} \sum_{i=0}^{n} y_i \cos kt_i, \quad k = 0, 1, \ldots, m \tag{6.133}$$

$$b_k = \frac{1}{m} \sum_{i=0}^{n} y_i \sin kt_i, \quad k = 1, 2, \ldots, m-1 \tag{6.134}$$

The trigonometric interpolant for the data is given by

$$G_n(t) = \frac{1}{2}(a_0 + a_m \cos mt) + \sum_{k=1}^{m-1} (b_k \sin kt + a_k \cos kt) \tag{6.135}$$

The inverse DFT gives back the data in terms of the Fourier coefficients as

$$y_i = G_n(t_i) = \frac{1}{2}(a_0 + a_m \cos mt_i) + \sum_{k=1}^{m-1} (b_k \sin kt_i + a_k \cos kt_i) \tag{6.136}$$

Remark If the data $\{(x_i, y_i)\}_{i=0}^n$ are generated by the periodic function $f(x)$ with the period $2L$ (instead of 2π) on the interval $[0, 2L]$, we can introduce the substitution $x = tL/\pi$ ($t = x\pi/L$) and consider the data $\{(t_i, y_i)\}_{i=0}^n$ generated by $F(t) = f(tL/\pi)$. In the formulas summarized above, we just need to replace t_i by $t_i = x_i\pi/L$ and t by $t = x\pi/L$ to achieve the interpolation of $f(x)$ by $g_n(x) = G_n(x\pi/L)$.

Remark Even though the trigonometric interpolant assumes periodicity, the DFT and inverse DFT can be used to transform back and forth between any data with equally spaced nodes and their Fourier coefficients.

Remark If the data generating function $f(x)$ is infinitely smooth and periodic, then the interpolation error $|f(x) - g_n(x)|$ decreases as n increases and decreases faster than n^{-p} for any integer p. The convergence is named spectral convergence. The accuracy of the approximation is called **spectral accuracy**.

Remark To carry out the DFT of $n + 1 = 2m$ data, each of the $2m$ Fourier coefficients needs to be computed with the summation. If we use the summation directly as in the above summary, we need n additions (assuming the basis vectors are calculated in advance and are ready to be used). So this naive approach requires $O(n^2)$ operations. However, a fast Fourier transform (FFT) algorithm can reduce this cost down to $O(n\log_2 n)$.

6.9 Exercises

Exercise 6.1 Find a polynomial $p(x)$ of the least degree such that $p(0) = 1$, $p'(0) = 0$, $p(1) = 0$ and $p'(1) = 1$.

Exercise 6.2 Interpolate the data in Table 6.6 by a degree-2 polynomial using (1) monomial interpolation, (2) Newton's interpolation and (3) Lagrange interpolation. (4) Show that the results in (1), (2) and (3) are the same. Why must they be the same?

x_i	-1	0	1
y_i	0	-1	0

TABLE 6.6
Data for interpolation.

Exercise 6.3 The data in Table 6.7 give the life expectation of the citizens in a region in a given year.
(1) Find an interpolating polynomial of degree 2 in Lagrange form that interpolates the data.
(2) Use your interpolating polynomial to estimate the life expectation in year 1995. Does your estimation seem reasonable?
(3) Use your interpolating polynomial to estimate the life expectation in year 1930 (i.e. extrapolating the data because 1930 is outside the range of the x-nodes of the data $[1980, 2000]$). Is your estimation reasonable?

Year	1980	1990	2000
Life expectation	60	61	66

TABLE 6.7
Life expectation vs. year.

Exercise 6.4 John measured the temperature outside his home one afternoon and obtained the data in Table 6.8.
(1) If $q_2(x) = c_0 + c_1 x + c_2 x^2$ interpolates the data, establish linear equations for c_0, c_1 and c_2 and find c_0, c_1 and c_2.
(2) Find a polynomial $p_2(x)$ of degree ≤ 2 in Lagrange form to interpolate the data. Simplify $p_2(x)$ to the standard form as $q_2(x)$.
(3) Explain why $p_2(x) = q_2(x)$.
(4) John forgot to measure the temperature at 3pm. Please estimate the temperature at 3pm for him. Is your estimation reasonable?
(5) If John wanted to use your simplified $p_2(x)$ to predict the temperature at 1pm in the next afternoon, what would be his predicted temperature? Is it reasonable? (Note: 1pm in the next afternoon is 24 hours after 1pm in the afternoon when John started the measurements.)

Exercise 6.5 There are two points $(0, 1)$ and $(1, 0)$.
(1) Find a polynomial $p_1(x)$ of degree 1 whose graph passes through the two points.
(2) Find a different polynomial of degree 2 whose graph passes through the same two points.
(3) Does the existence of such a different degree-2 polynomial violate the polynomial interpolant uniqueness theorem?

Time [pm]	1	2	4
Temperature [C]	30	32	30

TABLE 6.8
Temperature data.

Exercise 6.6 Interpolate the data in Table 6.9 by a Newton's interpolation polynomial. Then write the polynomial in nested form.

x_i	−1	1	2	3
y_i	7	11	28	63

TABLE 6.9
Data for interpolation.

Exercise 6.7 $p_4(x) = x^4 - x^3 + x^2 - x + 1$ interpolates the data in Table 6.10. If another data point $(3, 30)$ is added, find an interpolating polynomial $p_5(x)$ to interpolate the new data set using $p_4(x)$ and Newton's form.

x_i	−2	−1	0	1	2
y_i	31	5	1	1	11

TABLE 6.10
Data for interpolation.

Exercise 6.8 $p_3(x) = 3x^3 + 2x^2 + x + 10$ interpolates 4 data points at the nodes $x_0 = -2, x_1 = -1, x_2 = 1$ and $x_2 = 2$. If a new data point $(0, -2)$ is added, find a polynomial $p_4(x)$ to interpolate all the 5 data points.

Exercise 6.9 Let $L_i(x)$, $i = 0, 1, \ldots, n$, be a Lagrange basis function for the $(n+1)$ distinct nodes $x_0, x_1, \cdots, x_n$.
(1) Verify $L_0(x) + L_1(x) = 1$ for the case $n = 1$.
(2) Prove that in general (i.e. for any n) $L_0(x) + L_1(x) + \cdots + L_n(x) = 1$. (Hint: Interpolate the data $(x_0, 1), (x_1, 1), \ldots, (x_n, 1)$ using Lagrange interpolation. By observation, what polynomial interpolates the same data? Then make the argument that $L_0(x) + L_1(x) + \cdots + L_n(x)$ must be 1.)

Exercise 6.10 Let $p_5(x)$ be an interpolating polynomial that interpolates the data generated by $f(x) = e^{-x}$ at six equally spaced nodes between 0 and 1 (including 0 and 1). Find an upper bound of the interpolation error $|f(x) - p_5(x)|$ for $x \in [0, 1]$.

Exercise 6.11 Let $p_4(x)$ be an interpolating polynomial that interpolates the data generated by $f(x) = \sin x + \cos x$ at five equally spaced nodes between $-\pi/2$ and $\pi/2$ (including $-\pi/2$ and $\pi/2$). Find an upper bound of the interpolation error $|f(x) - p_4(x)|$ for $x \in [-\pi/2, \pi/2]$.

Exercise 6.12 Let $N_{n+1}(x) = (x - x_0)(x - x_1) \cdots (x - x_n)$. If $a = x_0 < x1 < \cdots < x_{n-1} < x_n = b$ and $x_0, x_1, \ldots, x_n$ are equally spaced, prove that $|N_{n+1}(x)| \leq \frac{n!}{4} h^{n+1}$, where spatial step $h = (b - a)/n$.

Exercise 6.13

$$S(x) = \begin{cases} Ax^3 + 3x^2 + x, & x \in [-1, 0] \\ Bx^3 + 3x^2 + x, & x \in [0, 1] \end{cases}$$

(1) Verify that $S(x)$ is a cubic spline on $[-1, 1]$. (2) If $S(x)$ is a natural cubic spline, find the values of A and B.

Exercise 6.14 Determine the values of a, b, c and d such that

$$S_{3,2}(x) = \begin{cases} s_1(x) = x^3, & x \in [0, 1] \\ s_2(x) = x^3 + bx^2 + cx + d, & x \in [1, 3] \end{cases}$$

is a cubic spline.

Exercise 6.15 Determine whether

$$S_{3,2}(x) = \begin{cases} s_1(x) = x^3 + 3x^2 + 7x - 5, & x \in [-1, 0] \\ s_2(x) = -x^3 + 3x^2 + 7x - 5 & x \in [0, 1] \end{cases}$$

is a natural cubic spline.

Exercise 6.16 Show that the natural cubic spline through the points $(0, 1)$, $(1, 2)$, $(2, 3)$ and $(3, 4)$ must be the same as the function $y = x + 1$.

Exercise 6.17 Interpolate the data in Table 6.11 by a natural cubic spline.

x_i	−1	0	1
y_i	0	2	−1

TABLE 6.11
Data for interpolation.

Exercise 6.18 Interpolate the data in Table 6.12 by a natural cubic spline.

Exercise 6.19 The data in Table 6.13 are interpolated by $g(x) = a_0 + a_1 \cos(\pi x) + b_1 \sin(\pi x)$. Find a_0, a_1 and b_1 using the interpolating conditions.

x_i	0.2	0.4	0.6	0.8
y_i	3.0	2.5	1.5	2.0

TABLE 6.12
Data for interpolation.

x_i	0	0.5	1
y_i	2	5	4

TABLE 6.13
Data for interpolation.

Exercise 6.20 Find the trigonometric interpolant $G_3(t)$ using DFT to interpolate the periodic function $f(t)$ at $t_0 = 0$, $t_1 = \pi/2$, $t_2 = \pi$ and $t_3 = 3\pi/2$, where $f(t)$ over the period $[0, 2\pi]$ is given as

$$f(t) = \begin{cases} t, & t \in [0, \pi] \\ 2\pi - t & t \in [\pi, 2\pi] \end{cases}$$

Sketch the graphs of $f(t)$ and $G_3(t)$ for $t \in [0, 2\pi]$ in one figure.

Exercise 6.21 Find the trigonometric interpolant $G_3(t)$ using DFT to interpolate the periodic function $f(t)$ at $t_0 = 0$, $t_1 = \pi/2$, $t_2 = \pi$ and $t_3 = 3\pi/2$, where the discontinuous $f(t)$ over the period $[0, 2\pi]$ is given as

$$f(t) = \begin{cases} 1, & t \in [0, \pi) \\ 0 & t \in [\pi, 2\pi) \end{cases}$$

Sketch the graphs of $f(t)$ and $G_3(t)$ for $t \in [0, 2\pi)$ in one figure.

6.10 Programming Problems

Problem 6.1 Write an MATLAB m-function

```
function y = Lagrange(xnodes,ynodes,x)
```

to construct and evaluate an interpolating polynomial in Lagrange form, where the *x*- and *y*-values of the data for interpolation are supplied in vectors xnodes and ynodes, respectively; the *x*-values at which the interpolating polynomial is evaluated are provided in the vector x; and the values of the polynomial are returned in the vector y.

Then write a MATLAB script `interpolationerror.m` to do the follows.

First, you are going to investigate how the degree of an interpolating polynomial affects interpolation error. Consider n data points (x_i, y_i) $(1 \leq i \leq n)$ generated by the function $f(x) = \sin(x)$ on the interval $[-5, 5]$, where n is chosen to be 11 and 21 respectively. Choose the nodes x_i to be equally spaced on the interval, including the end points -5 and 5. For each chosen value of n, interpolate the n data points by a polynomial $p_{n-1}(x)$ of degree $n - 1$. Plot this interpolating polynomial $p_{n-1}(x)$ and the data generating function $f(x)$ in the same figure by evaluating them at 401 equally spaced x-values on the interval $[-5, 5]$, and then plot the error $f(x) - p_{n-1}(x)$ in another figure. Now change your data generating function to the Runge function $f(x) = \frac{1}{1+x^2}$ and repeat the above process to produce two new figures. What have your observed in each case? What lessons have you learned?

Next, you are going to look at how the interpolation nodes affect interpolation errors. Repeat your work from the above paragraph, but this time change from using the equally spaced nodes to the Chebyshev nodes. The Chebyshev nodes are given in the vector `x_Chebyshev` defined by the following MATLAB statements (see Eqs. (6.75) and (6.76)):

```
c=-5; d=5;
k=0:n-1;
x_Chebyshev=(c+d)/2+((c-d)/2)*cos((2*k+1)*pi/(2*(n-1)+2));
```

How do these plots differ from those arising from equally-spaced nodes. Which set of interpolating nodes (the equally spaced nodes or the Chebyshev nodes) is better?

The error in polynomial interpolation is given by

$$f(x) - p_{n-1}(x) = \frac{f^{(n)}(\xi)}{n!} \omega(x), \tag{6.137}$$

where $\omega(x)$ is defined as

$$\omega(x) = (x - x_1)(x - x_2) \cdots (x - x_n). \tag{6.138}$$

Write a MATLAB m-function

```
function [w] = omega(xnodes,x)
```

to evaluate $\omega(x)$ defined in Eq. (6.146). The nodes $x_1, x_2, \ldots, x_n$ are passed to the m-function as the vector `xnodes`. The m-function receives the values of x as the input vector `x` and returns the output vector `w` (that has the same size as `x`) corresponding to the evaluations of $\omega(x)$. Clearly, $\omega(x)$ is important in determining the behavior of interpolation error. Using your m-function `omega`, plot the graphs of $\omega(x)$ for the above 11 equally distributed nodes and Chebyshev nodes in separate figures. Are your plots consistent with the error behaviors you have observed?

Problem 6.2 The daily average temperatures in some days of February 2021 in Dallas are provided below:

Date	Temperature (F)
5	46.6
7	44.6
8	49.8
9	31.4
11	25.3
13	23.5
14	14.5
15	7.3

We wish to use this data to estimate the average daily temperature for some other days and compare the estimations with the actual temperature. Using the interpolating polynomial function `Lagrange` in Problem 6.1 to do the following:

(1) Plot the interpolating polynomial through this data using 401 points in the interval $[5, 15]$ with a solid line. On the same plot, overlay the given data values as open circles.

(2) Using your interpolating polynomial, estimate the average daily temperature for the days: Feb. 1, 4, 6, 10, 12, 16, 18 and March 1.

(3) The actual average daily temperatures (F) for the those days are
   - Feb. 1: 44.3
   - Feb. 4: 57.3
   - Feb. 6: 45.6
   - Feb. 10: 28.6
   - Feb. 12: 23.2
   - Feb. 16: 17.5
   - Feb. 18: 25.3
   - March 1: 50.4

   Discuss the accuracy of your polynomial model for both interpolation (estimation for the days within the interval of the interpolating nodes) and extrapolation (estimation for the days outside the interval of the interpolating nodes).

(4) Extend your plot of the interpolating polynomial over the interval $[1, 29]$, again plotting the data with open circles and the interpolant with a solid line. How does this plot support your conclusions from part (3)?

(5) Use MATLAB built-in function `spline` to do (1) and (2) with cubic spline interpolation.

Problem 6.3 Suppose you are asked to interpolate n distinct data points (x_1, y_1), (x_2, y_2), ..., (x_n, y_n) by a polynomial of formal degree $n-1$ using Newton form. Here I denote the first data point as (x_1, y_1) instead of (x_0, y_0) to ease the conversion of the construction process into a MATLAB code (note that an array index in MATLAB starts from 1 instead of 0 by default). The interpolating polynomial in Newton form is

$$p_{n-1}(x) = a_1 + a_2(x - x_1) + a_3(x - x_1)(x - x_2) + \cdots$$
$$+ a_n(x - x_1)(x - x_2) \cdots (x - x_{n-1}). \tag{6.139}$$

The construction of Newton form is the process to determine the n unknown coefficients $a_1, a_2, \ldots, a_n$ using n interpolating conditions $p_{n-1}(x_1) = y_1$, $p_{n-1}(x_2) = y_2, \ldots, p_{n-1}(x_n) = y_n$. We know that if one more data point (x_{n+1}, y_{n+1}) joins the previous n data points, the new interpolating polynomial of degree n that interpolates the expanded data set (the n old points plus one new point) can be simply written as

$$p_n(x) = p_{n-1}(x) + a_{n+1}(x - x_1)(x - x_2) \cdots (x - x_n). \tag{6.140}$$

The extra coefficient a_{n+1} is determined from the added interpolating condition $p_n(x_{n+1}) = y_{n+1}$. It is

$$a_{n+1} = \frac{y_{n+1} - p_{n-1}(x_{n+1})}{(x_{n+1} - x_1)(x_{n+1} - x_2) \cdots (x_{n+1} - x_n)}. \tag{6.141}$$

This gives us a recursive formula to find coefficients $a_2, a_3, \ldots$ by starting with the constant polynomial (of degree 0) $p_0(x) = a_1 = y_1$ and letting n take 1, 2, $\ldots$ in the above formula. Here are more details. You start with $p_0(x) = a_1 = y_1$, which interpolates just the first one data point (x_1, y_1). Let $n = 1$ in Eq. (6.141), you obtain

$$a_2 = \frac{y_2 - p_0(x_2)}{(x_2 - x_1)}, \tag{6.142}$$

and the polynomial interpolating the first two data points (x_1, y_1) and (x_2, y_2):

$$p_1(x) = a_1 + a_2(x - x_1). \tag{6.143}$$

Let $n = 2$ in Eq. (6.141), you obtain

$$a_3 = \frac{y_3 - p_1(x_3)}{(x_3 - x_1)(x_3 - x_2)}, \tag{6.144}$$

and the polynomial interpolating the first three data points (x_1, y_1), (x_2, y_2) and (x_3, y_3):

$$p_2(x) = a_1 + a_2(x - x_1) + a_3(x - x_1)(x - x_2). \tag{6.145}$$

You can continue this process until you obtain all the coefficients in Newton form for any given data points.

As indicated by Eq. (6.141), when using Eq. (6.141) for the above recursion, you need to evaluate $p_{n-1}(x)$ at $x = x_{n+1}$ to obtain $p_{n-1}(x_{n+1})$, which can be done by nested multiplication; and you also need to compute $(x_{n+1} - x_1)(x_{n+1} - x_2) \cdots (x_{n+1} - x_n)$, which is to evaluate the function

$$\omega(x) = (x - x_1)(x - x_2) \cdots (x - x_n) \tag{6.146}$$

at $x = x_{n+1}$.

Now, you are asked to write three MATLAB m-functions to fulfill the above construction process. The first one is

```
function [w] = omega(xnodes,x)
```

This m-function is used to evaluate $\omega(x)$ defined in Eq. (6.146). The nodes $x_1, x_2, \ldots,$ x_n are passed to the m-function as the vector xnodes. The m-function receives the values of x as the input vector x and returns the evaluations as the output vector w. The second m-function is

```
function [p] = nestedform(a,xnodes,x)
```

This m-function is used to evaluate Newton form $p_{n-1}(x)$ defined in Eq. (6.139) using nested multiplication (Horner's method). The coefficients $a_1, a_2, \ldots, a_n$ are passed to the m-function as the vector a, the nodes $x_1, x_2, \ldots, x_{n-1}$ are passed to the m-function as the vector xnodes. The m-function receives the values of x as the input vector x and returns the evaluations as the output vector p. The third m-function is

```
function [a] = Newtonform(xnodes,ynodes)
```

It is used to construct Newton from of an interpolating polynomial. The input vectors xnodes and ynodes pass the x-values and y-values of the data points (x_1, y_1), (x_2, y_2), $\cdots$ into the m-function. The output vector a returns the coefficients $a_1, a_2, \ldots$ in Newton form. It should become clear to you that this function implements the recursion in Eq. (6.141) and needs to call the first two m-functions. To evaluate Newton form, you just use your m-function nestedform.

Last, you are asked to write a MATLAB script testNewtonform.m to test your m-functions by interpolating the data (x_i, y_i) $(1 \le i \le 5)$ generated by a polynomial function $f(x) = 3.1x^4 + 2.3x^3 - 6.6x^2 + 8.7x + 7.9$, where nodes x_i are $-2, -1, 0, 1,$ 2, and $y_i = f(x_i)$. After you construct the interpolating polynomial $p_4(x)$ (of formal degree 4) in Newton form, plot in one figure your interpolating polynomial $p_4(x)$ and the data generating function $f(x)$ by evaluating them at 201 equally spaced x-values on the interval $[-3, 3]$. In another figure, plot the error $f(x) - p(x)$ using the same 201 points. Make sure your plots are annotated.

Theoretically, do you expect that $p(x)$ and $f(x)$ are the same? Why? Do your plots verify your expectation? Can you comment on the amplitude of the computed error in your second plot?

Problem 6.4 write a program to find the trigonometric interpolant $G_n(t)$ (i.e. to determine the coefficients in $G_n(t)$) to interpolate the periodic function $f(t)$ at $n+1 = 2m$ equally spaced nodes over the period $[0, 2\pi)$.

Consider the hat function $f(t)$ given as

$$f(t) = \begin{cases} t, & t \in [0, \pi] \\ 2\pi - t & t \in [\pi, 2\pi] \end{cases}$$

For $m = 2, 4, 8, 16$ and 32, respectively, (1) plot the graphs of $f(t)$ and $G_n(t)$ using 500 equally spaced points t_k $(k = 1, 2, \ldots, 500)$ over $[0, 2\pi]$, and (2) find the maximum absolute error $\max_i |f(t_i) - G_n(t_i)|$ $(i = 1, 2, \ldots, 500)$. Plot the maximum absolute error versus m.

Now change $f(t)$ to the square wave function given as

$$f(t) = \begin{cases} 1, & t \in [0, \pi) \\ 0 & t \in [\pi, 2\pi) \end{cases}$$

and re-generate the above plots.

Finally, change $f(t)$ to

$$f(t) = 0.0001t^4(t - 2\pi)^4$$

and re-generate the above plots.

7

Numerical Integration

In this chapter, we start with a review of definite integrals and an overview of numerical integration. We then cover various numerical integration methods, including the midpoint rule, the trapezoidal rule, Simpson's rule and Gaussian quadrature rules. We give detailed error analysis for these rules. We also introduce degree of precision and change of intervals in numerical integration.

In probability theory, the cumulative distribution function of a continuous random variable X is

$$F(x) = P(X \leq x) = \int_{-\infty}^{x} g(t) \mathrm{d}t \qquad (7.1)$$

where $P(X \leq x)$ is the probability that the random variable X is less or equal to x, and $g(t)$ is the probability density function (PDF). The random variable X is normally distributed (described by the normal distribution) with mean (expected value) μ and variance σ^2 (standard deviation σ) if its probability density function $g(t)$ is a Gaussian function of the form

$$g(t) = \frac{1}{\sigma\sqrt{2\pi}} e^{-\frac{1}{2}\left(\frac{t-\mu}{\sigma}\right)^2} \qquad (7.2)$$

When the results of a series of measurements are described by the normal distribution, the probability that the error of a single measurement lies between $-z\sigma$ and $+z\sigma$ (the measurement is between $\mu - z\sigma$ and $\mu + z\sigma$), is given by $\mathrm{erf}(z/\sqrt{2})$, where $\mathrm{erf}(x)$ is called the error function and is defined as

$$\mathrm{erf}(x) = \frac{2}{\sqrt{\pi}} \int_{0}^{x} e^{-u^2} \mathrm{d}u \qquad (7.3)$$

The error function $\mathrm{erf}(x)$ cannot be evaluated in closed form in terms of elementary functions. We can use numerical integration to approximate the value of $\mathrm{erf}(x)$ for a given x with controllable accuracy to tabulate the values of $\mathrm{erf}(x)$. The error function $\mathrm{erf}(x)$ has already been extensively tabulated. In this chapter, we learn various so-called quadrature rules of numerical integration.

DOI: 10.1201/9781003201694-7

7.1 Definite Integrals

Let's first review the definition and properties of definite integrals before we approximate definite integrals.

In calculus, we introduced the definition of a definite integral when finding the **net area** A of a region that is formed between the curve $y = f(x)$ and the x-axis over the interval $[a, b]$ as shown in Fig. 7.1. The net area is the sum of positive areas of sub-regions above the x-axis and negative areas of sub-regions below the x-axis. For example, the net area of the shaded region in Fig. 7.1 is negative as the whole region is below the x-axis.

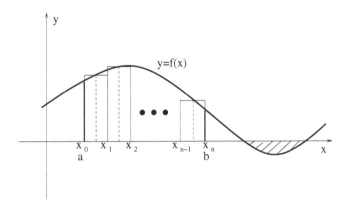

FIGURE 7.1
The area problem.

We first divide the interval $[a, b]$ into n subintervals using the $n + 1$ **partition points** $a = x_0 < x_1 < \cdots < x_n = b$ as in Fig. 7.1. We may denote the **partition** as P so that

$$P = \{a = x_0 < x_1 < \cdots < x_n = b\} \tag{7.4}$$

Let the length of the subinterval i, $[x_{i-1}, x_i]$, be $\Delta x_i = x_i - x_{i-1}$, $i = 1, 2, \ldots, n$. If the partition points are equally spaced so that

$$\Delta x_i = h = \frac{b - a}{n} \tag{7.5}$$

we may call this **uniform partition** as P_h.

Next, we draw a rectangle over each subinterval $[x_{i-1}, x_i]$ with the height $f((x_{i-1} + x_i)/2)$, $i = 1, 2, \ldots, n$, as illustrated in Fig. 7.1. Note that the rectangle with the base $x_i - x_{i-1} = h$ has the height $f((x_{i-1} + x_i)/2)$, which can be negative when the curve is below the x-axis. Its signed area can therefore be negative. We approximate the **net area** A of the region over $[a, b]$ by the sum of the signed areas of all these rectangles as

$$A \approx \sum_{i=1}^{n} f((x_{i-1} + x_i)/2)\Delta x_i \tag{7.6}$$

The sum is called a midpoint **Riemann sum**. Note that $(x_{i-1} + x_i)/2$ is the middle point of the interval $[x_{i-1}, x_i]$. Later, we recognize this sum as a quadrature rule called the composite midpoint rule for approximating a definite integral.

Remark The process of determining area is historically termed as quadrature. The word quadrature was initially used to mean squaring: the construction of a square using only compass and straightedge to have the same area as a given geometric shape.

By applying the limit process, the net area A is defined as

$$A = \lim_{\max \Delta x_i \to 0} \sum_{i=1}^{n} f((x_{i-1} + x_i)/2)\Delta x_i \tag{7.7}$$

which leads to the definition of a definite integral

$$\int_a^b f(x)\mathrm{d}x = \lim_{\max \Delta x_i \to 0} \sum_{i=1}^{n} f((x_{i-1} + x_i)/2)\Delta x_i \tag{7.8}$$

Remark The area $A = \int_a^b f(x)\mathrm{d}x$ is a net area.

Reasoned from the definition of a definite integral, definite integrals have the following properties.
- The variable x in $\int_a^b f(x)\mathrm{d}x$ is a just a place holder, and is called a **dummy variable**. It can be replaced by other symbols. For example $\int_a^b f(x)\mathrm{d}x = \int_a^b f(\theta)\mathrm{d}\theta$.
- **Linearity property**:

$$\int_a^b [\alpha f(x) + \beta g(x)]\mathrm{d}x = \alpha \int_a^b f(x)\mathrm{d}x + \beta \int_a^b g(x)\mathrm{d}x \tag{7.9}$$

- **Additive property**:

$$\int_a^b f(x)\mathrm{d}x = \int_a^c f(x)\mathrm{d}x + \int_c^b f(x)\mathrm{d}x \tag{7.10}$$

If we know an antiderivative of the integrand $f(x)$ in the definite integral $\int_a^b f(x)\mathrm{d}x$, then we can use the evaluation theorem (part of the **Fundamental Theorem of Calculus**) to find the true value of the definite integral. For example

$$\int_a^b x^n \mathrm{d}x = \frac{x^{n+1}}{n+1}\bigg|_a^b, \quad n \neq -1 \tag{7.11}$$

which can be used to find an antiderivative of a polynomial with the linearity property of definite integrals.

Finally, we present a generalized mean value theorem for definite integrals. This theorem will be used later for error analysis in numerical integration.

Theorem 7.1 (Generalized Mean Value Theorem for Integrals) *Let $f(x)$ and $g(x)$ be continuous real functions on the closed interval $[a, b]$. If the function $g(x)$ does not change sign on $[a, b]$ (i.e. $g(x) \geq 0$ on $[a, b]$ or $g(x) \leq 0$ on $[a, b]$), then there exists a number $c \in [a, b]$ such that*

$$\int_a^b f(x)g(x)\mathrm{d}x = f(c)\int_a^b g(x)\mathrm{d}x \tag{7.12}$$

Proof *We prove the case $g(x) \geq 0$ on $[a, b]$ but $g(x)$ is not a zero function. The case $g(x) \leq 0$ on $[a, b]$ is similar.*

Since $f(x)$ is continuous on $[a, b]$, it has a minimum value $L = f(x_L)$ at x_L and a maximum value $U = f(x_U)$ at x_U by the extreme value theorem and $L \leq f(x) \leq U$ on $[a, b]$.

Since continuous $g(x) \geq 0$ on $[a, b]$ and $g(x)$ is not a zero function, we then have

$$\int_a^b g(x)\mathrm{d}x \neq 0 \tag{7.13}$$

By the comparison theorem for definite integrals, we have

$$L\int_a^b g(x)\mathrm{d}x \leq \int_a^b f(x)g(x)\mathrm{d}x \leq U\int_a^b g(x)\mathrm{d}x \tag{7.14}$$

which gives

$$f(x_L) = L \leq \frac{\int_a^b f(x)g(x)\mathrm{d}x}{\int_a^b g(x)\mathrm{d}x} \leq U = f(x_U) \tag{7.15}$$

By the intermediate value theorem, there exists a value c between x_L and x_U ($c \in [a, b]$ as x_L and x_U are in $[a, b]$) such that

$$f(c) = \frac{\int_a^b f(x)g(x)\mathrm{d}x}{\int_a^b g(x)\mathrm{d}x} \Rightarrow \int_a^b f(x)g(x)\mathrm{d}x = f(c)\int_a^b g(x)\mathrm{d}x \tag{7.16}$$

∎

Remark It can be further shown that $c \in (a,b)$. In addition, if $g(x) = 1$, then the generalized mean value theorem for integrals is reduced the usual mean value theorem for integrals

$$\int_a^b f(x)\mathrm{d}x = f(c)(b-a) \tag{7.17}$$

7.2 Numerical Integration

In this section, we give an overview of what will come in this chapter to lay out a big picture for numerical integration.

We want to approximate the definite integral

$$I(f) = \int_a^b f(x)\mathrm{d}x \tag{7.18}$$

First we divide the interval $[a,b]$ into n subintervals by the partition P, where

$$P = \{a = x_0 < x_1 < \cdots < x_n = b\} \tag{7.19}$$

By the linearity property of a definite integral, we then have

$$\int_a^b f(x)\mathrm{d}x = \sum_{i=1}^n \int_{x_{i-1}}^{x_i} f(x)\mathrm{d}x \tag{7.20}$$

We may use a uniform partition P_h with $h = x_i - x_{i-1} = (b-a)/n$, $i = 1, 2, \ldots, n$.

Then the approximation of $I(f)$ can be obtained by approximating $\int_{x_{i-1}}^{x_i} f(x)\mathrm{d}x$ $(i = 1, 2, \ldots, n)$, a definite integral over each subinterval. The subinterval can be made very small (through a dense partition) so that the integrand $f(x)$ does not change much over the subinterval and can be well approximated by simple functions with known antiderivatives, say polynomials (again!).

So the approximation of $I(f)$ boils down to the approximation of $f(x)$ over a small subinterval $[\alpha, \beta]$, where $[\alpha, \beta]$ can be any subinterval $[x_{i-1}, x_i]$. We can approximate $f(x)$ by a polynomial $p(x)$ that interpolates data generated by $f(x)$ at some nodes in the subinterval. Accordingly we have

$$\int_\alpha^\beta f(x)\mathrm{d}x \approx \int_\alpha^\beta p(x)\mathrm{d}x \tag{7.21}$$

The integration of the interpolating polynomial $p(x)$ can be carried out in closed form, which produces the so-called **basic quadrature rule** $R(f)$:

$$R(f) = \int_\alpha^\beta p(x)\mathrm{d}x \tag{7.22}$$

Depending on what interpolation nodes in $[\alpha, \beta]$ are chosen to interpolate $f(x)$, different basic quadrature rules are derived. For example, if we use only one node $(\alpha + \beta)/2$, we obtain the **basic midpoint rule**, if we use α and β as the nodes, we obtain **basic trapezoidal rule**; if we use α, $(\alpha + \beta)/2$ and β as the nodes, we obtain **basic Simpson's rule**; and if we use a set of so-called Gauss nodes, we obtain a **basic Gaussian quadrature rule**.

Let's take the basic trapezoidal rule $R_T(f)$ as an example. Its interpolating polynomial $p(x)$ interpolates the data $(\alpha, f(\alpha))$ and $(\beta, f(\beta))$, and can be written in Lagrange form as

$$p(x) = f(\alpha)L_0(x) + f(\beta)L_1(x) \tag{7.23}$$

where $L_0(x) = (x - \beta)/(\alpha - \beta)$ and $L_1(x) = (x - \alpha)/(\beta - \alpha)$ are Lagrange basis functions. By the linearity of definite integrals, we obtain the basic trapezoidal rule $R_T(f)$ as

$$R_T(f) = \int_\alpha^\beta p(x)\mathrm{d}x = f(\alpha)\int_\alpha^\beta L_0(x)\mathrm{d}x + f(\beta)\int_\alpha^\beta L_1(x)\mathrm{d}x \tag{7.24}$$

With w_0 and w_1 defined as

$$w_0 = \int_\alpha^\beta L_0(x)\mathrm{d}x = \frac{\beta - \alpha}{2}, \quad w_1 = \int_\alpha^\beta L_1(x)\mathrm{d}x = \frac{\beta - \alpha}{2} \tag{7.25}$$

the form of the basic trapezoidal rule $R_T(f)$ is

$$R_T(f) = w_0 f(\alpha) + w_1 f(\beta) \tag{7.26}$$

which is a linear combination of the values of the integrand $f(x)$ at the interpolation nodes α and β using the corresponding weights w_0 and w_1. The weights are definite integrals of Lagrange basis functions over the subinterval $[\alpha, \beta]$. The other basic quadrature rules mentioned above can be derived similarly and have the similar forms.

The initial definite integral $I(f)$ is split with the partition P as

$$\int_a^b f(x)\mathrm{d}x = \sum_{i=1}^n \int_{x_{i-1}}^{x_i} f(x)\mathrm{d}x \tag{7.27}$$

With a basic quadrature rule $R(f)$ ready, we can apply $R(f)$ to approximate the definite integral over each subinterval and add the results to approximate $I(f)$. The sum of basic quadrature rules over all the subintervals is called the corresponding **composite quadrature rule** with the partition P, denoted as $R(f; P)$. For example, the basic midpoint rule $R_M(f)$ is

$$R_M(f) = \int_\alpha^\beta p(x)\mathrm{d}x = \int_\alpha^\beta f\left(\frac{\alpha + \beta}{2}\right)\mathrm{d}x = (\beta - \alpha)f\left(\frac{\alpha + \beta}{2}\right) \tag{7.28}$$

With $R_M(f)$ approximating $\int_{x_{i-1}}^{x_i} f(x)dx$, we obtain

$$
\begin{aligned}
\int_a^b f(x)dx &= \sum_{i=1}^n \int_{x_{i-1}}^{x_i} f(x)dx \\
&\approx \sum_{i=1}^n (x_i - x_{i-1}) f\left(\frac{x_{i-1}+x_i}{2}\right) = \sum_{i=1}^n f\left(\frac{x_{i-1}+x_i}{2}\right) \Delta x_i \quad (7.29)
\end{aligned}
$$

We then have the composite midpoint rule $R_M(f;P)$ with the partition P as

$$
R_M(f;P) = \sum_{i=1}^n f\left(\frac{x_{i-1}+x_i}{2}\right) \Delta x_i \tag{7.30}
$$

which approximates $I(f)$ as

$$
I(f) \approx R_M(f;P) \tag{7.31}
$$

Clearly the composite midpoint rule is just the midpoint Riemann sum introduced in the previous section.

The **quadrature error** of a basic quadrature rule $R(f) = \int_\alpha^\beta p(x)dx$ to approximate $\int_\alpha^\beta f(x)dx$ is defined as

$$
E(f) = \int_\alpha^\beta f(x)dx - R(f) = \int_\alpha^\beta f(x)dx - \int_\alpha^\beta p(x)dx = \int_\alpha^\beta [f(x) - p(x)]dx \tag{7.32}
$$

which can be analyzed by integrating the interpolation error $[f(x) - p(x)]$.

The quadrature error of a composite quadrature rule $R(f;P)$ to approximate $I(f) = \int_a^b f(x)dx$ is

$$
E(f;P) = I(f) - R(f;P) \tag{7.33}
$$

which is just an accumulation of the quadrature errors from basic quadrature rules that are applied to all the subintervals of the partition P.

7.2.1 Change of Intervals

It is much more convenient to derive and discuss a basic quadrature rule $R^*(F(t))$ that approximates $\int_{-1}^1 F(t)dt$, a definite integral over a canonical interval $[-1,1]$. So we need to know how to apply the rule $R^*(F(t))$ for the canonical interval of integration, $[-1,1]$, to approximate $\int_\alpha^\beta f(x)dx$, a definite integral over a general interval of integration, $[\alpha, \beta]$.

By the substitution rule for integrals, we have

$$
\int_{-1}^1 f(g(t))g'(t)dt = \int_{g(-1)}^{g(1)} f(x)dx \tag{7.34}
$$

We therefore seek a transformation $x = g(t)$ such that $x \in [\alpha, \beta]$ when $t \in [-1, 1]$, i.e. $\alpha = g(-1)$ and $\beta = g(1)$. It transforms the interval $[\alpha, \beta]$ to the canonical interval $[-1, 1]$ and transform $\int_\alpha^\beta f(x)dx$ to $\int_{-1}^1 F(t)dt$, where $F(t) = f(g(t))g'(t)$.

The transformation must satisfy $g(-1) = \alpha$ and $g(1) = \beta$. One choice is the linear function $g(t) = \frac{\beta - \alpha}{2}t + \frac{\beta + \alpha}{2}$ which interpolates the data $(t, x) = (-1, \alpha)$ and $(t, x) = (1, \beta)$. With this choice, we have

$$\int_{-1}^1 F(t)dt = \int_\alpha^\beta f(x)dx \tag{7.35}$$

where $F(t)$ is

$$F(t) = f(g(t))g'(t) = \frac{\beta - \alpha}{2} f\left(\frac{\beta - \alpha}{2}t + \frac{\beta + \alpha}{2}\right) \tag{7.36}$$

A basic quadrature rule $R^*(F)$ that approximates $\int_{-1}^1 F(t)dt$ has the form

$$R^*(F) = \sum_{j=0}^m A_j F(t_j)$$

which is a linear combination of the values of the integrand at the interpolation nodes t_j $(j = 0, 1, \ldots, m)$ in $[-1, 1]$ using the weights $A_j = \int_{-1}^1 L_j(t)dt$, where $L_j(t)$ is a Lagrange basis function.

If we want to apply the rule $R^*(F)$ to approximate $\int_\alpha^\beta f(x)dx$, we have

$$\int_\alpha^\beta f(x)dx = \int_{-1}^1 F(t)dt$$
$$\approx R^*(F) = \sum_{j=0}^m A_j F(t_j) = \sum_{j=0}^m A_j \frac{\beta - \alpha}{2} f\left(\frac{\beta - \alpha}{2}t_j + \frac{\beta + \alpha}{2}\right) \tag{7.37}$$

If we define

$$w_j = A_j \frac{\beta - \alpha}{2}, \quad x_j = \frac{\beta - \alpha}{2}t_j + \frac{\beta + \alpha}{2} \tag{7.38}$$

the rule that approximates $\int_\alpha^\beta f(x)dx$ is

$$R(f) = \sum_{j=0}^m w_j f(x_j) \tag{7.39}$$

whose weights w_j and nodes x_j $(x_j \in [\alpha, \beta])$ are transformed from the weights A_j and nodes t_j $(t_j \in [-1, 1])$ of $R^*(F)$.

Example 7.1 A basic Gaussian quadrature rule $R^*(F) = F(-\sqrt{1/3}) + F(\sqrt{1/3})$ is derived to approximate $\int_{-1}^1 F(t)dt$. Apply this rule to approximate $\int_1^5 x2^{-x^2}dx$. Find the error in the approximation using the true value of the definite integral.

[Solution:] The given rule has the following weights and nodes

$$A_0 = 1, \quad A_1 = 1$$

$$t_0 = -\sqrt{1/3}, \quad t_1 = \sqrt{1/3}$$

With change of intervals, the weights and nodes of the rule $R(f)$ from transformation for $\int_\alpha^\beta f(x)dx$ are

$$w_0 = A_0 \frac{\beta - \alpha}{2}, \quad w_1 = A_1 \frac{\beta - \alpha}{2}$$

$$x_0 = \frac{\beta - \alpha}{2}t_0 + \frac{\beta + \alpha}{2}, \quad x_0 = \frac{\beta - \alpha}{2}t_1 + \frac{\beta + \alpha}{2}$$

Here $\alpha = 1$ and $\beta = 5$, so we have $\frac{\beta - \alpha}{2} = 2$, $\frac{\beta + \alpha}{2} = 3$ and

$$w_0 = 2, \quad w_1 = 2$$

$$x_0 = 2(-\sqrt{1/3}) + 3, \quad x_0 = 2(\sqrt{1/3}) + 3$$

The rule from transformation for $\int_1^5 f(x)dx$ is

$$R(f) = 2f(-2\sqrt{1/3} + 3) + 2f(2\sqrt{1/3} + 3)$$

With the integrand $f(x) = x2^{-x^2}$, the definite integral $\int_1^5 x2^{-x^2}dx$ is approximated by

$$R(f) = 2(-2\sqrt{1/3} + 3)2^{-(-2\sqrt{1/3}+3)^2} + 2(2\sqrt{1/3} + 3)2^{-(2\sqrt{1/3}+3)^2} \approx 0.348$$

The true value of the definite integral in this example is available as

$$\int_1^5 x2^{-x^2}dx = -\frac{1}{2\ln 2}2^{-x^2}\Big|_1^5 = -\frac{1}{2\ln 2}(2^{-25} - 2^{-1}) \approx 0.361$$

The error of the approximation is

$$\int_1^5 x2^{x^2}dx - R(f) \approx 0.361 - 0.348 = 0.013$$

■

7.3 The Midpoint Rule

The **basic midpoint rule** $R_M(f)$ that approximates $\int_\alpha^\beta f(x)dx$ is

$$R_M(f) = (\beta - \alpha)f\left(\frac{\alpha + \beta}{2}\right) \tag{7.40}$$

which is derived by integrating over $[\alpha, \beta]$ a constant polynomial $p(x) = f((\alpha + \beta)/2)$ that interpolates the integrand $f(x)$ at the midpoint $(\alpha + \beta)/2$ of the interval $[\alpha, \beta]$. It has a geometric interpretation as illustrated in Fig. 7.2. The net area between the curve

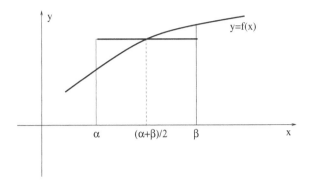

FIGURE 7.2
The basic midpoint rule.

$y = f(x)$ and the x axis over $[\alpha, \beta]$ is approximated by the net area of the rectangle, which is the product of the base $\beta - \alpha$ and the height $f((\alpha + \beta)/2)$.

The **composite midpoint rule** $R_M(f;P)$ with the partition $P = \{a = x_0 < x_1 < \cdots < x_n = b\}$ that approximates $I(f) = \int_a^b f(x)\mathrm{d}x = \sum_{i=1}^n \int_{x_{i-1}}^{x_i} f(x)\mathrm{d}x$ is

$$R_M(f;P) = \sum_{i=1}^n (x_i - x_{i-1})f((x_{i-1}+x_i)/2) \tag{7.41}$$

See Fig. 7.1 for the illustration of the approximation $I(f) \approx R_M(f;P)$, where $R_M(f;P)$ is a Riemann sum. For a uniform partition P_h with $h = (b-a)/n$, we have

$$R_M(f;P_h) = h \sum_{i=1}^n f(a+ih-h/2) \tag{7.42}$$

which can be easily coded. The pesudocode for $R_M(f;P_h)$ is given below.

Algorithm 12 The composite midpoint rule with a uniform partition for $\int_a^b f(x)\mathrm{d}x$

$h \leftarrow (b-a)/n$
$R_M \leftarrow 0$
for i from 1 to n **do**
$\quad x_M \leftarrow a+ih-h/2$
$\quad R_M \leftarrow R_M + f(x_M)$
end for
$R_M \leftarrow hR_M$

7.3.1   Degree of Precision (DOP)

The error of a basic quadrature rule is

$$I(f) - R(f) = \int_\alpha^\beta f(x)\mathrm{d}x - \int_\alpha^\beta p(x)\mathrm{d}x = \int_\alpha^\beta [f(x) - p(x)]\mathrm{d}x \tag{7.43}$$

which is the integration of the interpolation error $[f(x) - p(x)]$, where $p(x)$ is a polynomial that interpolates and approximates the integrand $f(x)$.

It is obvious that the basic midpoint rule $R_M(f)$ is exact with no error if the integrand $f(x)$ is a constant polynomial because $p(x) = f(x) = f((\alpha + \beta)/2)$ in this case. If $f(x)$ is a linear polynomial, $R_M(f)$ is also exact. This can be seen geometrically. However, if $f(x)$ is a quadratic polynomial, $R_M(f)$ is not exact anymore. We say the degree of precision (DOP) of the midpoint rule is 1, i.e. DOP $= 1$.

Definition 7.1 (DOP) *A basic quadrature rule $R(f)$ that approximates $\int_\alpha^\beta f(x)dx$ has DOP $= m$ if $R(f)$ is exact (i.e. $\int_\alpha^\beta f(x)dx = R(f)$ with no error) whenever $f(x)$ is a polynomial of degree $\leq m$, but not exact when $f(x)$ is a polynomial of degree $m + 1$.*

Obviously if a basic quadrature rule $R(f)$ is exact for any polynomial of degree $\leq m$, its corresponding composite quadrature rule $R(f; P)$ must also be exact for any polynomial of degree $\leq m$.

We can use the following theorem to test the DOP of a basic quadrature rule.

Theorem 7.2 (DOP) *A basic quadrature rule $R(f)$ that approximates $\int_\alpha^\beta f(x)dx$ has DOP $= m$ if*

$$\int_\alpha^\beta x^k dx = R(x^k), \quad k = 0, 1, \ldots, m \tag{7.44}$$

$$\int_\alpha^\beta x^{m+1} dx \neq R(x^{m+1}) \tag{7.45}$$

The proof of the above theorem is based on the linearity property of definite integrals and a quadrature rule $R(f)$ (see Exercises 7.5 and 7.6). The linearity property of $R(f)$ can be stated as

$$R(cf(x) + dg(x)) = cR(f(x)) + dR(g(x)) \tag{7.46}$$

We can even test the DOP of a basic rule by applying the above theorem to the case with $\alpha = -1$ and $\beta = 1$. This can be shown using the change of intervals.

Theorem 7.3 (DOP) *Denote the basic quadrature rule $R(f)$ as $R^*(F)$ when it* is applied to approximate $\int_{-1}^{1} F(t)dt$. The rule $R(f)$ has DOP $= m$ if

$$\int_{-1}^{1} t^k dt = R^*(t^k), \quad k = 0, 1, \ldots, m \tag{7.47}$$

$$\int_{-1}^{1} t^{m+1} dt \neq R^*(t^{m+1}) \tag{7.48}$$

The DOP of a basic quadrature rule $R(f) = \sum_{j=0}^{M} A_j f(x_j)$ must be $\geq M$ by the polynomial interpolant existence and uniqueness theorem.

Example 7.2 Find the DOP of the basic midpoint rule $R_M(f) = (\beta - \alpha)f((\alpha + \beta)/2)$ that approximates $\int_{\alpha}^{\beta} f(x)dx$.

[Solution:] We only need to consider the basic quadrature rule for $\int_{-1}^{1} F(t)dt$ with $\alpha = -1$ and $\beta = 1$. The rule becomes $R_M^*(F) = 2F(0)$. Create a table to list the values $\int_{-1}^{1} t^k dt$ and $R^*(t^k)$ for $k = 0, 1, \ldots$ until the values do not match (Table 7.1).

k	t^k	$\int_{-1}^{1} t^k dt$	$R^*(t^k)$
0	1	2	2
1	t	0	0
2	t^2	2/3	0

TABLE 7.1
Finding DOP.

It can be seen from the table that the midpoint rule $R_M(f)$ has DOP $= 1$. ∎

We can derive a basic quadrature rule by maximizing its DOP. Below is an example.

Example 7.3 Suppose the quadrature rule

$$R^*(F) = A_0 F(-1) + A_1 F(0) + A_2 F(1)$$

approximates $\int_{-1}^{1} F(t)dt$. Find the weights A_0, A_1 and A_2 by maximizing the DOP of the rule. What is the DOP of this rule?

[Solution:] Create a table to list the values $\int_{-1}^{1} t^k dt$ and $R^*(t^k)$ for $k = 0, 1, \ldots$. To find the 3 weights, we match the values for $k = 0, 1, 2$ to obtain 3 equations for the 3 weights. After the weights are determined, we continue the table (for $k = 3, \ldots$) with the known weights until the values do not match to find the DOP of the rule (Table 7.2).

The three equations for the 3 weights are

$$2 = A_0 + A_1 + A_2$$

$$0 = -A_0 + A_2$$

k	t^k	$\int_{-1}^{1} t^k \mathrm{d}t$	$R^*(t^k)$
0	1	2	$A_0 + A_1 + A_2$
1	t	0	$-A_0 + A_2$
2	t^2	2/3	$A_0 + A_2$
3	t^3	0	$-A_0 + A_2 = 0$
4	t^4	2/5	$A_0 + A_2 = 2/3$

TABLE 7.2
Maximizing DOP to derive a quadrature rule.

$$2/3 = A_0 + A_2$$

which can be solved to give $A_0 = A_2 = 1/3$ and $A_1 = 4/3$. So the rule is

$$R^*(F) = \frac{1}{3}[F(-1) + 4F(0) + F(1)]$$

Its DOP is 3. Later we will recognize this rule as the basic Simpson's rule applied to the canonical interval $[-1, 1]$. ■

Remark The rule in the above example is called the basic Simpson's rule for the canonical interval $[-1, 1]$. Using change of intervals, we can obtain the basic Simpson's rule $R_S(f)$ to approximate $\int_{\alpha}^{\beta} f(x)\mathrm{d}x$ as

$$R_S(f) = \frac{\beta - \alpha}{6}\left[f(\alpha) + 4f\left(\frac{\alpha + \beta}{2}\right) + f(\beta)\right] \qquad (7.49)$$

Note that if a basic quadrature rule is exact for a function, then its corresponding composite quadrature rule is also exact for the same function because the composite rule is a repeated application of the basic rule.

7.3.2   Error of the Midpoint Rule

Theorem 7.4 (Midpoint Rule Theorem) *The error to approximate the definite integral*

$$I(f) = \int_a^b f(x)\mathrm{d}x \qquad (7.50)$$

where $f(x) \in C_{[a,b]}^2$, by the composite midpoint rule

$$R_M(f; P_h) = h\sum_{i=1}^{n} f(a + ih - h/2) \qquad (7.51)$$

where P_h is a uniform partition with $h = (b-a)/n$, is

$$E_M(f;P_h) = I(f) - R_M(f;P_h) = \frac{f''(\eta)}{24}(b-a)h^2 \qquad (7.52)$$

where η is some number in (a,b). If h is small enough (n is large enough), we have an asymptotic estimate of the error as

$$E_M(f;P_h) \approx \frac{h^2}{24}[f'(b) - f'(a)] \qquad (7.53)$$

Eq. (7.52) is a useful form of the error as it can be used to
- estimate the error upper bound for a given partition P_h
- determine the interval length h (or the number of subintervals n) for a given error tolerance

Example 7.4 If the composite midpoint rule

$$R_M(f;P_h) = h\sum_{i=1}^{n} f(a+ih-h/2)$$

is applied to approximate

$$I(f) = \int_{-1}^{2} \sin(x^2)dx$$

with absolute error at most 10^{-6}, how many subintervals are needed?
[Solution:] The absolute error of the approximation is

$$|E_M(f;P_h)| = |I(f) - R_M(f;P_h)| = \frac{|f''(\eta)|}{24}(b-a)h^2$$

Here $f(x) = \sin(x^2)$, $a = -1$, $b = 2$, $\eta \in (a,b) = (-1,2)$ and $h = (b-a)/n = 3/n$, where n is the number of subintervals.
    We have $f'(x) = 2x\cos(x^2)$ and $f''(x) = 2\cos(x^2) - 4x^2\sin(x^2)$. For $-1 < \eta < 2$, we use the triangle inequality to get

$$|f''(\eta)| = |2\cos(\eta^2) - 4\eta^2\sin(\eta^2)| \leq |2\cos(\eta^2)| + |4\eta^2\sin(\eta^2)| < 2 + 4(2)^2 = 18$$

So we have the upper bound of the absolute error in terms of n as

$$|E_M(f;P_h)| = \frac{|f''(\eta)|}{24}(b-a)h^2 < \frac{18}{24}3(3/n)^2 = \frac{81}{4n^2}$$

We let the upper bound be at most 10^{-6} to assure the error is at most 10^{-6}. So we have

$$\frac{81}{4n^2} \leq 10^{-6} \Rightarrow n \geq \sqrt{\frac{81}{4 \times 10^{-6}}} = 4500$$

    The answer is that we use at least 4500 subintervals. ∎

Because the asymptotic error estimate does not involve the unknown number η, it is easy to compute. The asymptotic error estimate indicates that

$$\frac{E_M(f;P_{2h})}{E_M(f;P_h)} = \frac{I(f) - R_M(f;P_{2h})}{I(f) - R_M(f;P_h)} \approx \frac{(2h)^2}{h^2} = 4$$

when h is small enough, meaning that the absolute error increases by a factor of 4 if a small h is doubled. In general, if the error of a composite quadrature rule $E(f,P_h)$ with a uniform partition P_h is of $O(h^p)$, where p is a positive integer, then for small h we have the ratio

$$\frac{E(f;P_{2h})}{E(f;P_h)} \approx \frac{(2h)^p}{h^p} = 2^p$$

and the value of p can be estimated by

$$p \approx \log_2\left[\frac{E(f;P_{2h})}{E(f;P_h)}\right]$$

Example 7.5 If the composite midpoint rule

$$R_M(f;P_h) = h\sum_{i=1}^{n} f(a + ih - h/2)$$

is applied to approximate

$$I(f) = \int_{-1}^{2} \sin(x^2)\,dx$$

with $h = (b-a)/n = 3/4500$ (the value from the previous example), estimate the absolute error of the approximation. What is the estimate of the absolute error if h is reduced by half?
[Solution:] Since $h = 3/4500 = 1/1500$ is quite small, we may use the asymptotic error estimate

$$E_M(f;P_h) \approx \frac{h^2}{24}[f'(b) - f'(a)]$$

Here $f(x) = \sin(x^2)$, $f'(x) = 2x\cos(x^2)$, $a = -1$, $b = 2$ and $h = 1/1500$. We have

$$|E_M(f;P_h)| \approx \frac{|4\cos 4 + 2\cos 1|}{24(1500)^2} \approx 2.8407 \times 10^{-8}$$

If h is reduced by half, as $E_M(f;P_h) = O(h^2)$, the error is reduced by a factor of $2^2 = 4$ to about $\frac{1}{4} \times 2.8407 \times 10^{-8} \approx 7.1018 \times 10^{-9}$. ∎

Remark If the integrand $f(x)$ in the integral $I(f) = \int_a^b f(x)\,dx$ is periodic with $f'(a) = f'(b)$, then the asymptotic error estimate $E_M(f;P_h) \approx \frac{h^2}{24}[f'(b) - f'(a)] = 0$ for the composite middle point rule $R_M(f;P_h)$ indicates that the error at very small h may be much smaller than non-periodic case because of the periodicity $f'(a) = f'(b)$.

The quadrature error of a basic quadrature rule $R(f) = \int_\alpha^\beta p(x)\mathrm{d}x$ to approximate $\int_\alpha^\beta f(x)\mathrm{d}x$ is

$$E(f) = \int_\alpha^\beta f(x)\mathrm{d}x - R(f) = \int_\alpha^\beta [f(x) - p(x)]\mathrm{d}x \tag{7.54}$$

which is an integration of the interpolation error $[f(x) - p(x)]$.

Remark   In the basic midpoint rule, $p(x)$ is a polynomial of degree $n = 0$ (a constant polynomial) that interpolates the $f(x)$ at the middle point $(\alpha + \beta)/2$. By the interpolation error theorem, we have

$$f(x) - p(x) = f'(\xi)\left(x - \frac{\alpha + \beta}{2}\right) \tag{7.55}$$

where $\xi = \xi(x)$ depends on x. So the error of the basic midpoint rule is

$$\int_\alpha^\beta f'(\xi(x))\left(x - \frac{\alpha + \beta}{2}\right)\mathrm{d}x \tag{7.56}$$

whose integrand is the product of the two functions $f'(\xi(x))$ and $N_1(x) = (x - (\alpha + \beta)/2)$. We may be tempted to apply the generalized mean value theorem for integrals to simplify the above error formula into a more useful form, but we cannot because we have no information about the sign of $f'(\xi(x))$, and $N_1(x)$ changes signs in the interval $[\alpha, \beta]$ and satisfies $\int_\alpha^\beta N_1(x)\mathrm{d}x = 0$.

More generally, if $M + 1$ interpolation nodes x_j, $j = 0, 1, \ldots, M$, are used in deriving a basic quadrature rule $R(f)$, and the rule has DOP larger than M, then

$$\int_\alpha^\beta N_{M+1}(x)\mathrm{d}x = \int_\alpha^\beta (x - x_0)(x - x_1)\cdots(x - x_M)\mathrm{d}x = 0 \tag{7.57}$$

because $\int_\alpha^\beta N_{M+1}(x)\mathrm{d}x = R(N_{M+1}(x))$ when DOP of $R(f)$ is larger than M, and $R(N_{M+1}(x)) = 0$ ($N_{M+1}(x)$ is zero at the interpolation nodes).

A more general approach to derive the error of a basic quadrature rule is to relate the error with the DOP of the rule. Below we give a full proof of the Theorem 7.4 about the error of the midpoint rule. Some ideas about quadrature error analysis can be seen in the proof.

Proof   *The basic midpoint rule $R_M(f)$ has* DOP $= 1$, *meaning it is exact for polynomial integrand of degree* ≤ 1. *Let's apply Taylor's theorem to write the integrand $f(x)$ as*

$$f(x) = T_1(x) + R_1(x) \tag{7.58}$$

where

$$T_1(x) = f(a) + f'(a)(x-a) \tag{7.59}$$

is the Taylor polynomial of degree 1 for $f(x)$ about the midpoint $a = (\alpha + \beta)/2$ and

$$R_1(x) = f''(c(x))(x-a)^2/2!, \quad c(x) \in [\alpha, \beta] \tag{7.60}$$

is Taylor's remainder. By the linearity properties for definite integrals and quadrature rules, we have

$$\int_\alpha^\beta f(x)dx = \int_\alpha^\beta T_1(x)dx + \int_\alpha^\beta R_1(x)dx \tag{7.61}$$

$$R_M(f) = R_M(T_1) + R_M(R_1) \tag{7.62}$$

Since $R_M(f)$ is exact for polynomial integrand of degree ≤ 1, we have

$$R_M(T_1) = \int_\alpha^\beta T_1(x)dx \tag{7.63}$$

We also have

$$R_M(R_1) = (\beta - \alpha)R_1((\alpha+\beta)/2) = (\beta-\alpha)R_1(a) = 0 \tag{7.64}$$

So we have $R_M(f) = R_M(T_1)$ by Eq. (7.62) and get the error of $R_M(f)$ as

$$E_M(f) = \int_\alpha^\beta f(x)dx - R_M(f) = \int_\alpha^\beta [f(x) - T_1(x)]dx$$

$$= \int_\alpha^\beta R_1(x)dx = \int_\alpha^\beta \frac{f''(c(x))}{2!}(x-a)^2 dx \tag{7.65}$$

Now we can apply the generalized mean value theorem for integrals because $(x-a)^2 \geq 0$ ($= 0$ only when $x = a$). So we have the error of the midpoint rule $R_M(f)$

$$E_M(f) = \int_\alpha^\beta \frac{f''(c(x))}{2!}(x-a)^2 dx = \frac{f''(\xi)}{2!}\int_\alpha^\beta (x-a)^2 dx = \frac{f''(\xi)}{24}(\beta - \alpha)^3 \tag{7.66}$$

where ξ is a number in (α, β).

The quadrature error of a composite quadrature rule $R(f;P)$ to approximate $I(f) = \int_a^b f(x)dx$ is

$$E(f;P) = I(f) - R(f;P) \tag{7.67}$$

which is the accumulation of the quadrature errors as the basic quadrature rule $R(f)$ is applied to all the subintervals of the partition P. The error of the composite midpoint rule $R_M(f;P)$ is

$$E_M(f;P) = I(f) - R_M(f;P) = \sum_{i=1}^n \frac{f''(\xi_i)}{24}(x_i - x_{i-1})^3 \tag{7.68}$$

where ξ_i is a number in (x_{i-1}, x_i). For a uniform partition P_h with $h = x_i - x_{i-1} = (b-a)/n$ $(i = 1, 2, \ldots, n)$, we have

$$E_M(f; P_h) = I(f) - R_M(f; P_h) = \frac{h^3}{24} \sum_{i=1}^{n} f''(\xi_i) \tag{7.69}$$

If the integrand $f(x)$ has continuous $f''(x)$ on $[a, b]$, then $f''(x)$ has a minimum value L and a maximum value U by the extreme value theorem and $L \leq f''(x) \leq U$ on $[a, b]$. So we have

$$nL \leq \sum_{i=1}^{n} f''(\xi_i) \leq nU \Rightarrow L \leq \frac{\sum_{i=1}^{n} f''(\xi_i)}{n} \leq U \tag{7.70}$$

By the intermediate value theorem, there exists a number $\eta \in (a, b)$ such that

$$f''(\eta) = \frac{\sum_{i=1}^{n} f''(\xi_i)}{n} \Rightarrow \sum_{i=1}^{n} f''(\xi_i) = nf''(\eta) \tag{7.71}$$

Finally we have

$$E_M(f; P_h) = I(f) - R_M(f; P_h) = \frac{h^3}{24} nf''(\eta) = \frac{f''(\eta)}{24}(b-a)h^2 \tag{7.72}$$

Realize that

$$\sum_{i=1}^{n} f''(\xi_i)h$$

in Eq. (7.69) is a Riemann sum for the definite integral $\int_a^b f''(x)dx$. When h is small enough, we have

$$\sum_{i=1}^{n} f''(\xi_i)h \approx \int_a^b f''(x)dx = f'(b) - f'(a) \tag{7.73}$$

and we obtain the asymptotic estimate of the error as

$$E_M(f; P_h) \approx \frac{h^2}{24}[f'(b) - f'(a)] \tag{7.74}$$

∎

7.4   The Trapezoidal Rule

The **basic trapezoidal rule** $R_T(f)$ that approximates $\int_\alpha^\beta f(x)dx$ is

$$R_T(f) = \frac{\beta - \alpha}{2}[f(\alpha) + f(\beta)] \tag{7.75}$$

which is derived by integrating from α to β the polynomial $p(x)$ that interpolates $f(x)$ at the nodes α and β. The geometric interpretation of the basic trapezoidal rule $R_T(f)$ is illustrated in Fig. 7.3, where the net area between the curve $y = f(x)$ and the x-axis is approximated by the net area of the trapezoid, which is half of the product of the height $\beta - \alpha$ with the sum of two signed bases $f(\alpha) + f(\beta)$. We know that the DOP

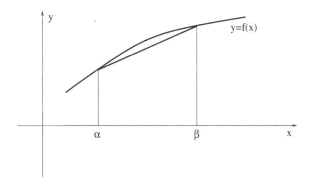

FIGURE 7.3
The basic trapezoidal rule.

of the basic trapezoidal rule is at least 1. It turns out that its DOP is equal to 1.

The **composite trapezoidal rule** $R_T(f;P)$ with the partition P that approximates $I(f) = \int_a^b f(x)dx$ is

$$R_M(f;P) = \sum_{i=1}^{n} \frac{x_i - x_{i-1}}{2}[f(x_{i-1}) + f(x_i)] \tag{7.76}$$

For a uniform partition P_h with $h = (b-a)/n$, we have

$$R_T(f;P_h) = h\left[\frac{f(a) + f(b)}{2} + \sum_{i=1}^{n-1} f(x_i)\right] \tag{7.77}$$

where $x_i = a + ih$ $(i = 1, 2, \ldots, n-1)$ is an interior partition point shared by two subintervals. Below is its pseudocode.

Algorithm 13 The composite trapezoidal rule with a uniform partition for $\int_a^b f(x)dx$

$h \leftarrow (b-a)/n$
$R_T \leftarrow (f(a) + f(b))/2$
for i from 1 to $n-1$ **do**
$\quad x_i \leftarrow a + ih$
$\quad R_T \leftarrow R_T + f(x_i)$
end for
$R_T \leftarrow hR_T$

Example 7.6 Approximate $I(f) = \int_0^1 \frac{1}{1+x} dx$ by $R_T(f;P)$, where P is a nonuniform partition with the partition points $x_0 = 0$, $x_1 = 0.4$ and $x_2 = 1$.

[Solution:] There are $n = 2$ subintervals: $[x_0, x_1] = [0, 0.4]$ and $[x_1, x_2] = [0.4, 1]$. With $f(x) = 1/(1+x)$, we have

$$R_T(f;P) = \sum_{i=1}^{n} \frac{x_i - x_{i-1}}{2} [f(x_{i-1}) + f(x_i)]$$

$$= \frac{x_1 - x_0}{2} [f(x_0) + f(x_1)] + \frac{x_2 - x_1}{2} [f(x_1) + f(x_2)]$$

$$= \frac{0.4 - 0}{2} \left[\frac{1}{1+0} + \frac{1}{1+0.4} \right] + \frac{1 - 0.4}{2} \left[\frac{1}{1+0.4} + \frac{1}{1+1} \right] \approx 0.70714$$

∎

Example 7.7 Approximate $I(f) = \int_0^1 \frac{1}{1+x} dx$ by $R_T(f;P_h)$ with $h = 0.25$.

[Solution:] Here $h = (1-0)/n = 0.25$, so $n = 4$. The partition points of P_h are $x_i = a + ih = 0 + 0.25i$, $i = 0, 1, 2, 3, 4$, i.e. $x_0 = a = 0$, $x_1 = 0.25$, $x_2 = 0.5$, $x_3 = 0.75$ and $x_4 = b = 1$. For $f(x) = 1/(1+x)$, we have

$$R_T(f;P_h) = h \left[\frac{f(a) + f(b)}{2} + \sum_{i=1}^{n-1} f(x_i) \right]$$

$$= 0.25 \left[\frac{1}{2} \left(\frac{1}{1+0} + \frac{1}{1+1} \right) + \frac{1}{1+0.25} + \frac{1}{1+0.5} + \frac{1}{1+0.75} \right] \approx 0.69702$$

∎

> **Remark** In the previous example, the true value of $I(f) = \ln 2 \approx 0.69315$. The error of the approximation is $E_T(f;P_h) \approx 0.69315 - 0.69702 = -0.00387$, which is negative. The composite trapezoidal rule overestimates $I(f)$ because the graph of $f(x)$ is concave up.

The error of the basic trapezoidal rule is the integration of the interpolation error. We have

$$E_T(f) = \int_\alpha^\beta f(x) dx - R_T(f) = \int_\alpha^\beta \frac{f''(\xi)}{2!} (x - \alpha)(x - \beta) dx$$

Here $N_2(x) = (x - \alpha)(x - \beta) \leq 0$ for $x \in [\alpha, \beta]$, and we can apply the generalized mean value theorem for integrals to obtain

$$E_T(f) = \int_\alpha^\beta \frac{f''(\xi(x))}{2!} N_1(x) dx = \frac{f''(\eta)}{2} \int_\alpha^\beta N_1(x) dx = -\frac{f''(\eta)}{12} (\beta - \alpha)^3$$

where η is a number in $[\alpha, \beta]$. We then can obtain the following theorem.

Theorem 7.5 (Trapezoidal Rule Theorem) *The error to approximate the definite integral*

$$I(f) = \int_b^a f(x)dx$$

where $f(x) \in C_{[a,b]}^2$, by the composite trapezoidal rule

$$R_T(f;P_h) = h \left[\frac{f(a) + f(b)}{2} + \sum_{i=1}^{n-1} f(x_i) \right]$$

where P_h is a uniform partition with $h = (b-a)/n$ and $x_i = a + ih$, is

$$E_T(f;P_h) = I(f) - R_T(f;P_h) = -\frac{f''(\zeta)}{12}(b-a)h^2$$

where ζ is some number in (a,b). If h is small enough (n is large enough), we have an asymptotic estimate of the error as

$$E_T(f;P_h) \approx -\frac{h^2}{12}[f'(b) - f'(a)]$$

Example 7.8 If the composite trapezoidal rule $R_T(f;P_h)$ is used to approximate $I(f) = \int_0^1 e^{-x^2}dx = \frac{\sqrt{\pi}}{2}\text{erf}(x)$ with an absolute error at most $10^{-4}/2$, how small should h be?

[Solution:] The absolute error of the approximation is

$$|E_T(f;P_h)| = |I(f) - R_T(f;P_h)| = \frac{|f''(\zeta)|}{12}(b-a)h^2$$

Here $f(x) = e^{-x^2}$, $a = 0$, $b = 1$ and $\zeta \in [0,1]$.
    We have

$$f'(x) = -2xe^{-x^2}, \quad f''(x) = (4x^2 - 2)e^{-x^2}$$

For $\zeta \in [0,1]$, by the triangle inequality, we have

$$|f''(\zeta)| = |4\zeta^2 - 2|e^{-\zeta^2} \le (|4\zeta^2| + |2|)e^{-\zeta^2} \le 6$$

So we have

$$|E_T(f;P_h)| = \frac{|f''(\zeta)|}{12}(b-a)h^2 \le \frac{6}{12}(1-0)h^2 = \frac{h^2}{2}$$

Let the error upper bound be at most the error tolerance. We have

$$\frac{h^2}{2} \le \frac{10^{-4}}{2} \Rightarrow h \le 10^{-2}$$

So we use h at most 0.01. ∎

Example 7.9 Use the composite trapezoidal rule $R_T(f;P_h)$ to approximate $I(f) = \int_0^1 xe^{x^2}dx$ with the number of the subintervals $n = 64, 128, 256, 512, 1024$ and 2048, respectively.

[Solution:] The true value of the definite integral is $I(f) = (e-1)/2$. The table generated by MATLAB below lists the value of h, the error $E_T(f;P_h)$ and the error ratio $|E_T(f;P_{2h})|/|E_T(f;P_h)|$ for each value of n.

n	h	E_h	E_2h|/|E_h|
64	0.015625	-0.000145558	3.999336e+00
128	0.0078125	-3.63909e-05	3.999834e+00
256	0.00390625	-9.09783e-06	3.999959e+00
512	0.00195312	-2.27446e-06	3.999990e+00
1024	0.000976562	-5.68616e-07	3.999997e+00
2048	0.000488281	-1.42154e-07	3.999999e+00

The table clearly shows that the error ratio $|E_T(f;P_{2h})|/|E_T(f;P_h)| \approx 4$, which implies $E_T(f;P_h) = O(h^2)$.

We have

$$y \approx Cx^p \Rightarrow \log_{10}y| = p\log_{10}x + \log_{10}C$$

which indicates that the slope of the loglog plot of y versus x is the power p.

So the second-order accuracy of the composite trapezoidal rule $R_T(f;P_h)$ (i.e. $E_T(f;P_h) = O(h^2)$) can also be reflected by a loglog plot of $|E_T(f;Ph)|$ versus h as shown in Fig. 7.4. The approximate slope 2 implies that $|E_T(f;P_h)| \approx Ch^2$.

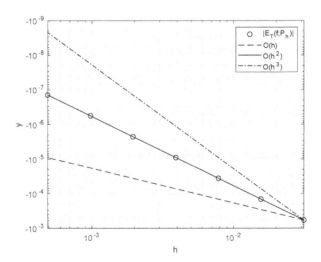

FIGURE 7.4

The order of accuracy of the composite trapezoidal rule $R_T(f;P_h)$ revealed by the loglog plot of the error $E_T(f;P_h)$ versus h.

■

7.5 Simpson's Rule

The **basic Simposon's rule** $R_S(f)$ that approximates $\int_\alpha^\beta f(x)dx$ is

$$R_S(f) = \frac{\beta - \alpha}{6}\left[f(\alpha) + 4f\left(\frac{\alpha + \beta}{2}\right) + f(\beta)\right] \tag{7.78}$$

which can be derived by integrating from α to β the polynomial $p(x)$ that interpolates $f(x)$ at the three nodes α, $(\alpha + \beta)/2$ and β in the integration interval $[\alpha, \beta]$. The geometric interpretation of basic Simpson's rule $R_S(f)$ is approximation of the net area between the curve $y = f(x)$ and the x-axis by the net area between the x-axis and the parabola through the three points on the curve $y = f(x)$ at α, β and the middle point $(\alpha + \beta)/2$.

An easier way to derive the rule $R_S(f)$ is to determine the weights w_0, w_1 and w_2 in the form of the rule

$$R_S(f) = w_0 f(\alpha) + w_1\left(\frac{\alpha + \beta}{2}\right) + w_2 f(\beta) \tag{7.79}$$

by maximizing the DOP of the rule. We can derive the rule for the canonical interval $[-1, 1]$ (see Example 7.3) and then apply change of intervals to derive the rule for the interval $[\alpha, \beta]$. We have known that the DOP of the basic Simpson's rule is 3.

If we apply the basic Simpson's rule to each subinterval $[x_{x-1}, x_i]$, $i = 1, 2, \ldots, n$ in the partition $P = \{a = x_0 < x_1 < \cdots < x_n = b\}$ and sum the results, we obtain the following **composite Simpson's rule** $R_S(f; P)$ that approximates $I(f) = \int_a^b f(x)dx$:

$$R_S(f; P) = \sum_{i=1}^{n} \frac{x_i - x_{i-1}}{6}\left[f(x_{i-1}) + 4f\left(\frac{x_{i-1} + x_i}{2}\right) + f(x_i)\right] \tag{7.80}$$

For a uniform partition P_h with $h = (b - a)/n$, we have

$$R_S(f; P_h) = \frac{h}{6}\left[f(a) + f(b) + 2\sum_{i=1}^{n-1} f(x_i) + 4\sum_{i=1}^{n} f\left(\frac{x_{i-1} + x_i}{2}\right)\right] \tag{7.81}$$

where $x_i = a + ih$ and $(x_{i-1} + x_i)/2 = a + ih - h/2$, $i = 1, 2, \ldots, n$. Below is the pesudocode to implement the rule $R_S(f; P_h)$.

The error of a basic quadrature rule can be derived using Peano's theorem that relates the error of the rule with the DOP of the rule.

Algorithm 14 The composite Simpson's rule with a uniform partition for $\int_a^b f(x)\mathrm{d}x$

$h \leftarrow (b-a)/n$
$S_2 \leftarrow 0$
for i from 1 to $n-1$ **do**
$\quad x_i \leftarrow a + ih$
$\quad S_2 \leftarrow S_2 + f(x_i)$
end for
$S_4 \leftarrow 0$
for i from 1 to n **do**
$\quad x_M \leftarrow a + ih - h/2$
$\quad S_4 \leftarrow S_4 + f(x_M)$
end for
$R_S \leftarrow (h/6)(f(a) + f(b) + 2S_2 + 4S_4)$

Theorem 7.6 (Peano Kernel Theorem) *Let $R(f)$ be a basic quadrature rule with* $\mathrm{DOP} = m$ *that approximates* $I(f) = \int_\alpha^\beta f(x)\mathrm{d}x$, *where* $f \in C^{m+1}[\alpha, \beta]$. *Then there exists a function* $K(x)$, *called the* **Peano kernel**, *that does not depend on* $f(x)$ *and its derivatives such that*

$$E(f) = I(f) - R(f) = \int_\alpha^\beta K(x)f^{(m+1)}(x)\mathrm{d}x$$

When $K(x)$ does not change sign on $[\alpha, \beta]$, by the generalized mean value theorem for integrals, we have

$$E(f) = I(f) - R(f) = \kappa f^{(m+1)}(\eta)$$

where κ is the Peano constant that does not depend on $f(x)$ and its derivatives, and η is a number in (α, β). The Peano constant κ can be easily determined by applying the rule to $f(x) = x^{m+1}$ because $f^{(m+1)}(\eta) = (m+1)!$ is a constant now.

Remark The proof of the theorem (omitted here) uses the integral form of Taylor's remainder and shares some similarity to our derivation of the error for the basic Simpson's rule below.

By applying Peano kernel theorem to the basic Simpson's rule, the error of the basic Simpson's rule is established as

$$E_S(f) = \int_\alpha^\beta f(x)\mathrm{d}x - R_S(f) = -\frac{f^{(4)}(\eta)}{2880}(\beta - \alpha)^5 \tag{7.82}$$

Below we derive a different form of the error. The beginning part of the derivation is very similar to that for the basic midpoint rule.

Since the DOP of Simpson's rule is 3, let's apply Taylor's theorem to write the integrand $f(x)$ as

$$f(x) = T_3(x) + R_3(x) \tag{7.83}$$

where

$$T_3(x) = f(\alpha) + f'(\alpha)(x - \alpha) + \frac{f''(\alpha)}{2!}(x - \alpha)^2 + \frac{f'''(\alpha)}{3!}(x - \alpha)^3 \tag{7.84}$$

is the Taylor polynomial of degree 3 for $f(x)$ about α (the lower limit of the definite integral), and

$$R_3(x) = \frac{f^{(4)}(c(x))}{4!}(x - \alpha)^4, \quad c(x) \in [\alpha, \beta] \tag{7.85}$$

is Taylor's remainder. By the linearity properties for definite integrals and quadrature rules, we have

$$\int_\alpha^\beta f(x)dx = \int_\alpha^\beta T_3(x)dx + \int_\alpha^\beta R_3(x)dx \tag{7.86}$$

$$R_S(f) = R_S(T_3) + R_S(R_3) \tag{7.87}$$

Since $R_S(f)$ is exact for polynomial integrand of degree ≤ 3, we have

$$R_S(T_3) = \int_\alpha^\beta T_3(x)dx \tag{7.88}$$

So we get the error of $R_S(f)$ as

$$E_S(f) = \int_\alpha^\beta f(x)dx - R_S(f) = \int_\alpha^\beta R_3(x)dx - R_S(R_3)$$

$$= \int_\alpha^\beta \frac{f^{(4)}(c(x))}{4!}(x - \alpha)^4 dx - R_S(R_3) \tag{7.89}$$

By the generalized mean value theorem for integrals, we have

$$\int_\alpha^\beta \frac{f^{(4)}(c(x))}{4!}(x - \alpha)^4 dx = \frac{f^{(4)}(\xi)}{4!} \int_\alpha^\beta (x - \alpha)^4 dx = \frac{f^{(4)}(\xi)}{120}(\beta - \alpha)^5 \tag{7.90}$$

where ξ is a number in (α, β).

Next we need to find $R_S(R_3)$. We have

$$R_S(R_3) = \frac{\beta - \alpha}{6} \left[R_3(\alpha) + 4R_3\left(\frac{\alpha + \beta}{2}\right) + R_3(\beta) \right]$$

$$= \frac{\beta - \alpha}{6} \left[0 + 4\frac{f^{(4)}(c(\frac{\alpha+\beta}{2}))}{4!}\left(\frac{\alpha + \beta}{2} - \alpha\right)^4 + \frac{f^{(4)}(c(\beta))}{4!}(\beta - \alpha)^4 \right]$$

$$= \frac{(\beta - \alpha)^5}{6 \cdot 4 \cdot 4!} \left[f^{(4)}\left(c\left(\frac{\alpha + \beta}{2}\right)\right) + 4f^{(4)}(c(\beta)) \right] \tag{7.91}$$

where $c((\alpha + \beta)/2)$ and $c(\beta)$ are in $[\alpha, \beta]$. If $f^{(4)}(x) \in C[\alpha, \beta]$, by the intermediate value theorem, we have

$$f^{(4)}\left(c\left(\frac{\alpha + \beta}{2}\right)\right) + 4f^{(4)}(c(\beta)) = 5f^{(4)}(\eta) \tag{7.92}$$

for some $\eta \in [\alpha, \beta]$. So we have

$$R_S(R_3) = \frac{5f^{(4)}(\eta)}{24 \cdot 4!}(\beta - \alpha)^5 \tag{7.93}$$

$$E_S(f) = \frac{f^{(4)}(\xi)}{120}(\beta - \alpha)^5 - \frac{5f^{(4)}(\eta)}{24 \cdot 4!}(\beta - \alpha)^5$$

$$= \frac{(\beta - \alpha)^5}{24 \cdot 120}(24f^{(4)}(\xi) - 25f^{(4)}(\eta))$$

$$= \frac{(\beta - \alpha)^5}{24 \cdot 120}[25(f^{(4)}(\xi) - f^{(4)}(\eta)) - f^{(4)}(\xi)] \tag{7.94}$$

Using the mean value theorem, we can write the error as

$$E_S(f) = \frac{25f^{(5)}(\zeta)}{2880}(\xi - \eta)(\beta - \alpha)^5 - \frac{f^{(4)}(\xi)}{2880}(\beta - \alpha)^5 \tag{7.95}$$

Remark Note that $|\xi - \eta| \leq (\beta - \alpha)$. If $(\beta - \alpha)$ is small, we have

$$E_S(f) \approx -\frac{f^{(4)}(\xi)}{2880}(\beta - \alpha)^5 \tag{7.96}$$

which is in the same form as the error derived from Peano kernel theorem.

Theorem 7.7 (Simpson's Rule Theorem) *The error to approximate the definite integral*

$$I(f) = \int_b^a f(x)\mathrm{d}x$$

where $f(x) \in C^4_{[a,b]}$, by the composite Simpson's rule

$$R_S(f; P_h) = \frac{h}{6}\left[f(a) + f(b) + 2\sum_{i=1}^{n-1} f(x_i) + 4\sum_{i=1}^{n} f\left(\frac{x_{i-1} + x_i}{2}\right)\right] \tag{7.97}$$

where P_h is a uniform partition with $h = (b - a)/n$ and $x_i = a + ih$, is

$$E_S(f; P_h) = I(f) - R_S(f; P_h) = -\frac{f^{(4)}(\zeta)}{2880}(b - a)h^4 \tag{7.98}$$

> where ζ is some number in (a,b). If h is small enough (n is large enough), we
> have an asymptotic estimate of the error as
>
> $$E_T(f;P_h) \approx -\frac{h^4}{2880}[f'''(b) - f'''(a)] \qquad (7.99)$$

The theorem can be easily proved by using the error formula for the basic Simpson's rule and the intermediate value theorem.

Remark   A corrected Simpson's rule can be defined as

$$R_{CS}(f;P_h) = R_S(f;P_h) - \frac{h^4}{2880}[f'''(b) - f'''(a)] \qquad (7.100)$$

which is usually more accurate than $R_S(f;P_h)$.

7.6   Newton-Cotes Rules

The midpoint rule, the trapezoidal rule and Simpson's rule belong to Newton-Cotes quadrature rules. Newton-Cotes rules are interpolatory quadrature rules with equally spaced nodes. They are closed when the nodes include the ending points of the interval of integration. Otherwise, they are open. So the trapezoidal and Simpson's rule are closed Newton-Cotes rule, while the midpoint rule is an open Newton-Cotes rule.

A higher-order Newton-Cotes rule use more equally spaced nodes, but it may suffer from catastrophic Runge's phenomenon where the error grows for large number of nodes. We prefer to use a composite Newton-Cotes rule that is a composition of a low-order basic Newton-Cotes rule rather than a high-order basic rule over the entire interval of integration. An open Newton-cotes rule does not use the values of the integrand at the ending points of the interval of integration. So it can be used to integrate a function with singularities at the ending points.

7.7   Gaussian Quadrature Rules

On the canonical interval $[-1,1]$, the Legendre polynomial $P_{m+1}(t)$ $(m = 0,1,2,\ldots)$ has $(m+1)$ zeros $t_0, t_1, \ldots, t_m$, as illustrated in Fig. 7.5. These zeros are called **Gauss**

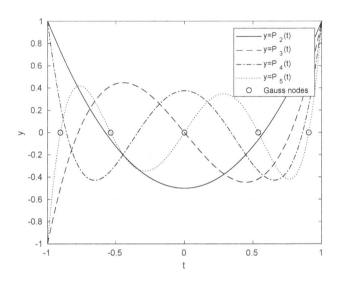

FIGURE 7.5
The zeros of Legendre polynomials.

nodes. We can derive a quadrature rule to approximate $\int_{-1}^{1} F(t)\,dt$ by integrating a polynomial that interpolates $F(t)$ at these Gauss nodes. As usual, the quadrature rule has the form

$$R_G^*(F) = A_0 F(t_0) + A_1 F(t_1) + \cdots + A_m F(t_m) = \sum_{j=0}^{m} A_j F(t_j) \qquad (7.101)$$

where the weights $A_j = \int_{-1}^{1} L_j(t)\,dt$ $(j = 0, 1, \ldots, m)$ are the integration of the Lagrangian basis functions associated with the Gauss nodes, and they are called **Gauss weights**. The quadrature rule $R_G^*(F)$ uses the Gauss nodes and the Gauss weights and is called a basic Gaussian quadrature rule. Table 7.3 lists the $(m + 1)$ Gauss nodes and the corresponding weights for $m = 0, 1, 2, 3, 4$, respectively.

> **Remark** If the Gauss nodes are known in a Gaussian quadrature rule, the corresponding Gauss weights can be derived by maximizing the DOP of the rule instead of integrating the associated Lagrange basis functions. We can also derive both the Gauss nodes and the Gauss weights of a Gaussian quadrature rule together by maximizing the DOP of the rule, which requires solving a nonlinear system.

The reason to use Gauss nodes, the zeros of a Legendre polynomial, to derive a Gaussian quadrature rule is the orthogonality of Legendre polynomials. A Legendre

m	$m+1$ Gauss nodes t_j	$m+1$ Gauss weights A_j
0	0	2
1	$-\sqrt{1/3}$	1
	$+\sqrt{1/3}$	1
2	$-\sqrt{3/5}$	5/9
	0	8/9
	$+\sqrt{3/5}$	5/9
3	$-\sqrt{(3-4\sqrt{0.3})/7}$	$(6+\sqrt{10/3})/12$
	$-\sqrt{(3+4\sqrt{0.3})/7}$	$(6-\sqrt{10/3})/12$
	$+\sqrt{(3-4\sqrt{0.3})/7}$	$(6+\sqrt{10/3})/12$
	$+\sqrt{(3+4\sqrt{0.3})/7}$	$(6-\sqrt{10/3})/12$
4	$-\sqrt{(5-2\sqrt{10/7})/9}$	$0.3(-0.7+5\sqrt{0.7})/(-2+5\sqrt{0.7})$
	$-\sqrt{(5+2\sqrt{10/7})/9}$	$0.3(0.7+5\sqrt{0.7})/(2+5\sqrt{0.7})$
	0	128/225
	$+\sqrt{(5-2\sqrt{10/7})/9}$	$0.3(-0.7+5\sqrt{0.7})/(-2+5\sqrt{0.7})$
	$+\sqrt{(5+2\sqrt{10/7})/9}$	$0.3(0.7+5\sqrt{0.7})/(2+5\sqrt{0.7})$

TABLE 7.3
Gauss nodes and weights.

polynomial $P_{m+1}(t)$ of degree $(m+1)$ is orthogonal to any polynomials of degree $\le m$ on $[-1,1]$, that is

$$\int_{-1}^{1} P_{m+1}(t)q_m(t)dt$$

where $q_m(t)$ is a polynomial in t of degree $\le m$. This fact makes the DOP of a basic Gaussian quadrature rule using $(m+1)$ Gauss nodes be as high as $(2m+1)$.

Theorem 7.8 (Gaussian Quadrature Theorem: Part 1) *Gaussian quadrature rule $R_G^*(F) = \sum_{j=0}^{m} A_j F(t_j)$ using $(m+1)$ Gauss nodes $t_0, t_1, \ldots, t_m$ (and corresponding weights $A_0, A_1, \ldots, A_m$) is exact for $\int_{1}^{-1} F(t)dt$ (i.e. $\int_{-1}^{1} F(t)dt = R_G^*(F)$) if $F(t)$ is a polynomial of degree $\le (2m+1)$. In other words, the DOP of $R_G^*(F)$ is $(2m+1)$.*

Remark The statement in the theorem is a remarkable result. Using $(m+1)$ Gauss nodes, a Gaussian quadrature rule can integrate all polynomials of degrees from 0 to $(2m+1)$ exactly. For example, a Gaussian quadrature rule with 5 Gauss nodes can integrate all polynomials of degrees from 0 to 9 exactly.

Proof *Let $F(t)$ be a polynomial of degree $\leq (2m+1)$. Let $P_{m+1}(t)$ be the Legendre polynomial of degree $(m+1)$. Then $F(t)$ can be decomposed as*

$$F(t) = P_{m+1}(t)q_m(t) + r_m(t) \tag{7.102}$$

where both the quotient $q_m(t)$ and the residue $r_m(t)$ are polynomials of degree $\leq m$. By the orthogonality of Legendre polynomials, we have

$$\int_{-1}^{1} F(t)\mathrm{d}t = \int_{-1}^{1} P_{m+1}(t)q_m(t)\mathrm{d}t + \int_{-1}^{1} r_m(t)\mathrm{d}t = \int_{-1}^{1} r_m(t)\mathrm{d}t \tag{7.103}$$

Since the Gauss nodes t_j, $j = 0,1,\ldots,m$, are the zeros of the Legendre polynomial $P_{m+1}(t)$, we therefore have

$$F(t_j) = P_{m+1}(t_j)q_m(t_j) + r_m(t_j) = r_m(t_j) \tag{7.104}$$

Therefore $r_m(t)$ is a polynomial of degree $\leq m$ that interpolates $F(t)$ at the $(m+1)$ Gauss nodes, and in Lagrange form it is

$$r_m(t) = \sum_{j=0}^{m} F(t_j)L_j(t) \tag{7.105}$$

We thus have

$$\int_{-1}^{1} r_m(t)\mathrm{d}t = R_G^*(F) \tag{7.106}$$

Finally we have proved that

$$\int_{-1}^{1} F(t)\mathrm{d}t = \int_{-1}^{1} r_m(t)\mathrm{d}t = R_G^*(F) \tag{7.107}$$

when $F(t)$ is a polynomial of degree $\leq (2m+1)$. ∎

Remark The Gauss weights $A_j = \int_{-1}^{1} L_j(t)\mathrm{d}t$ associated with the Gauss nodes t_j in the Gaussian quadrature rule $R_G^*(F) = \sum_{j=0}^{m} A_j F(t_j)$ are positive.

Proof *The Lagrange basis function*

$$L_i(t) = \frac{(t-t_0)\cdots(t-t_{i-1})(t-i+1)\cdots(t-t_m)}{(t_i-t_0)\cdots(t_i-t_{i-1})(t_i-i+1)\cdots(t_i-t_m)} \tag{7.108}$$

is a polynomial of degree m satisfying $L_i(t_j) = \delta_{ij}$, where $i = 0, 1, \ldots, m$ and $j = 0, 1, \ldots, m$. So $[L_i(t)]^2$ is a polynomial of degree 2m. Since the DOP of the Gaussian quadrature rule $R_G^*(F)$ is $(2m+1)$, we have

$$\int_{-1}^{1} [L_i(t)]^2 \mathrm{d}t = R_G^*([L_i(t)]^2) \tag{7.109}$$

Note that

$$R_G^*([L_i(t)]^2) = A_j [L_i(t_j)]^2 = A_j \delta_{ij}^2 = A_i \tag{7.110}$$

So we have

$$A_i = \int_{-1}^{1} [L_i(t)]^2 \mathrm{d}t > 0 \tag{7.111}$$

■

Use change of intervals, we can apply a basic Gaussian quadrature rule $R_G^*(F)$ over the interval $[-1, 1]$ to the definite integral $\int_\alpha^\beta f(x)\mathrm{d}x$ as

$$\int_\alpha^\beta f(x)\mathrm{d}x \approx R_G(f) = \frac{\beta - \alpha}{2} \sum_{j=0}^{m} A_j f\left(\frac{\beta - \alpha}{2} t_j + \frac{\beta + \alpha}{2}\right) \tag{7.112}$$

We can then construct a **composite Gaussian quadrature** rule $R_G(f;P)$ for the partition $P = \{a = x_0 < x_1 < \cdots x_n = b\}$ as follows.

$$I(f) = \int_a^b f(x)\mathrm{d}x = \sum_{i=1}^{n} \int_{x_{i-1}}^{x_i} f(x)\mathrm{d}x$$

$$\approx \sum_{i=1}^{n} \left[\frac{x_i - x_{i-1}}{2} \sum_{j=0}^{m} A_j f\left(\frac{x_i - x_{i-1}}{2} t_j + \frac{x_i + x_{i-1}}{2}\right) \right] := R_G(f;P) \tag{7.113}$$

If the partition $P = P_h$ is uniform with $h = x_i - x_{i-1}$, $i = 1, 2, \ldots, n$ and $x_i = a + ih$ $(i = 0, 1, \ldots, n)$, then we have

$$R_G(f;P_h) = \frac{h}{2} \sum_{i=1}^{n} \sum_{j=0}^{m} A_j f\left(\frac{h}{2} t_j + a + ih - \frac{h}{2}\right) \tag{7.114}$$

Below is the pseudocode to implement $R_G(f;P_h)$.

Algorithm 15 The composite Gaussian quadrature rule with a uniform partition for $\int_a^b f(x)dx$

$h \leftarrow (b-a)/n$
$R_G \leftarrow 0$
for i from 1 to n **do**
    for j from 0 to m **do**
        $x_M \leftarrow a + ih - h/2$
        $R_G \leftarrow R_G + A_j f((h/2)t_j + x_M)$
    end for
end for
$R_G \leftarrow (h/2)R_G$

Theorem 7.9 (Gaussian Quadrature Theorem: Part 2) *The composite Gaussian quadrature rule* $R_G(f;P_h) = \frac{h}{2}\sum_{i=1}^{n}\sum_{j=0}^{m}A_j f\left(\frac{h}{2}t_j + a + ih - \frac{h}{2}\right)$ *that approximates* $\int_a^b f(x)dx$ *has the error*

$$E_G(f;P_h) = \frac{[(m+1)!]^4}{(2m+3)![(2m+2)!]^2} f^{(2m+2)}(\xi)(b-a)h^{2m+2} \qquad (7.115)$$

The proof of the theorem is out of the scope of this textbook. Below we establish a weaker result for the basic Gaussian quadrature rule in a similar way as we did for Simpson's rule.

Since the DOP of the basic Gaussian quadrature rule $R_G(f)$ (with $m+1$ nodes) is $(2m+1)$, let's apply Taylor's theorem to write the integrand $f(x)$ as

$$f(x) = T_{2m+1}(x) + R_{2m+1}(x) \qquad (7.116)$$

where $T_{2m+1}(x)$ is the Taylor polynomial of degree $(2m+1)$ for $f(x)$ about α (the lower limit of the definite integral $\int_\alpha^\beta f(x)dx$), and the Taylor's remainder is

$$R_{2m+1}(x) = \frac{f^{(2m+2)}(c(x))}{(2m+2)!}(x-\alpha)^{2m+2}, \quad c(x) \in [\alpha, \beta] \qquad (7.117)$$

By the linearity properties for definite integrals and quadrature rules, we have

$$\int_\alpha^\beta f(x)dx = \int_\alpha^\beta T_{2m+1}(x)dx + \int_\alpha^\beta R_{2m+1}(x)dx \qquad (7.118)$$

$$R_G(f) = R_G(T_{2m+1}) + R_G(R_{2m+1}) \qquad (7.119)$$

where $R_G(f)$ is the basic Gaussian quadrature rule (with $m+1$ Gauss nodes) applied over $[\alpha, \beta]$. The Gaussian quadrature rule $R_G(f)$ is exact for polynomial integrand of degree $\leq (2m+1)$, so we have

$$R_G(T_{2m+1}) = \int_\alpha^\beta T_{2m+1}(x)dx \qquad (7.120)$$

So we get the error of $R_G(f)$ as

$$E_G(f) = \int_\alpha^\beta f(x)dx - R_G(f) = \int_\alpha^\beta R_{2m+1}(x)dx - R_G(R_{2m+1})$$

$$= \int_\alpha^\beta \frac{f^{(2m+2)}(c(x))}{(2m+2)!}(x-\alpha)^{2m+2}dx - R_G(R_{2m+1}) \qquad (7.121)$$

By the generalized mean value theorem for integrals, we have

$$\int_\alpha^\beta \frac{f^{(2m+2)}(c(x))}{(2m+2)!}(x-\alpha)^{2m+2}dx = \frac{f^{(2m+2)}(\xi)}{(2m+2)!} \int_\alpha^\beta (x-\alpha)^{2m+2}dx$$

$$= \frac{f^{(2m+2)}(\xi)}{(2m+3)(2m+2)!}(\beta-\alpha)^{2m+3} \qquad (7.122)$$

where ξ is a number in (α, β).

Next we need to find $R_G(R_{2m+1})$. Let

$$x_j = \frac{\beta-\alpha}{2}t_j + \frac{\beta+\alpha}{2} \qquad (7.123)$$

By change of intervals, we have

$$R_G(R_{2m+1}) = \frac{\beta-\alpha}{2}\sum_{j=0}^m A_j R_{2m+1}(x_j) = \frac{\beta-\alpha}{2}\sum_{j=0}^m A_j \frac{f^{(2m+2)}[c(x_j)]}{(2m+2)!}(x_j-\alpha)^{2m+2}$$

$$= \left(\frac{\beta-\alpha}{2}\right)^{2m+3}\sum_{j=0}^m A_j \frac{f^{(2m+2)}[c(x_j)]}{(2m+2)!}(t_j+1)^{2m+2} \qquad (7.124)$$

Assume $f^{(2m+2)}(x) \in C_{[\alpha,\beta]}$. With $A_j > 0$ and $(t_j+1)^{2m+2} > 0$, we have

$$\sum_{j=0}^m A_j(t_j+1)^{2m+2}f^{(2m+2)}[c(x_j)] = f^{(2m+2)}(\eta)\sum_{j=0}^m A_j(t_j+1)^{2m+2} \qquad (7.125)$$

for some $\eta \in [\alpha, \beta]$ by the intermediate value theorem. Let $C_G = \sum_{j=0}^m A_j(t_j+1)^{2m+2}$. We then have

$$R_G(R_{2m+1}) = \frac{C_G f^{(2m+2)}(\eta)}{2^{2m+3} \cdot (2m+2)!}(\beta-\alpha)^{2m+3} \qquad (7.126)$$

$$E_G(f) = \frac{f^{(2m+2)}(\xi)}{(2m+3)(2m+2)!}(\beta-\alpha)^{2m+3} - \frac{C_G f^{(2m+2)}(\eta)}{2^{2m+3} \cdot (2m+2)!}(\beta-\alpha)^{2m+3}$$

$$= \frac{(\beta-\alpha)^{2m+3}}{(2m+3)!}\left(f^{(2m+2)}(\xi) - \frac{(2m+3)C_G}{2^{2m+3}}f^{(2m+2)}(\eta)\right) \qquad (7.127)$$

Let $C = \frac{(2m+3)C_G}{2^{2m+3}} - 1$, then

$$E_G(f) = \frac{(\beta-\alpha)^{2m+3}}{(2m+3)!}(f^{(2m+2)}(\xi) - f^{(2m+2)}(\eta) - Cf^{(2m+2)}(\eta)) \qquad (7.128)$$

Using the mean value theorem, we can then write the error as

$$E_G(f) = \frac{f^{(2m+3)}(\zeta)}{(2m+3)!}(\xi - \eta)(\beta - \alpha)^{2m+3} - \frac{Cf^{(2m+2)}(\eta)}{(2m+3)!}(\beta - \alpha)^{2m+3} \quad (7.129)$$

Remark Note that $|\xi - \eta| \leq (\beta - \alpha)$. If $(\beta - \alpha)$ is small, we have

$$E_G(f) \approx -\frac{Cf^{(2m+2)}(\eta)}{(2m+3)!}(\beta - \alpha)^{2m+3} \quad (7.130)$$

Using the error formula $E_G(f)$ for the basic Gaussian quadrature rule $R_G(f)$ and the intermediate value theorem, we can establish the error formulas for the composite Gaussian quadrature rule $R_G(f; P_h)$ as

$$E_G(f; P_h) = O(h^{2m+2}) \quad (7.131)$$

Remark There are other types of Gaussian quadrature rules such as Gauss-Radau quadrature ($t_m = 1$ and the other nodes are chosen to maximize the DOP) and Gauss-Lobatto quadrature ($t_0 = -1$, $t_m = 1$ and the other nodes are chosen to maximize the DOP).

7.8 Other Numerical Integration Techniques

Shown in Fig. 7.6 is an illustration of an integrand $f(x)$ which has a singularity at the ending point a ($f(a)$ is unbounded) or rapid oscillatory change over a portion of the interval of integration $[a, b]$. In addition, the ending point b may be ∞. Below we briefly describe some techniques to handle such integrals.

7.8.1 Integration with Singularities

Consider the integral

$$\int_{-1}^{1} \frac{1}{\sqrt{1 - x^2}} dx = \pi$$

Suppose we want to numerically integrate the integral to approximate the value of π. We cannot apply the composite trapezoidal rule or composite Simpson's rule to do so because the integrand is singular at the ending points -1 and 1. However, we can apply open composite Newton-Cotes rules or composite Gaussian quadrature rules.

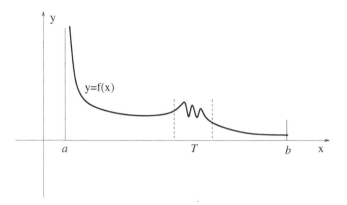

FIGURE 7.6
A special integrand.

Sometimes, we can use the substitution rule to transform an improper definite integral (with a singularity at an ending point or infinite lower or upper limit) to a proper integral with respect a new variable. For example, we can use $x = \sin t$, $t \in [-\pi/2, \pi/2]$ to transform $\int_{-1}^{1} \frac{1}{\sqrt{1-x^2}} dx = \pi$ to $\int_{-\pi/2}^{\pi/2} dt$. Some useful transformations include

$$x = \frac{1}{t}, \quad x = -\ln t, \quad x = \frac{t}{1-t}, \quad x = \tan t, \quad x = \sqrt{\frac{1+t}{1-t}} \qquad (7.132)$$

7.8.2 Adaptive Integration

As illustrated in Fig. 7.6, an integrand of a definite integral may vary slowly somewhere but rapidly elsewhere over the interval of integration. To approximate such an integral, a composite quadrature rule with uniform partition P_h needs a very small value of h to capture the rapid change of the integrand for good accuracy. Such small h is not necessary for slowly-varying portion of the interval and requires a lot of evaluations of the integrand. So it is more efficient to use adaptive integration that distributes more partition points for rapidly-changing portions and less for slowly-varying portions. We can use error estimate to test which portions need more partition points. For the integral of f over the subinterval T in the interval of integration, the error estimate may be given as

$$|I(f; P_h^T) - R_1(f; P_h^T)| \approx |R_2(f; P_{h^*}^T) - R_1(f; P_h^T)|$$

where P_h^T is a uniform partition of T, and $R_2(f; P_{h^*}^T)$ is rule much more accurate than $R_1(f; P_h^T)$ with either higher DOP or finer h^*. If the error estimate is larger than the error tolerance, we split the portion T and refine the partition.

> **Remark** MATLAB provides the function `integral` for numerical integration. The syntax to call the function is
>
> ```
> q = integral(fun,xmin,xmax)
> ```
>
> which numerically integrates function `fun` from `xmin` to `xmax` using global adaptive quadrature and default error tolerance.

7.9 Exercises

Exercise 7.1 (1) Derive a basic quadrature rule $R(f)$ to approximate $\int_{\alpha}^{\beta} f(x)dx$ by integrating an interpolating polynomial $p(x)$ of degree 0 that interpolates one data point generated by $f(x)$ at the node β. (2) Give a geometric interpretation of the rule and then re-derive the rule using the geometric interpretation.

Exercise 7.2 (1) Find the degree-1 polynomial $p_1(x)$ to interpolate the data generated by $f(t)$ at $t = -1$ and $t = 1$.
(2) Derive a quadrature rule $R(f)$ to approximate $I(f) = \int_{-1}^{1} f(t)dt$ by integrating $p_1(x)$.
(3) Check that your derived rule is correct using a geometric argument.
(4) If your rule $R(f)$ is used to approximate $\int_{2}^{8} f(x)dx$, write down the approximation.

Exercise 7.3 (1) Derive a basic quadrature rule $R(F)$ to approximate $\int_{-1}^{1} F(t)dt$ by integrating an interpolating polynomial $p(t)$ of degree 2 that interpolates the three data points generated by $F(t)$ at the nodes $-1, 0$ and 1. (2) Apply the quadrature rule $R(F)$ to approximate $\int_{a}^{b} f(x)dx$ using change of intervals.

Exercise 7.4 (1) Find the weights A_0 and A_1 in a quadrature rule $R_G(F) = A_0 f\left(-\sqrt{\dfrac{1}{3}}\right) + A_1 f\left(\sqrt{\dfrac{1}{3}}\right)$ to approximate $\int_{-1}^{1} F(t)dt$ by maximizing the DOP of the rule. (2) What is the DOP of the derived rule?

Exercise 7.5 Prove that if a quadrature rule $R(f) = A_0 F(t_0) + A_1 F(t_1)$ is exact (with no error) for $I(1) = \int_{-1}^{1} 1dt$ and $I(t) = \int_{-1}^{1} tdt$ then it is exact for $I(2 + 3t) = \int_{-1}^{1} (2 + 3t)dt$.

Exercise 7.6 A basic quadrature rule $R(f)$ for approximating $\int_{-\alpha}^{\beta} f(x)dx$. is exact for $f(x) = 1$, $f(x) = x$ and $f(x) = x^2$. Show that the rule is exact (i.e. having no error) for any polynomial integrand $f(x) = c_0 + c_1 x + c_2 x^2$ of degree 2, where the coefficients c_0, c_1 and c_2 are constants.

Exercise 7.7 (1) Approximate $I(f) = \int_0^2 4^x dx$ using the composite trapezoidal rule with uniform partition $R_T(f; P_h)$, where $h = 1/2$. (2) Find the true value of the definite integral using the antiderivative of $f(x) = 4^x$. (3) Does your approximation overestimate or underestimate the true value? Use the graph of $f(x)$ to explain the overestimation or underestimation geometrically.

Exercise 7.8

$$I(f) = \int_0^3 x^2 dx$$

(1) Find the exact value of $I(f)$.
(2) Let P be a nonuniform partition of the interval $[0,3]$ by the 3 partition points $x_0 = 0$, $x_1 = 2$ and $x_2 = 3$. Find the approximation of $I(f)$ by the composite trapezoidal rule $R_T(f; P)$.
(3) Let P_h be a uniform partition of the interval $[0,3]$ with $h = 1$. Find the approximation of $I(f)$ by the composite trapezoidal rule $R_T(f; P_h)$.
(4) If the integrand is changed to $f(x) = \pi x - \sqrt{2}$, the composite trapezoidal rule $R_T(f; P)$ is exact for any partition P. Why?

Exercise 7.9 The composite trapezoidal rule $R_T(f; P_h)$ with the uniform partition P_h is used to approximate the definite integral $I(f) = \int_0^1 \sin(\sin x)dx$. If the error is at most $\dfrac{10^{-6}}{6}$, how many partition points are needed? (Hint: You may use the triangle inequality $|a \pm b| \leq |a| + |b|$.)

Exercise 7.10 $I_f = \int_{-1}^1 \cos(x^2)dx$ is to be approximated by a composite Trapezoidal rule $R_T(f; P_h)$, where the uniform partition P_h divides the interval $[-1, 1]$ into equally spaced subintervals of the length h. Determine the values of h such that the error $|I_f - R_T(f; P_h)| \leq 10^{-10}$.

Exercise 7.11

$$I(f) = \int_0^2 x^3 dx$$

(1) Find the exact value of $I(f)$.
(2) Let P_h be a uniform partition of the interval $[0,2]$ with 2 subintervals. Find the approximation of $I(f)$ by the composite Simpson's rule $R_S(f; P_h)$. What is the error of the approximation?
(3) If the integrand is changed to $f(x) = \pi x^3 + \sqrt{2}x^2 - 2021x + e$, what is the error in the composite Simpson's rule $R_S(f; P)$ for any partition P? Why?

Exercise 7.12

$$I(f) = \int_0^6 x^4 dx$$

(1) Find the approximation of $I(f)$ by the composite Simpson's rule $R_S(f;P)$, where the partition P has the partition points $x_0 = 0$, $x_1 = 4$, $x_2 = 6$.
(3) If the absolute error in the approximation $R_S(g;P_h)$ of another definite integral $I(g)$ is 10^{-10}, where $R_S(g;P_h)$ is the composite Simpson's rule with the uniform partition P_h, what is an estimate of the absolute error in the approximation $R_S(g;P_{2h})$ of $I(g)$?

Exercise 7.13 Apply the composite Simpson's rule $R_S(f;P_h)$ using the uniform partition P_h with $h = 0.5$ to approximate the definite integral $I(f) = \int_0^2 4^x dx$. Find the exact value of $I(f)$ and use it to compute the error of the approximation.

Exercise 7.14 A definite integral $I(f) = \int_1^4 \ln(x)dx$ is to be approximated by a composite Simpson's rule $R_S(f;P_h)$ with a uniform partition of the interval $[1,2]$. Determine how many subintervals are needed in the partition P_h such that the error $|I(f) - R_S(f;P_h)| \leq 10^{-8}$.

Exercise 7.15

$$I(f) = \int_0^2 \sin(\pi x/2)dx$$

P_h is a uniform partition of the interval $[0,2]$ with $h=1$.
(1) Find the exact value of $I(f)$.
(2) Approximate $I(f)$ by the composite midpoint rule $R_M(f;P_h)$.
(3) Approximate $I(f)$ by the composite trapezoidal rule $R_T(f;P_h)$.
(4) Approximate $I(f)$ by the composite Simpson's rule $R_S(f;P_h)$.
(5) Approximate $I(f)$ by the composite Gaussian quadrature rule $R_G(f;P_h)$ in which the basic Gaussian quadrature rule on each subinterval uses two Gauss nodes.
(6) Does your result $R_T(f;P_h)$ underestimate or overestimate $I(f)$? Use the graph of the integrand to explain the reason.

Exercise 7.16 $R(F) = \dfrac{5}{9}F\left(-\sqrt{\dfrac{3}{5}}\right) + \dfrac{8}{9}F(0) + \dfrac{5}{9}F\left(\sqrt{\dfrac{3}{5}}\right)$ is a basic Gaussian

quadrature rule derived using the Gauss nodes $-\sqrt{\dfrac{3}{5}}$, 0 and $\sqrt{\dfrac{3}{5}}$ for $\int_{-1}^1 F(t)dt$.

Apply this basic rule to approximate $\int_{-5}^5 \sin\left(\dfrac{\pi x^2}{30}\right)dx$.

Exercise 7.17 A Gauss quadrature rule for $I(f) = \int_{-1}^1 F(t)dt$ has $n+1 = 2$ nodes t_0 and t_1 with corresponding weights A_0 and A_1.
(1) List all monomials t^k for which the rule is exact.
(2) It is known that $t_1 = \sqrt{1/3}$ and $A_1 = 1$. Determine the values of the node t_0 and the weight A_0 by maximizing the DOP of the rule.

Exercise 7.18 Let P_h be a uniform partition with the subinterval length h for the definite integral $I(f)$. Let P_{2h} be a uniform partition with subinterval length $2h$ (double that of P_h).

(1) If the absolute error in the approximation $R_T(f; P_h)$ of the definite integral $I(f)$ is 10^{-10}, where $R_T(f; P_h)$ is the composite trapezoidal rule with P_h, what is an estimate of the absolute error in the approximation $R_T(f; P_{2h})$ of $I(f)$?

(2) If the absolute error in the approximation $R_S(f; P_h)$ of the definite integral $I(f)$ is 10^{-10}, where $R_S(f; P_h)$ is the composite Simpson's rule with P_h, what is an estimate of the absolute error in the approximation $R_S(f; P_{2h})$ of $I(f)$?

(3) A composite quadrature rule $R(f; P_h)$ has the absolute error of $O(h^p)$. If the absolute error in the approximation $R(f; P_h)$ of $I(f)$ is 10^{-10}, and the absolute error in the approximation $R(f; P_{2h})$ of $I(f)$ is 3.2×10^{-9}, what is the value of p?

7.10 Programming Problems

Problem 7.1 Write a MATLAB m-function `Trapezoidal.m` that implements $R_T(f; P_h)$, the composite trapezoidal rule with a uniform partition. Write an m-script `testTrapezoidal.m` that tests `Trapezoidal` by approximating the definite integral

$$I(f) = \int_{-1}^{1} \sin(\pi x) \exp(x)\, dx = \frac{\pi \left(e - \frac{1}{e} \right)}{1 + \pi^2}$$

using uniform partition P_h with $n = 10, 20, \ldots, 1280$ subintervals. To avoid repetitions in your code, define an array for the values of n and write a for-loop. Display your results in a table such that each row of the table consists of the values of h, the approximation $R_T(f; P_h)$, the error $E_h^T = I(f) - R_T S(f; P_h)$, and the ratio $|E_h^T / E_{2h}^T|$ (Note that the ratio for the row for $n = 10$ is not available, so leave it blank). Discuss the results.

Problem 7.2 Write a MATLAB m-function `Simpson.m` that implements $R_S(f; P_h)$, the composite Simpson's rule with a uniform partition. Write an m-script `testSimpson.m` that tests `Simpson` by approximating the definite integral

$$I(f) = \int_{-1}^{1} \sin(\pi x) \exp(x)\, dx = \frac{\pi \left(e - \frac{1}{e} \right)}{1 + \pi^2}$$

using uniform partition P_h with $n = 10, 20, \ldots, 320$ subintervals. To avoid repetitions in your code, define an array for the values of n and write a for-loop. Display your results in a table such that each row of the table consists of the values of h, the approximation $R_S(f; P_h)$, the error $E_h^S = I(f) - R_S(f; P_h)$ and the ratio $|E_h^S / E_{2h}^S|$ (Note that the ratio for the row for $n = 10$ is not available, so leave it blank). Discuss the results.

Problem 7.3 Write a MATLAB m-function `Gauss2.m` that implements $R_G(f; P_h)$, the composite Gaussian quadrature rule with a uniform partition. The basic Gaussian quadrature rule for each subinterval uses 2 Gauss nodes. Write an m-script `testGauss2.m` that tests `Gauss2` by approximating the definite integral

$$I(f) = \int_{-1}^{1} \sin(\pi x) \exp(x) \, dx = \frac{\pi \left(e - \frac{1}{e}\right)}{1 + \pi^2}$$

using uniform partition P_h with $n = 10, 20, \ldots, 160$ subintervals. To avoid repetitions in your code, define an array for the values of n and write a for-loop. Display your results in a table such that each row of the table consists of the values of h, the approximation $R_G(f; P_h)$, the error $E_h^G = I(f) - R_G(f; P_h)$ and the ratio $|E_h^G / E_{2h}^G|$ (Note that the ratio for the row for $n = 10$ is not available, so leave it blank). Discuss the results.

Problem 7.4 Use the composite Gaussian quadrature rule $R_G(f; P_h)$ with a uniform partition P_h to approximate the value of π by approximating the definite integral

$$\int_0^1 \frac{4}{1 + x^2} \, dx = \pi$$

with ten subintervals over $[0, 1]$ and the basic Gaussian quadrature rule of 5 Gauss nodes for each subinterval.

Problem 7.5 Consider the heat conduction in a semi-infinite solid. Suppose at time $t = 0$, the initial temperature of the solid is the constant T_i, and we heat the flat surface of the solid instantly to temperature T_s. The temperature $T(x, t)$ of the solid, where x is the distance from a point in the solid to the flat surface of the solid, is given by

$$\frac{T(x, t) - T_i}{T_s - T_i} = \text{erfc}\left(\frac{x}{\sqrt{4\alpha t}}\right), \tag{7.133}$$

where α is a physics constant called the heat diffusivity, and the function $\text{erfc}(z)$ is defined as

$$\text{erfc}(z) = 1 - \frac{2}{\sqrt{\pi}} \int_0^z e^{-s^2} \, ds. \tag{7.134}$$

In this project, we will consider a solid with heat diffusivity $\alpha = 3.39 \times 10^{-5} \, [m^2/s]$ and the case with the initial temperature $T_i = 20$ and the surface temperature $T_s = 100$.

First, write an m-function named `errfunc.m` with the first line

```
ef = errfunc(z,n)
```

that evaluates the error function defined in Eq. (7.134) for a given value of z using a composite quadrature rule with n equally spaced intervals. You can define an anonymous function for the integrand in Eq. (7.134) prior to calling your composite quadrature rule.

Then write an m-function named `solidtemp.m` with the first line

```
T = solidtemp(x,t)
```

that finds the temperature of solid defined in Eq. (7.133) for given values of x and t by calling the first m-function `errfunc` with large enough n, say $n = 1000$. Be careful how you set your upper limit of integration.

Now, write a MATLAB script named `heatconduction.m` that will create a single figure with the following plots overlaid on one another (use `hold on` and `hold off` in MATLAB):

- Plot $T(x, 360)$, the temperature distribution at 0.1 hour, for x ranging from $0[m]$ to $1[m]$, using at least 400 points.
- Repeat the above plot for 0.5, 2 and 10 [hours] (Note: 1 hour=3600 seconds).

Your plots should be appropriately annotated. Briefly discuss the plots.

8

Numerical Differentiation

In this chapter, we present different numerical differentiation techniques. We first introduce the method of undetermined coefficients. We then describe differentiation by interpolation, including differentiation using DFT. Finally, we present Richardson extrapolation.

To investigate the dynamics of an object freely moving in a fluid, we may use a high-speed camera to track the center of mass (CM) of the object. Suppose the vertical coordinates y of the CM at different time t are recorded in the form as in Table 8.1

t	t_0	t_1	$\cdots$	t_n
y	y_0	y_1	$\cdots$	y_n

TABLE 8.1
Vertical coordinates y at different time t.

The time step $h = t_i - t_{i-1}$ ($i = 1, 2, \ldots, n$) is a small constant (determined by the frame rate of the high-speed camera). We want to use the data to estimate the instantaneous vertical velocity v and acceleration a at each time instant t_i ($i = 0, 1, \ldots, n$) to understand the dynamics of the object. The vertical coordinate y is a function $y = f(t)$ of the time t, the velocity $v = f'(t)$ is its first derivative and the acceleration $a = f''(t)$ is its second derivative. However in this experiment, we have only discrete data about $f(t)$ instead of an expression for $f(t)$. Numerical differentiation can be used to approximate the derivatives of $f(t)$ using only discrete values of $f(t)$. Numerical differentiation is also widely used to solve differential equations (see examples in Chapter 9).

8.1 Differentiation Using Taylor's Theorem

The derivative of a function $f(x)$ at $x = a$ is defined as

$$f'(a) = \lim_{\Delta x \to 0} \frac{f(a + \Delta x) - f(a)}{\Delta x} \tag{8.1}$$

DOI: 10.1201/9781003201694-8

if the limit exists. The derivative $f'(a)$ can be interpreted as the slope of the tangent line to the curve $y = f(x)$ at $x = a$, as illustrated in Fig. 8.1. We have a numerical differentiation scheme

$$f'(a) \approx \frac{f(a + \Delta x) - f(a)}{\Delta x} \tag{8.2}$$

if we use a finite but small Δx to compute the quotient at the right-hand side of Eq. (8.1), which can be interpreted as the slope of the secant line through the two points at $x = a$ and $x = a + \Delta x$ on the curve $y = f(x)$. If $\Delta = h > 0$, we call

$$\frac{f(a + h) - f(a)}{h} \tag{8.3}$$

the **forward finite difference** approximation of $f'(a)$. If $\Delta x = -h < 0$, we call

$$\frac{f(a) - f(a - h)}{h} \tag{8.4}$$

the **backward finite difference** approximation of $f'(a)$. The average of the forward and backward finite difference approximations (i.e. the average of the slopes of two secant lines) gives another two-point approximation

$$\frac{f(a + h) - f(a - h)}{2h} \tag{8.5}$$

which is called the **centered finite difference** approximation (the slope of the secant line through the two points at $x = a - h$ and $x = a + h$ on the curve $y = f(x)$). Fig. 8.1 shows the geometric interpretations of $f'(a)$ and its approximations.

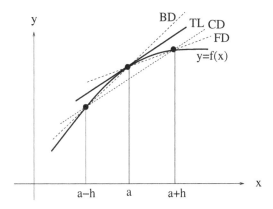

FIGURE 8.1
The derivative $f'(a)$ and its finite difference approximations (FD: forward difference, BD: backward difference, CD: centered difference) are the slopes of the tangent (TL) and secant lines.

In Example 2.5 in Chapter 2, we use Taylor's theorem to find the dependence of the absolute error of the forward finite difference approximation on h as

$$\left| f'(a) - \frac{f(a+h) - f(a)}{h} \right| = \frac{|f''(c)|}{2!} h = O(h) \tag{8.6}$$

Similarly, we have

$$\left| f'(a) - \frac{f(a) - f(a-h)}{h} \right| = O(h) \tag{8.7}$$

Below we show that the error of the centered finite difference approximation depends on h as $O(h^2)$ instead of $O(h)$ due to fortune cancellation in averaging. By Taylor's theorem, we have

$$f(a+h) = f(a) + f'(a)h + \frac{f''(a)}{2!} h^2 + \frac{f'''(a)}{3!} h^3 + \cdots \tag{8.8}$$

$$f(a-h) = f(a) - f'(a)h + \frac{f''(a)}{2!} h^2 - \frac{f'''(a)}{3!} h^3 + \cdots \tag{8.9}$$

We see the cancellation of the term $\frac{f''(a)}{2!} h^2$ in

$$f(a+h) - f(a-h) = 2f'(a)h + 2\frac{f'''(a)}{3!} h^3 + \cdots \tag{8.10}$$

So we have

$$\frac{f(a+h) - f(a-h)}{2h} = f'(a) + \frac{f'''(a)}{3!} h^2 + \cdots \tag{8.11}$$

and the centered finite difference approximation has the error of $O(h^2)$.

The error in a finite difference approximation is called **truncation error** (due to truncation of a convergent Taylor series to a finite number of terms). It is mathematical approximation error due to finite h and truncation. It is present even in exact arithmetic. Mathematically the truncation error approaches 0 as $h \to 0$. However, computationally we should not use a too small h because of roundoff error and catastrophic cancellation, as seen in Problem 3.1 in Chapter 3.

8.1.1 The Method of Undetermined Coefficients

A finite difference approximation given in the previous section approximates $f'(a)$ as a linear combination of function values $f(a-h)$, $f(a)$ and $f(a+h)$. Since a Taylor's series expansion expresses a function value $f(x)$ in terms of $f(a)$, $f'(a)$, $f''(a)$, ... We can use Taylor's series expansions to determine a linear combination of function values to approximate the derivative $f^{(k)}(a)$ ($k = 0, 1, ...$). The process is called the method of undetermined coefficients, as demonstrated in the following example.

Example 8.1 Determine a finite difference approximation of $f''(a)$ using the function values $f(a-h)$, $f(a)$ and $f(a+h)$ and find the dependence of the error on h in the big-O notation.

[Solution:] We write the approximation as a linear combination of the function values as

$$f''(a) \approx C_1 f(a-h) + C_2 f(a) + C_3 f(a+h) \tag{8.12}$$

which has the three undetermined coefficients C_1, C_2 and C_3. To determine these coefficients, we use the following Taylor series expansions

$$f(a+h) = f(a) + f'(a)h + \frac{f''(a)}{2!}h^2 + \frac{f'''(a)}{3!}h^3 + \frac{f^{(4)}(a)}{4!}h^4 + \frac{f^{(5)}(a)}{5!}h^5 \ldots \tag{8.13}$$

$$f(a-h) = f(a) - f'(a)h + \frac{f''(a)}{2!}h^2 - \frac{f'''(a)}{3!}h^3 + \frac{f^{(4)}(a)}{4!}h^4 - \frac{f^{(5)}(a)}{5!}h^5 \ldots \tag{8.14}$$

We have

$$C_1 f(a-h) + C_2 f(a) + C_3 f(a+h) =$$

$$(C_1 + C_2 + C_3)f(a) + (C_1 - C_3)f'(a) + (C_1 + C3)\frac{f''(a)}{2!}h^2$$

$$+(C_1 - C_3)\frac{f'''(a)}{3!}h^3 + (C_1 + C3)\frac{f^{(4)}(a)}{4!}h^4 + (C_1 - C3)\frac{f^{(5)}(a)}{5!}h^5 + \cdots \tag{8.15}$$

To approximate $f''(a)$ by $C_1 f(a-h) + C_2 f(a) + C_3 f(a+h)$, we require that the coefficient in front of $f''(a)$ at the right-hand side be 1 and try to get rid of as many lower powers of h as possible for better accuracy (since h is assumed to be a small quantity). Since we have three unknown coefficients, we can write down the three equations

$$C_1 + C_2 + C_3 = 0 \tag{8.16}$$

$$C_1 - C_3 = 0 \tag{8.17}$$

$$(C_1 + C_3)\frac{h^2}{2!} = 1 \tag{8.18}$$

which can be solved to give $C_1 = C_3 = 1/h^2$ and $C_2 = -2/h^2$. So the finite difference approximation of $f''(a)$ is

$$f''(a) \approx \frac{f(a-h) - 2f(a) + f(a+h)}{h^2}$$

which is called the **centered finite difference** approximation of the second derivative. By Eq. (8.15), we have

$$\frac{f(a-h) - 2f(a) + f(a+h)}{h^2} = f''(a) + \frac{f^{(4)}(a)}{12}h^2 + \cdots \tag{8.19}$$

Note that the term $(C_1 - C_3)\frac{f'''(a)}{3!}h^3$ vanishes too (as $C_1 = C_3$) even though it is not included in the three equations, and the omitted other nonzero terms have powers of h higher than 2. So the error of the centered finite difference approximation of $f''(a)$ is of $O(h^2)$. We say the approximation is of second order. ∎

Remark If n function values are combined to approximate a derivative, there are n coefficients to be determined. We may want to write down the first $n+2$ terms in each Taylor series expansion so that we can write down n equations for the unknown coefficients from the first n terms, allow the possibility of the automatic cancellation of the $(n+1)$-th term, and has the $(n+1)$-th or $(n+2)$-term to derive the error in big-O notation.

8.2 Differentiation Using Interpolation

A finite difference approximation approximates a derivative of a function $f(x)$ using function values. We can derive such an approximation using interpolation. In interpolation, we can use function values to interpolate the function $f(x)$, leading to an approximation of $f(x)$ by the interpolant $p(x)$. Accordingly we can approximate the derivative of $f(x)$ by the derivative of $p(x)$. Below is an example.

Example 8.2 Derive a finite difference approximation of $f''(a)$ in the form of $C_1 f(a-h) + C_2 f(a) + C_3 f(a+h)$ by approximating $f''(a)$ with $p_2''(a)$, where $p_2(x)$ is a polynomial of degree ≤ 2 that interpolates $f(x)$ at the nodes $a-h$, a and $a+h$.
[Solution:] The interpolating polynomial $p_2(x)$ in Lagrange form is

$$p_2(x) = f(a-h)L_0(x) + f(a)L_1(x) + f(a+h)L_2(x)$$

where

$$L_0(x) = \frac{[x-a][x-(a+h)]}{[(a-h)-a][(a-h)-(a+h)]} = \frac{[x-a][x-(a+h)]}{2h^2}]$$

$$L_1(x) = \frac{[x-(a-h)][(x-(a+h)]}{[a-(a-h)][a-(a+h)]} = \frac{[x-(a-h)][(x-(a+h)]}{-h^2}$$

$$L_2(x) = \frac{[x-(a-h)][(x-a)]}{[(a+h)-(a-h)][(a+h)-a]} = \frac{[x-(a-h)][(x-a)]}{2h^2}$$

We have

$$p_2''(x) = f(a-h)L_0''(x) + f(a)L_1''(x) + f(a+h)L_2''(x)$$

and obtain the centered finite difference approximation again

$$f''(a) \approx p_2''(a) = \frac{f(a-h) - 2f(a) + f(a+h)}{h^2}$$

with

$$C_1 = L_0'(a) = 1/h^2, \quad C_2 = L_1'(a) = -2/h^2, C_3 = L_2'(a) = 1/h^2$$

∎

We know the interpolation error in polynomial interpolation is given by

$$f(x) - p_n(x) = \frac{f^{(n+1)}(\xi)}{(n+1)!} N_{n+1}(x), \quad N_{n+1}(x) = (x - x_0)(x - x_1) \cdots (x - x_n) \quad (8.20)$$

where $p_n(x)$ is the polynomial of degree $\leq n$ that interpolates the $n+1$ data points $\{(x_i, f(x_i))\}_{i=0}^{n}$ generated by $f(x)$. We can use the interpolation error to find the truncation error in a finite difference approximation that is derived using interpolation. Below is an example.

Example 8.3 In the previous example, we use polynomial interpolation to derive a finite difference approximation of $f''(a)$ as

$$f''(a) \approx p_2''(a) = \frac{f(a-h) - 2f(a) + f(a+h)}{h^2}$$

Derive the truncation error in this approximation.
[Solution:] The interpolation error is

$$f(x) - p_2(x) = \frac{f'''(\xi)}{3!} N_3(x), \quad N_3(x) = [x - (a-h)][x - a][x - (a+h)]$$

where ξ depends on x and is a function $\xi(x)$ of x. We have

$$f''(x) - p_2''(x) = \left[\frac{f'''(\xi)}{3!}\right]'' N_3(x) + 2\left[\frac{f'''(\xi)}{3!}\right]' N_3'(x) + \left[\frac{f'''(\xi)}{3!}\right] N_3''(x)$$

where

$$N_3'(x) = [x - (a-h)][x - a] + [x - (a-h)][x - (a+h)] + [x - a][x - (a+h)]$$

$$N_3''(x) = 2[x - (a-h)] + 2[x - a] + 2[x - (a+h)]$$

Note that $N_3(a) = 0$, $N_3'(a) = -h^2$ and $N_3''(a) = 0$. We then have

$$f''(a) - p_2''(a) = -2\left[\frac{f'''(\xi)}{3!}\right]' h^2$$

that is

$$f''(a) = \frac{f(a-h) - 2f(a) + f(a+h)}{h^2} - 2\left[\frac{f'''(\xi)}{3!}\right]' h^2$$

The error term is $-2\left[\frac{f'''(\xi)}{3!}\right]' h^2$, which is of $O(h^2)$. ∎

8.2.1 Differentiation Using DFT

Given the data $\{(t_i, y_i)\}_{i=0}^n$ generated by $F(t)$ at the following equally spaced nodes

$$t_i = ih, \quad h = \frac{2\pi}{n+1} = \frac{\pi}{m}; \quad i = 0, 1, \dots, n, \quad n+1 = 2m \tag{8.21}$$

the trigonometric interpolant for the data (see Section 6.8) is given by

$$G_n(t) = \frac{1}{2}(a_0 + a_m \cos mt) + \sum_{k=1}^{m-1}(b_k \sin kt + a_k \cos kt) \tag{8.22}$$

where the $2m$ Fourier coefficients a_k and b_k are obtained by the discrete Fourier transform (DFT) of the data as

$$a_k = \frac{1}{m}\sum_{i=0}^{n} y_i \cos kt_i, \quad k = 0, 1, \dots, m \tag{8.23}$$

$$b_k = \frac{1}{m}\sum_{i=0}^{n} y_i \sin kt_i, \quad k = 1, 2, \dots, m-1 \tag{8.24}$$

The inverse DFT gives back the data in terms of the Fourier coefficients as

$$y_i = G_n(t_i) = \frac{1}{2}(a_0 + a_m \cos mt_i) + \sum_{k=1}^{m-1}(b_k \sin kt_i + a_k \cos kt_i) \tag{8.25}$$

As $G_n(t)$ approximates $F(t)$, we can approximate the derivative of $F(t)$ by the derivative of $G_n(t)$, i.e. $F'(t) \approx G_n'(t)$. We have

$$G_n'(t) = \frac{1}{2}(-ma_m \sin mt) + \sum_{k=1}^{m-1}(kb_k \cos kt - ka_k \sin kt) \tag{8.26}$$

At the interpolating nodes t_i, $\sin mt_i = \sin i\pi = 0$ and

$$G_n'(t_i) = \sum_{k=1}^{m-1}(kb_k \cos kt_i - ka_k \sin kt_i) \tag{8.27}$$

Finally we obtain the approximation

$$F'(t_i) \approx G_n'(t_i) \tag{8.28}$$

where $G_n'(t_i)$ can be written in the form of the inverse DFT

$$G_n'(t_i) = \frac{1}{2}(\bar{a}_0 + \bar{a}_m \cos mt_i) + \sum_{k=1}^{m-1}(\bar{b}_k \sin kt_i + \bar{a}_k \cos kt_i) \tag{8.29}$$

with the Fourier coefficients

$$\bar{a}_0 = 0, \quad \bar{a}_m = 0, \quad \bar{b}_k = -ka_k, \quad \bar{a}_k = kb_k \tag{8.30}$$

Similarly, we have $G_n''(t_i)$ in the form of the inverse DFT as

$$G_n'(t_i) = \frac{1}{2}(\bar{\bar{a}}_0 + \bar{\bar{a}}_m \cos m t_i) + \sum_{k=1}^{m-1}(\bar{\bar{b}}_k \sin k t_i + \bar{\bar{a}}_k \cos k t_i) \tag{8.31}$$

with the Fourier coefficients

$$\bar{\bar{a}}_0 = 0, \quad \bar{\bar{a}}_m = 0, \quad \bar{\bar{b}}_k = -k\bar{a}_k = -k^2 b_k, \quad \bar{\bar{a}}_k = k\bar{b}_k = -k^2 a_k \tag{8.32}$$

Remark If the data $\{(x_i, y_i)\}_{i=0}^n$ are generated by the periodic function $f(x)$ with the period $2L$ (instead of 2π) on the interval $[0, 2L]$, we can introduce the substitution $x = tL/\pi$ ($t = x\pi/L$) and consider the data $\{(t_i, y_i)\}_{i=0}^n$ generated by $F(t) = f(tL/\pi)$. With $t_i = x_i\pi/L$ and $t = x\pi/L$, we achieve the interpolation of $f(x)$ as $g_n(x) = G_n(x\pi/L)$. So $f'(x_i)$ can be approximated by $g_n'(x_i) = G_n'(t_i)\pi/L$, which amounts to

$$g_n'(x_i) = \frac{1}{2}\left[\bar{A}_0 + \bar{A}_m \cos\left(\frac{m\pi x_i}{L}\right)\right] + \sum_{k=1}^{m-1}\left[\bar{B}_k \sin\left(\frac{k\pi x_i}{L}\right) + \bar{A}_k \cos\left(\frac{k\pi x_i}{L}\right)\right] \tag{8.33}$$

with the Fourier coefficients

$$\bar{A}_0 = 0, \quad \bar{A}_m = 0, \quad \bar{B}_k = -(k\pi/L)a_k, \quad \bar{A}_k = (k\pi/L)b_k \tag{8.34}$$

8.3 Richardson Extrapolation

In this section, we demonstrate a strategy to improve the accuracy of an approximation of a derivative. The strategy is called Richardson extrapolation. Below is a demonstration of Richardson extrapolation works.

By Taylor's series expansion, we have

$$f(a+h) = f(a) + f'(a)h + \frac{f''(a)}{2!}h^2 + \frac{f'''(a)}{3!}h^3 + \frac{f^{(4)}(a)}{4!}h^4 + \cdots \tag{8.35}$$

from which we obtain

$$\frac{f(a+h) - f(a)}{h} = f'(a) + \frac{f''(a)}{2!}h + \frac{f'''(a)}{3!}h^2 + \frac{f^{(4)}(a)}{4!}h^3 + \cdots \tag{8.36}$$

We denote the left-hand side as $D(h)$, i.e.

$$D(h) = \frac{f(a+h) - f(a)}{h} \tag{8.37}$$

which is a finite difference approximation (forward difference if $h > 0$ and backward difference if $h < 0$) of $f'(a)$ with the accuracy of $O(h)$.

Suppose we have computed the finite difference approximations $D(h)$ (the finite difference approximation with h) and $D(h/2)$ (the finite difference approximation with $h/2$). We have the following expansions of the approximations

$$D(h) = f'(a) + \frac{f''(a)}{2!}h + \frac{f'''(a)}{3!}h^2 + \frac{f^{(4)}(a)}{4!}h^3 + \cdots \tag{8.38}$$

$$D\left(\frac{h}{2}\right) = f'(a) + \frac{f''(a)}{2!}\left(\frac{h}{2}\right) + \frac{f'''(a)}{3!}\left(\frac{h}{2}\right)^2 + \frac{f^{(4)}(a)}{4!}\left(\frac{h}{2}\right)^3 + \cdots \tag{8.39}$$

We can eliminate the dominant error terms of $O(h)$ in the above equations as

$$2D(h/2) - D(h) = f'(a) - \frac{f'''(a)}{3!}\frac{h^2}{2} - \frac{f^{(4)}(a)}{4!}\frac{3h^3}{4} - \cdots \tag{8.40}$$

If we denote the left-hand side of the previous equation as

$$R_1(h) = 2D(h/2) - D(h) \tag{8.41}$$

then $R_1(h)$ (called an extrapolation) is an improved approximation of $f'(a)$ with the accuracy $O(h^2)$. The accuracy is improved from $O(h)$ to $O(h^2)$ by the extrapolation $R_1(h)$.

If we want, we can use

$$R_1(h) = f'(a) - \frac{f'''(a)}{3!}\frac{h^2}{2} - \frac{f^{(4)}(a)}{4!}\frac{3h^3}{4} - \cdots \tag{8.42}$$

to form the extrapolation $R_2(h)$ by combining $R_1(h)$ and $R_1(h/2)$ to eliminate the dominant error terms of $O(h^2)$ and obtain the accuracy $O(h^3)$. We can repeatedly apply the process to further improve the accuracy.

8.4   Exercises

Exercise 8.1 Use (1) forward, (2) backward and (3) centered finite difference approximations with $h = 0.1$ to approximate the first derivative of $f(x) = e^x$ at $a = 0$. Calculate the absolute error of each approximation.

Exercise 8.2 Use centered finite difference approximations with $h = 0.1$ to approximate the first derivative of $f(x) = x^2$ at $a = 1$. Why the absolute error of the approximation is zero in this case.

Exercise 8.3 Suppose the vertical coordinates y of the center of mass (CM) of a falling object at different time t are recorded as in Table 8.2. Estimate the vertical velocity and acceleration of the object at time $t = 0.2$.

t [s]	0.1	0.2	0.3
y [m]	0.05	0.21	0.44

TABLE 8.2
Vertical coordinates y at different time t.

Exercise 8.4 What is wrong in the following analysis that uses Taylor's theorem.
   We have
$$f(a+h) = f(a) + f'(a)h + \frac{f''(c)}{2!}h^2$$

$$f(a-h) = f(a) - f'(a)h + \frac{f''(c)}{2!}h^2$$

By subtracting the above two formulas, we obtain
$$f(a+h) - f(a-h) = 2f'(a)h$$

Exercise 8.5 If the centered finite difference approximation
$$A = \frac{f(a+h) - 2f(a) + f(a-h)}{h^2}$$

is used to approximate $T = f''(a)$. Find out how the absolute error of the approximation depend on h. Write the error in the big-O notation.

Exercise 8.6 If the centered finite difference approximation
$$A = \frac{f(a-2h) - 8f(a-h) + 8f(a+h) - f(a+2h)}{12h}$$

is used to approximate $T = f'(a)$. Find out how the absolute error of the approximation depend on h. Write the error in the big-O notation.

Exercise 8.7 Use the method of undetermined coefficients to find a finite difference approximation of $f'(a)$ using the function values $f(a)$, $f(a+h)$ and $f(a+2h)$. The approximation is called a one-sided finite difference approximation as it uses function values at only one side of a. Find the dependence of the error on h in the big-O notation.

Exercise 8.8 Use the method of undetermined coefficients to find a one-sided finite difference approximation of $f''(a)$ using the function values $f(a)$, $f(a+h)$ and $f(a+2h)$. Find the dependence of the error on h in the big-O notation.

Exercise 8.9 Use the method of undetermined coefficients to find a one-sided finite difference approximation of $f'(a)$ using the function values $f(a)$, $f(a+h)$, $f(a+2h)$ and $f(a+3h)$. Find the dependence of the error on h in the big-O notation.

Exercise 8.10 Use the method of undetermined coefficients to determine α and β in the following so-called compact finite difference approximation

$$\alpha f'(a-h) + f'(a) + \alpha f'(a+h) \approx \beta f(a-h) - \beta f(a-h)$$

such that the approximation is as accurate as possible.

Exercise 8.11 Let $p_2(x)$ be a polynomial of degree ≤ 2 that interpolates $f(x)$ at a, $a+h$ and $a+2h$. Use $p_2(x)$ to find a finite difference approximation of $f'(a)$. Use the interpolation error to find the dependence of the error on h in the big-O notation.

Exercise 8.12 Let $p_2(x)$ be a polynomial of degree ≤ 2 that interpolates $f(x)$ at a, $a+h$ and $a+2h$. Use $p_2(x)$ to find a finite difference approximation of $f''(a)$.

Exercise 8.13 Let A_h be an approximation of the true value T, where A_h depends on a small parameter h. Suppose the error $T - A_h$ satisfies $T - A_h \approx Ch$. How can you use A_h and $A_{h/2}$, where $A_{h/2}$ is an approximation computed with $h/2$, to estimate the true value T?

Exercise 8.14 Let A_h be an approximation of the true value T, where A_h depends on a small parameter h. Suppose the error $T - A_h$ satisfies $T - A_h \approx Ch$, where C is a constant. How can you use A_h and A_{2h}, where A_{2h} is an approximation computed with $2h$, to estimate of the true value T?

Exercise 8.15 Let A_h be an approximation of the true value T, where A_h depends on a small parameter h. Suppose the error $T - A_h$ satisfies

$$T - A_h = C_2 h^2 + C_3 h^3 + \cdots$$

where C_2 and C_3 are known constants. How can you use A_h and A_{2h}, where A_{2h} is an approximation computed with $2h$, to estimate the true value T? What is the error of your estimation in terms of h in big-O notation?

8.5 Programming Problems

Problem 8.1 Write a code to approximate the derivative $f'(a)$ for $f(x) = e^x$ at $a = 0$ using centered finite difference approximations with $h = 0.2, 0.1, 0.05, 0.025$ and 0.0125, respectively. Let E_h denote the error of the approximation with h. Plot the error E_h versus h in loglog scale. Compute the ratio E_{2h}/E_h for each value of h except the largest value $h = 0.2$. Comment on your results.

Problem 8.2 Write a code to approximate the derivative $f''(a)$ for $f(x) = e^x$ at $a = 0$ using centered finite difference approximation with $h = 0.2, 0.1, 0.05, 0.025$ and 0.0125, respectively. Let E_h denote the error of the approximation with h. Plot the error E_h versus h in loglog scale. Compute the ratio E_{2h}/E_h for each value of h except the largest value $h = 0.2$.Comment on your results.

Problem 8.3 Redo problem 3.1.

Problem 8.4 write a program to find the trigonometric interpolant $G_n(t)$ (i.e. to determine the coefficients in $G_n(t)$) to interpolate the periodic function $f(t) = 0.0001t^4(t - 2\pi)^4$ at $n + 1 = 2m$ equally spaced nodes over the period $[0, 2\pi)$.

Then for $m = 2, 4, 8, 16$ and 32, respectively, (1) plot the graphs of $f'(t)$ and $G'_n(t)$ using 500 equally spaced points t_k $(k = 1, 2, \ldots, 500)$ over $[0, 2\pi]$, and (2) find the maximum absolute error $\max_i |f'(t_i) - G'_n(t_i)|$ $(i = 1, 2, \ldots, 500)$. Plot the maximum absolute error versus m.

9

Initial Value Problems and Boundary Value Problems

In this chapter, we describe some basic numerical methods for solving initial value problems (IVPs) and boundary value problems (BVPs). We use these methods to introduce some important concepts such as local truncation error, global error, consistency, convergence, stability and the difference between explicit and implicit methods.

What is an IVP? To give an example, let's consider the cooling of a small cup of hot coffee in an air-conditioned room. We assume the temperature u of the coffee is uniform and a function of the time t only. By Newton's law of cooling, the rate of change of the temperature $u(t)$ is proportional to the difference between $u(t)$ and the temperature of the surrounding air, i.e. $u(t)$ satisfies the following ordinary differential equation (ODE)

$$u'(t) = -k(u(t) - a(t)) \tag{9.1}$$

where k is a positive constant, and $a(t)$ is the temperature of the surrounding air. Suppose the initial temperature $u(0)$ (the initial value of $u(t)$) of the coffee is η and the controlled air temperature $a(t)$ is a known function of the time t. The temperature of the coffee as a function of the time can be determined from the following IVP

$$u'(t) = -k(u(t) - a(t)), \quad t > 0 \tag{9.2}$$
$$u(0) = \eta \tag{9.3}$$

where $u(0) = \eta$ specifies the initial temperature and is called the **initial condition**.

What is a BVP? To give an example, let's consider the steady (not changing with time) temperature distribution in a thin straight rod along the x-axis. The temperature u along the rod is a function of the coordinate x only. By Fourier's law of heat conduction, the heat flux in the x-direction is proportional to $u'(x)$. By energy conservation, the rate at which the heat flux changes with respect to x balances the heating/cooling intensity along the rod. Accordingly, the temperature distribution $u(x)$ satisfies the following ODE

$$u''(x) = -kf(x) \tag{9.4}$$

where k is a positive constant, and $f(x)$ is the heating ($f(x) > 0$) /cooling ($f(x) < 0$) intensity. Suppose the temperature of the two ends of the rod at $x = 0$ and $x = 1$ are fixed to the constants α and β, respectively. The conditions $u(0) = \alpha$ and $u(1) = \beta$

DOI: 10.1201/9781003201694-9

are called **boundary conditions**. Here the boundary conditions specify the values of the solution at the two boundaries and are called **Dirichlet boundary conditions** (A boundary condition is called **a Neumann boundary condition** if it specifies the value of the derivative of the solution at a boundary). The temperature distribution in the rod as a function of the position can be determined from the following BVP

$$u''(x) = -kf(x), \quad 0 < x < 1 \tag{9.5}$$

$$u(0) = \alpha, \quad u(1) = \beta \tag{9.6}$$

9.1   Initial Value Problems (IVPs)

An IVP takes the form

$$u'(t) = f(u(t), t)), \quad t > t_0 \tag{9.7}$$

$$u(t_0) = c \tag{9.8}$$

where c is the initial data at $t = t_0$, and often $t_0 = 0$ for simplicity. From now on, we treat t as time (which is not necessarily the case in real applications).

Remark   The IVP above has a unique solution if f satisfies the Lipschitz condition

$$|f(u,t) - f(v,t)| \leq L|u - v| \tag{9.9}$$

for all $u \in \mathbb{R}$, $v \in \mathbb{R}$ and $t \geq t_0$, where L is called a Lipschitz constant that is independent of u and v. A stronger requirement for the uniqueness of the solution is the continuity of $f(u,t)$ and $\frac{\partial}{\partial u} f(u,t)$.

An IVP can include a system of ODEs and initial data as

$$\mathbf{u}(t) = \mathbf{f}(\mathbf{u}(t), t), \quad t > t_0$$

$$\mathbf{u}(t_0) = \mathbf{c}$$

where $\mathbf{u}$, $\mathbf{f}$ and are vectors with multiple components, and each component of $\mathbf{f}$ is a function of the components of $\mathbf{u}$ and t.

An IVP with an ODE of high order can be reduced to a system of first order ODEs, as demonstrated in the following example.

Example 9.1   Consider the motion of a particle along a straight line. Denote the position function of the particle as $u(t)$. Then the velocity and acceleration of the particle are $u'(t)$ and $u''(t)$, respectively. Suppose the resultant time-dependent force applied to the particle is $f(t)$, and the initial position and velocity of the particle are $u(0) = s_0$ and $u'(0) = v_0$, respectively. By Newton's second law, the position function is determined from the following IVP

$$u''(t) = f(t)/m, \quad t > 0 \tag{9.10}$$
$$u(0) = s_0, \quad u'(0) = v_0 \tag{9.11}$$

where m is the mass of the particle.

By introducing $v(t) = u'(t)$, we can rewrite the IVP as

$$\mathbf{u}(t) = \mathbf{f}(\mathbf{u}(t), t), \quad t > 0 \tag{9.12}$$
$$\mathbf{u}(0) = \mathbf{c} \tag{9.13}$$

where

$$\mathbf{u}(t) = \begin{bmatrix} u(t) \\ v(t) \end{bmatrix}, \quad \mathbf{f}(\mathbf{u}(t), t) = \begin{bmatrix} v(t) \\ f(t)/m \end{bmatrix}, \quad \mathbf{c} = \begin{bmatrix} s_0 \\ v_0 \end{bmatrix} \tag{9.14}$$

■

Below we introduce some basic numerical methods for solving the IVP for a scalar function $u(t)$

$$u'(t) = f(u(t), t)), \quad t > t_0 \tag{9.15}$$
$$u(t_0) = c \tag{9.16}$$

In these methods, we start from the initial data at $t = t_0$ to march forward in time (as the underlying physics process does) to compute approximations of the function values at successive time $t_1, t_2, \ldots$. For simplicity, we consider evenly spaced time with the **time step size** Δt, that is $t_{n+1} - t_n = \Delta t$ for $n = 0, 1, \ldots$. We denote the approximation of the true function value $u(t_n)$ at t_n as $u^{(n)}$, i.e.

$$u^{(n)} \approx u(t_n), \quad n = 1, 2, \ldots \tag{9.17}$$

Note that we are given the initial data $u^{(0)} = u(t_0) = c$, and we want to compute the approximations $u^{(1)}, u^{(2)}, \ldots$ successively in time. So these methods are called **time-marching methods**.

9.1.1 Euler's Method

Euler's method is seldom used in practice, but it is the simplest method that can be easily analyzed to illustrate some basic ideas and concepts in solving IVPs. Below we show two ways to derive this method. One uses numerical differentiation, and the other numerical integration.

In Chapter 2, we have used Taylor's theorem to derive the following result

$$\frac{f(a+h) - f(a)}{h} = f'(a) + \frac{f''(c)}{2!}h \tag{9.18}$$

where $h > 0$, $c \in (a, a+h)$, and the left-hand side is called the forward finite difference approximation of $f'(a)$ (with the error of $O(h)$) in Chapter 8. Applying this result to the ODE $u'(t) = f(u(t), t)$ in the IVP at time $t = t_n$ ($n = 0, 1, 2, \ldots$), we have

$$\frac{u(t_{n+1}) - u(t_n)}{\Delta t} = f(u(t_n), t_n) + \frac{u''(c_n)}{2!}\Delta t \tag{9.19}$$

where $c_n \in (t_n, t_{n+1})$. If we drop the last term $\frac{u''(c_n)}{2!}\Delta t$ (which is small when Δt is small), we have the approximation

$$\frac{u(t_{n+1}) - u(t_n)}{\Delta t} \approx f(u(t_n), t_n) \tag{9.20}$$

The corresponding equation

$$\frac{u^{(n+1)} - u^{(n)}}{\Delta t} = f(u^{(n)}, t_n) \tag{9.21}$$

is called the **disretized equation** for the ODE in the IVP. Note that the discretized equation is satisfied by $u^{(n)}$ and $u^{(n+1)}$, which are approximations of $u(t_n)$ and $u(t_{n+1})$, respectively.

The discretized equation gives us a time-marching method

$$u^{(n+1)} = u^{(n)} + \Delta t f(u^{(n)}, t_n), \quad n = 0, 1, 2, \ldots \tag{9.22}$$

which allows us to start with the initial data $u^{(0)} = u(t_0) = c$ and compute $u^{(1)}$, $u^{(2)}$, $\cdots$ successively. The method is called Euler's method. Below is its pseudocode.

Algorithm 16 Euler's method for the IVP $u'(t) = f(u(t), t))$, $t_0 < t < T$; $u(t_0) = c$

$n_{\text{end}} \leftarrow \lfloor (T - t_0)/\Delta t \rfloor$
$u^{(0)} \leftarrow c$
for n from 0 to $n_{\text{end}} - 1$ **do**
$\quad u^{(n+1)} \leftarrow u^{(n)} + \Delta t f(u^{(n)}, t_n)$
$\quad t_{n+1} \leftarrow t_n + \Delta t$
end for

Example 9.2 Apply Euler's method to solve the IVP

$$u'(t) = u(t), \ 0 < t < 1; \quad u(0) = 1$$

with $\Delta t = 0.2$. The exact solution is $u(t) = e^t$.
Solution: Starting with $u^{(0)} = 1$, we use

$$u^{(n+1)} = u^{(n)} + \Delta t u^{(n)}, \quad n = 0, 1, 2, \ldots$$

to compute $u^{(1)}$, $u^{(2)}$, $\ldots$ successively and organize the results in the following table.

The last column in the table give the absolute error at each time. The exact and computed solutions are compared in Fig. 9.1. ∎

t_n	$u^{(n)}$	$u(t_n) = e^{t_n}$	$\left| u(t_n) - u^{(n)} \right|$
0	1.0000	1.0000	0
0.2000	1.2000	1.2214	0.0214
0.4000	1.4400	1.4918	0.0518
0.6000	1.7280	1.8221	0.0941
0.8000	2.0736	2.2255	0.1519

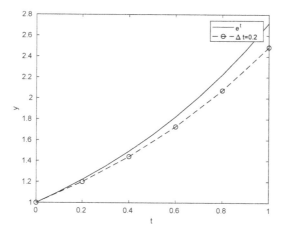

FIGURE 9.1

Comparison between the exact solution and the approximate solution computed by Euler's method for an IVP.

Remark In the above example, the right-hand side of the ODE $f(u,t) = u$ is an explicit function of only u and does not depend on t explicitly (f depends on t implicitly as u is a function of t). We say such an ODE is an **autonomous ODE**.

Solving the IVP can be related to numerical integration. If we integrate the ODE in the IVP from t_n to t_{n+1}, we have

$$u(t_{n+1}) - u(t_n) = \int_{t_n}^{t_{n+1}} f(u(t),t)\,dt \tag{9.23}$$

The definite integral at the right-hand side can be approximated by a quadrature rule. If it is approximated by the left-endpoint rule $R_L(f) = f(u(t_n),t_n)(t_{n+1} - t_n)$, we have

$$u(t_{n+1}) - u(t_n) \approx f(u(t_n),t_n)(t_{n+1} - t_n) \tag{9.24}$$

which can be written as

$$\frac{u(t_{n+1}) - u(t_n)}{\Delta t} \approx f(u(t_n), t_n) \tag{9.25}$$

from which we can again derive Euler's method (as we have done so using finite difference).

> **Remark** Numerical differentiation and numerical integration can be used to develop other methods for solving IVPs.

9.1.1.1   Local Truncation Error and Global Error

The derivation of Euler's method above clearly indicate that the exact solution $u(t)$ does not satisfy the discretized equation (as the discretized equation comes from the ODE after a small term is dropped)

$$\frac{u^{(n+1)} - u^{(n)}}{\Delta t} - f(u^{(n)}, t_n) = 0 \tag{9.26}$$

Instead $u(t)$ satisfies

$$\frac{u(t_{n+1}) - u(t_n)}{\Delta t} - f(u(t_n), t_n) = \frac{u''(c_n)}{2!}\Delta t \tag{9.27}$$

We call the term $\frac{u''(c_n)}{2!}\Delta t$ the local truncation error (LTE) and denote it as τ_n. That is

$$\tau_n = \frac{u(t_{n+1}) - u(t_n)}{\Delta t} - f(u(t_n), t_n) = \frac{u''(c_n)}{2!}\Delta t \tag{9.28}$$

It is called a truncation error because it is the term we drop to obtain the discretized equation and is the discrepancy that the exact solution fails to satisfy the discretized equation. It is local because it is the discrepancy at the moment $t = t_n$. In general, a method has order of accuracy p if its LTE is of $O(h^p)$. Euler's method has LTE of $\mathcal{O}(h)$ and is first order accurate.

If we use the exact value $u(t_n)$ (instead of the approximation $u^{(n)}$) to compute the next approximation, denoted as $u_*^{(n+1)}$ (not the same as $u^{(n+1)}$), in Euler's method (as in the first time marching step of Euler's method where the exact initial condition $u(t_0)$ is used to compute $u^{(1)}$), we have

$$u_*^{(n+1)} = u(t_n) + \Delta t f(u(t_n), t_n) = u(t_n) + u'(t_n)\Delta t \tag{9.29}$$

which is the tangent line approximation (linear approximation) of $u(t_{n+1})$, as illustrated in Fig. 9.2. We then have

$$u_*^{(n+1)} - u(t_n) - \Delta t f(u(t_n), t_n) = 0 \tag{9.30}$$

$$[u_*^{(n+1)} - u(t_{n+1})] + [u(t_{n+1}) - u(t_n) - \Delta t f(u(t_n), t_n)] = 0 \tag{9.31}$$

By definition of LTE, we have

$$\tau_n \Delta t = u(t_{n+1}) - u(t_n) - \Delta t f(u(t_n), t_n) \tag{9.32}$$

So we obtain

$$u(t_{n+1}) - u_*^{(n+1)} = \tau_n \Delta t \tag{9.33}$$

In Fig. 9.2, the length of a vertical line segment between a solid circle (exact solution) and a cross-mark (linear approximation) therefore represents a LTE.

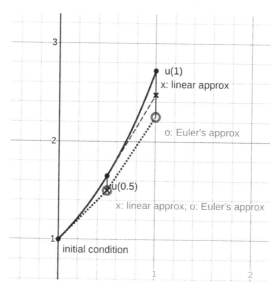

FIGURE 9.2
Geometric interpretation of Euler's method, LTE and global error.

The error in the approximation $u^{(n+1)}$ of $u(t_{n+1})$ in Euler's method is defined as

$$e_{n+1} = u(t_{n+1}) - u^{(n+1)} \tag{9.34}$$

which can be written as

$$[u(t_{n+1}) - u_*^{(n+1)}] + [u_*^{(n+1)} - u^{(n+1)}] = \tau_n \Delta t + [u_*^{(n+1)} - u^{(n+1)}] \tag{9.35}$$

Except for $u^{(1)}$, this is not just due to the LTE because we use the approximate value $u^{(n)}$ (instead of the true value $u(t_n)$) to compute $u^{(n+1)}$. Geometrically speaking, we start from a wrong point $(t_n, u^{(n)})$ and use a wrong slope $f(u^{(n)}, t_n)$ to draw the tangent line to obtain the next approximation, as illustrated in Fig. 9.2. So the error in $u^{(n+1)}$ has contributions from both the LTE and the error in the previous approximation $u^{(n)}$. The latter has accumulation effect. So the error e_{n+1} is called global error. In Fig. 9.2, a global error is represented by a line segment between a solid circle (exact solution) and an open circle (Euler's approximation).

Remark The definitions and discussions about the LTE and global error of Euler's method can be extended to other methods for IVPs.

9.1.1.2 Consistency, Convergence and Stability

The LTE of Euler's method satisfies

$$|\tau_n| = \left|\frac{u''(c_n)}{2!}\Delta t\right| \to 0, \quad \text{as} \quad \Delta t \to 0 \tag{9.36}$$

Since the LTE is the discrepancy between the discretized equation and the ODE, we say the discretized equation is consistent with the ODE if the LTE $\tau_n \to 0$ as $\Delta t \to 0$. Euler's method is consistent.

As seen in the previous section, the global error is different from the LTE. The LTE approaches zero does not necessarily imply the global error must approach zero. We say a method is convergent if its global error approaches zero too as $\Delta t \to 0$:

$$|e_n| = |u(t_n) - u^{(n)}| \to 0, \quad \text{as} \quad \Delta t \to 0 \tag{9.37}$$

i.e. the approximate solution approaches the exact solution as $\Delta t \to 0$:

$$\lim_{\Delta t \to 0} u^{(n)} = u(t_n) \tag{9.38}$$

Theorem 9.1 (Convergence of Euler's Method) *If $f(u,t)$ satisfies the Lipschitz condition*

$$|f(u,t) - f(v,t)| \le L|u - v| \tag{9.39}$$

for any $u \in \mathbb{R}$, $v \in \mathbb{R}$ and $t \ge t_0$, and $|u''(t)| \le M$ is bounded for $t > t_0$, then Euler's method is convergent and its global error is of $O(\Delta t)$.

Proof *The time marching process in Euler's method is*

$$u^{(n+1)} = u^{(n)} + \Delta t f(u^{(n)}, t_n), \quad n = 0, 1, 2, \dots \tag{9.40}$$

By definition, the LTE τ_n satisfies

$$u(t_{n+1}) = u(t_n) + \Delta t f(u(t_n), t_n) + \tau_n \Delta t \tag{9.41}$$

Subtracting the above two equations, we obtain

$$e_{n+1} = e_n + \Delta t [f(u(t_n), t_n) - f(u^{(n)}, t_n)] + \tau_n \Delta t \tag{9.42}$$

which establishes the relation between the two consecutive global errors e_{n+1} and e_n. By the triangle inequality, We then have

$$|e_{n+1}| \le |e_n| + \Delta t |f(u(t_n), t_n) - f(u^{(n)}, t_n)| + |\tau_n|\Delta t \tag{9.43}$$

Applying the Lipschitz condition

$$|f(u(t_n),t_n) - f(u^{(n)},t_n)| \leq L|u(t_n) - u^{(n)}| = L|e_n| \tag{9.44}$$

we obtain

$$|e_{n+1}| \leq (1 + L\Delta t)|e_n| + |\tau_n|\Delta t \tag{9.45}$$

We know that

$$|\tau_n| = \frac{|u''(c_n)|}{2!}\Delta t \tag{9.46}$$

where $|u''(c_n)|$ is bounded as

$$|u''(t)| \leq M \tag{9.47}$$

So we finally obtain

$$|e_{n+1}| \leq (1 + L\Delta t)|e_n| + M(\Delta t)^2/2 \tag{9.48}$$

Let $\rho = 1 + L\Delta t$ and $d = M(\Delta t)^2/2$. They are constants. We have

$$|e_{n+1}| \leq \rho|e_n| + d, \quad n = 0, 1, 2, \ldots \tag{9.49}$$

We can apply this inequality recursively as follows

$$|e_{n+1}| \leq \rho|e_n| + d \leq \rho(\rho|e_{n-1}| + d) + d = \rho^2|e_{n-1}| + \rho d + d$$
$$\leq \rho^2(\rho|e_{n-2}| + d) + \rho d + d = \rho^3|e_{n-2}| + \rho^2 d + \rho d + d$$
$$\leq \cdots\cdots$$
$$\leq \rho^{n+1}|e_0| + \rho^n d + \rho^{n-1}d + \cdots + \rho^2 d + \rho d + d$$
$$= \rho^{n+1}|e_0| + \frac{d(1-\rho^{n+1})}{1-\rho} = \frac{d(1-\rho^{n+1})}{1-\rho} \tag{9.50}$$

where we apply $e_0 = 0$ in the last step. As $\Delta t \to 0$, we have $d = M(\Delta t)^2/2 \to 0$. We then have $|e_{n+1}| \to 0$ as $\Delta t \to 0$, which proves that Euler's method is convergent.
Note that the Maclaurin series of the natural exponential function gives

$$e^{L\Delta t} = 1 + L\Delta t + \frac{(L\Delta t)^2}{2!} + \cdots \geq 1 + L\Delta t = \rho \tag{9.51}$$

We have

$$\rho^{n+1} \leq [e^{L\Delta t}]^{n+1} = e^{L(t_{n+1}-t_0)} \tag{9.52}$$

So with $d/(\rho - 1) = M\Delta t/(2L)$, we obtain

$$|e_{n+1}| \leq \frac{d(\rho^{n+1}) - 1}{\rho - 1} \leq \frac{M[e^{L(t_{n+1}-t_0)} - 1]}{2L}\Delta t \tag{9.53}$$

which proves that the global error in Euler's method is of $O(\Delta t)$. ∎

Let's now consider a simple test IVP

$$u'(t) = \lambda u(t), \quad u(0) = c \tag{9.54}$$

where λ is a constant. The true solution of the test problem is

$$u(t) = ce^{\lambda t} \tag{9.55}$$

If $\lambda < 0$, then the true solution decays with t. When a method with a finite Δt is applied to the test problem, it is required at the least that the approximate solution should not grow, i.e.

$$|u^{(n+1)}| \le |u^{(n)}| \tag{9.56}$$

If a method satisfies this condition, we say the method is **absolutely stable**. When we apply Euler's method to the test problem, we have

$$u^{(n+1)} = u^{(n)} + \Delta t[\lambda u^{(n)}] = (1 + \lambda \Delta t)u^{(n)} \tag{9.57}$$

So Euler's method is absolutely stable if

$$|1 + \lambda \Delta t| \le 1 \tag{9.58}$$

Let $z = \lambda \Delta t$. The interval of absolute stability for Euler's method is $-2 \le z \le 0$.

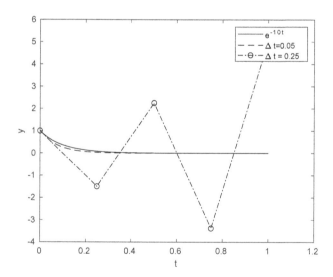

FIGURE 9.3
Test of the absolute stability for Euler's method.

Example 9.3 Apply Euler's method to solve the test problem $u'(t) = -10u(t), u(0) = 1$ with $\Delta t = 0.05$ and $\Delta t = 0.25$, respectively. Fig. 9.3 indicates that the method is stable with $\Delta t = 0.05$ as $z = \lambda \Delta t = -0.5 \in [-2, 0]$, while unstable with $\Delta t = 0.25$ as $z = \lambda \Delta t = -2.5 \notin [-2, 0]$. ■

Remark There are many forms of stability to guarantee convergence of a numerical method for an IVP, including zero-stability, absolute stability, A-stability, A(α)-stability and L-stability, which can be seen with details in a course on numerical analysis of ODEs.

9.1.1.3 Explicit and Implicit Methods

Euler's method uses the known approximation $u^{(n)}$ at the current time t_n (time step n) to explicitly compute the next approximation $u^{(n+1)}$ at the next time t_{n+1} (time step $n+1$) as

$$u^{(n+1)} = u^{(n)} + \Delta t f(u^{(n)}, t_n), \quad n = 0, 1, 2, \ldots \tag{9.59}$$

So Euler's method is also called **forward Euler method** (marches forward in time) and an explicit method.

Consider the ODE in the IVP at t_{n+1}

$$u'(t_{n+1}) = f(u(t_{n+1}), t_{n+1}) \tag{9.60}$$

If we approximate $u'(t_{n+1})$ by the backward finite difference approximation (instead of forward finite difference approximation in forward Euler method)

$$u'(t_{n+1}) \approx \frac{u^{(n+1)} - u^{(n)}}{\Delta t} \tag{9.61}$$

We end up with the so-called **backward Euler method**

$$u^{(n+1)} = u^{(n)} + \Delta t f(u^{(n+1)}, t_{n+1}) \tag{9.62}$$

which is substantially different from forward Euler method in its time-marching process. Now f is determined by the next unknown approximation $u^{(n+1)}$ so that $u^{(n+1)}$ is defined implicitly by an equation. So Backward Euler method is called an implicit method. If $f(u, t)$ is a nonlinear function of u, the equation for $u^{(n+1)}$ is nonlinear, and a nonlinear root finding problem needs to be solved at every time step.

Remark Explicit methods are easier to implement than implicit methods. An explicit method generally uses less computational time in one step than an implicit method. However, explicit methods may suffer from numerical instability or need very small time steps for stable computation. For many applications, implicit methods are superior to explicit methods.

9.1.2 Taylor Series Methods

The Taylor series expansion of $u(t_{n+1})$ about $u(t_n)$ is

$$u(t_{n+1}) = u(t_n) + u'(t_n)\Delta t + \frac{u''(t_n)}{2!}(\Delta t)^2 + \cdots \tag{9.63}$$

The forward Euler method can be derived by dropping all terms of $O((\Delta t)^2)$ and higher and replacing $u'(t_n)$ by $f(u(t_n), t_n)$ according to the ODE $u'(t) = f(u,t)$. We thus obtain

$$u(t_{n+1}) \approx u(t_n) + f(u(t_n), t_n)\Delta t \tag{9.64}$$

which suggests the method

$$u^{(n+1)} = u^{(n)} + f(u^{(n)}, t_n)\Delta t \tag{9.65}$$

So the forward Euler method is also a Taylor series method. It is a Taylor series method of only order 1 as the local truncation error (LTE) is of $O(\Delta t)$ (i.e. $O((\Delta t)^2)/\Delta t$ as compared with the ODE).

A Taylor series method of higher accuracy can be derived by keeping more terms in the Taylor series above. If we keep the term of $O((\Delta t)^2)$, we have

$$u(t_{n+1}) \approx u(t_n) + u'(t_n)\Delta t + \frac{u''(t_n)}{2!}(\Delta t)^2 \tag{9.66}$$

The ODE tells us that $u'(t_n) = f(u(t_n), t_n)$. We now need to compute the second derivative $u''(t_n)$ so that the approximated equation gives us a Taylor series method to compute $u^{(n+1)}$ from the known $u^{(n)}$. We can compute the higher derivatives by repeatedly differentiating the ODE. We can get $u''(t)$ as

$$u''(t) = \frac{d}{dt}f(u,t) = \frac{\partial f(u,t)}{\partial u}u'(t) + \frac{\partial f(u,t)}{\partial t}$$
$$= \frac{\partial f(u,t)}{\partial u}f(u,t) + \frac{\partial f(u,t)}{\partial t} \tag{9.67}$$

Finally we have

$$u(t_{n+1}) \approx u(t_n) + f(u(t_n), t_n)\Delta t + \frac{1}{2}\left[\frac{\partial f(u,t)}{\partial u}f(u,t) + \frac{\partial f(u,t)}{\partial t}\right](\Delta t)^2 \tag{9.68}$$

which give a Taylor series method of order 2

$$u^{(n+1)} = u^{(n)} + f(u^{(n)}, t_n)\Delta t + \frac{1}{2}\left[\frac{\partial f(u^{(n)}, t_n)}{\partial u}f(u^{(n)}, t_n) + \frac{\partial f(u^{(n)}, t_n)}{\partial t}\right](\Delta t)^2 \tag{9.69}$$

The local truncation error (LTE) of this Taylor series method is of order $O((\Delta t)^2)$. So it is a method of order 2. We can implement the method by computing the follows in each marching step in the shown order

$$P_1 = f(u^{(n)}, t_n) \tag{9.70}$$

$$P_2 = \frac{\partial f(u^{(n)}, t_n)}{\partial u} P_1 + \frac{\partial f(u^{(n)}, t_n)}{\partial t} \tag{9.71}$$

$$u^{(n+1)} = u^{(n)} + P_1 \Delta t + \frac{P_2}{2}(\Delta t)^2 \tag{9.72}$$

The corresponding pesudocode is

Algorithm 17 The Taylor series method of order 2 for the IVP $u'(t) = f(u(t), t))$, $t_0 < t < T$; $u(t_0) = c$

$n_{\text{end}} \leftarrow \lfloor (T - t_0)/\Delta t \rfloor$
$u^{(0)} \leftarrow c$
for n from 0 to $n_{\text{end}} - 1$ **do**
$\quad t_{n+1} \leftarrow t_n + \Delta t$
$\quad P_1 \leftarrow f(u^{(n)}, t_n)$
$\quad P_2 \leftarrow \frac{\partial f(u^{(n)}, t_n)}{\partial u} P_1 + \frac{\partial f(u^{(n)}, t_n)}{\partial t}$
$\quad u^{(n+1)} \leftarrow u^{(n)} + P_1 \Delta t + \frac{P_2}{2}(\Delta t)^2$
end for

9.1.3   Runge-Kutta (RK) Methods

As shown before, we can derive a method for solving IVPs by numerical integration. Integrating the ODE $u'(t) = f(u, t)$ from t_n to t_{n+1}, we have

$$u(t_{n+1}) - u(t_n) = \int_{t_n}^{t_{n+1}} f(u(t), t) \mathrm{d}t \tag{9.73}$$

The definite integral at the right-hand side can be approximated by a quadrature rule.

If we apply the midpoint rule to approximate the definite integral, we obtain from Eq. (9.73) the following

$$u(t_{n+1}) - u(t_n) \approx \Delta t f(u(t_{n+1/2}), t_{n+1/2}) \tag{9.74}$$

where $t_{n+1/2} = t_n + \Delta t/2$. However, we need $u(t_{n+1/2})$. We can approximate $u(t_{n+1/2})$ by U using Euler's method as

$$U = u^{(n)} + \frac{\Delta t}{2} f(u^{(n)}, t_n) \tag{9.75}$$

With this approximation, we obtain the following two-stage method

$$U = u^{(n)} + \frac{\Delta t}{2} f(u^{(n)}, t_n) \tag{9.76}$$

$$u^{(n+1)} = u^{(n)} + \Delta t f(U, t_{n+1/2}) \tag{9.77}$$

This method is called a two-stage explicit Runge-Kutta (RK) method.

Remark By replacing approximate values $u^{(n)}$ and $u^{(n+1)}$ with the true values $u(t_n)$ and $u(t_{n+1})$ in this two-stage RK method, the local truncation error of the method is defined as

$$\tau_n = \frac{u(t_{n+1}) - u(t_n)}{\Delta t} - f\left(u(t_n) + \frac{\Delta t}{2} f(u(t_n), t_n), t_{n+1/2}\right) \qquad (9.78)$$

By applying Taylor's expansion of $u(t_n + \Delta t)$ about t_n and Taylor's expansion of $f(u(t_n) + \Delta t u'(t_n)/2, t_n + \Delta t/2)$ about $(u(t_n), t_n)$ (see Chapter 5 for Taylor's expansions in two dimensions) and noticing that $\frac{\partial f}{\partial u} u' + \frac{\partial f}{\partial t} = \frac{df}{dt} = u''$, we can obtain $\tau_n = O((\Delta t)^2)$ We therefore say the method is of order 2.

If we apply Simpson's rule to approximate the definite integral, we obtain from Eq. (9.73) the following

$$u(t_{n+1}) - u(t_n) \approx \frac{\Delta t}{6}[f(u(t_n), t_n) + 4f(u(t_{n+1/2}), t_{n+1/2}) + f(u(t_{n+1}), t_{n+1})] \quad (9.79)$$

Let $U_1 = u^{(n)}$. We approximate $u(t_{n+1/2})$ by

$$U_2 = u^{(n)} + \frac{\Delta t}{2} f(U_1, t_n) \qquad (9.80)$$

which applies the forward Euler method, and by

$$U_3 = u^{(n)} + \frac{\Delta t}{2} f(U_2, t_{n+1/2}) \qquad (9.81)$$

which looks like the application of the backward Euler method (except that U_2 instead of U_3 is used to evaluate f to make the approximation explicit). We then approximate $u(t_{n+1})$ by

$$U_4 = u^{(n)} + \Delta t f(U_3, t_{n+1/2}) \qquad (9.82)$$

Finally we have a four-stage explicit method

$$U_1 = u^{(n)} \qquad (9.83)$$

$$U_2 = u^{(n)} + \frac{\Delta t}{2} f(U_1, t_n) \qquad (9.84)$$

$$U_3 = u^{(n)} + \frac{\Delta t}{2} f(U_2, t_{n+1/2}) \qquad (9.85)$$

$$U_4 = u^{(n)} + \Delta t f(U_3, t_{n+1/2}) \qquad (9.86)$$

$$u^{(n+1)} = u^{(n)} + \frac{\Delta t}{6}[f(U_1, t_n) + 2f(U_2, t_{n+1/2}) + 2f(U_3, t_{n+1/2}) + f(U_4, t_{n+1})]$$

$$(9.87)$$

which is called the classical Runge-Kutta (RK) method of order 4. Its local truncation error (LTE) is of $O((\Delta t)^4)$. If we introduce

$$K_1 = f(U_1, t_n) = f(u^{(n)}, t_n) \tag{9.88}$$

$$K_2 = f(U_2, t_{n+1/2}) = f\left(u^{(n)} + \frac{\Delta t}{2} f(U_1, t_n), t_{n+1/2}\right) = f\left(u^{(n)} + \frac{\Delta t}{2} K_1, t_{n+1/2}\right) \tag{9.89}$$

$$K_3 = f(U_3, t_{n+1/2}) = f\left(u^{(n)} + \frac{\Delta t}{2} f(U_2, t_{n+1/2}), t_{n+1/2}\right) = f\left(u^{(n)} + \frac{\Delta t}{2} K_2, t_{n+1/2}\right) \tag{9.90}$$

$$K_4 = f(U_4, t_{n+1}) = f(u^{(n)} + \Delta t f(U_3, t_{n+1/2}), t_{n+1}) = f(u^{(n)} + \Delta t K_3, t_{n+1}) \tag{9.91}$$

we can write the method as

$$K_1 = f(u^{(n)}, t_n) \tag{9.92}$$

$$K_2 = f\left(u^{(n)} + \frac{\Delta t}{2} K_1, t_{n+1/2}\right) \tag{9.93}$$

$$K_3 = f\left(u^{(n)} + \frac{\Delta t}{2} K_2, t_{n+1/2}\right) \tag{9.94}$$

$$K_4 = f(u^{(n)} + \Delta t K_3, t_{n+1}) \tag{9.95}$$

$$u^{(n+1)} = u^{(n)} + \frac{\Delta t}{6}(K_1 + 2K_2 + 2K_3 + K_4) \tag{9.96}$$

The corresponding pesudocode is

Algorithm 18 The classical RK method for the IVP $u'(t) = f(u(t), t))$, $t_0 < t < T$; $u(t_0) = c$

$n_{end} \leftarrow \lfloor (T - t_0)/\Delta t \rfloor$
$u^{(0)} \leftarrow c$
for n from 0 to $n_{end} - 1$ **do**
$\quad t_{n+1/2} \leftarrow t_n + \frac{\Delta t}{2}$
$\quad t_{n+1} \leftarrow t_n + \Delta t$
$\quad K_1 \leftarrow f(u^{(n)}, t_n)$
$\quad K_2 \leftarrow f(u^{(n)} + \frac{\Delta t}{2} K_1, t_{n+1/2})$
$\quad K_3 \leftarrow f(u^{(n)} + \frac{\Delta t}{2} K_2, t_{n+1/2})$
$\quad K_4 \leftarrow f(u^{(n)} + \Delta t K_3, t_{n+1})$
$\quad u^{(n+1)} \leftarrow u^{(n)} + \frac{\Delta t}{6}(K_1 + 2K_2 + 2K_3 + K_4)$
end for

Example 9.4 Apply the classical RK method of order 4 to solve the IVP

$$u'(t) = u(t), \ 0 < t < 1; \quad u(0) = 1$$

with $\Delta t = 0.1$. The exact solution is $u(t) = e^t$.

Solution: Here $f(u,t) = u$, $\Delta t = 0.1$, $t_n = n\Delta t$ and $t_{n+1/2} = (n+1/2)\Delta t$. Starting with the IC $u^{(0)} = 1$, we use

$$K_1 = f(u^{(n)}, t_n) = u^{(n)}$$

$$K_2 = f\left(u^{(n)} + \frac{\Delta t}{2}K_1, t_{n+1/2}\right) = u^{(n)} + \frac{\Delta t}{2}K_1$$

$$K_3 = f\left(u^{(n)} + \frac{\Delta t}{2}K_2, t_{n+1/2}\right) = u^{(n)} + \frac{\Delta t}{2}K_2$$

$$K_4 = f(u^{(n)} + \Delta t K_3, t_{n+1}) = u^{(n)} + \Delta t K_3$$

$$u^{(n+1)} = u^{(n)} + \frac{\Delta t}{6}(K_1 + 2K_2 + 2K_3 + K_4)$$

for $n = 0, 1, 2, \ldots, 9$ to compute $u^{(1)}$, $u^{(2)}$, $\ldots$, $u^{(10)}$ successively in MATLAB and organize the results in the following table.

```
tn    un       u(tn)     |u(tn)-un|
---------------------------------
0.1  1.10517  1.10517  8.47423e-08
0.2  1.22140  1.22140  1.87309e-07
0.3  1.34986  1.34986  3.10513e-07
0.4  1.49182  1.49182  4.57561e-07
0.5  1.64872  1.64872  6.32103e-07
0.6  1.82212  1.82212  8.38299e-07
0.7  2.01375  2.01375  1.08087e-06
0.8  2.22554  2.22554  1.36520e-06
0.9  2.45960  2.45960  1.69738e-06
1.0  2.71828  2.71828  2.08432e-06
```

The last column in the table give the absolute error at each time. The exact and computed solutions are compared in Fig. 9.4. ∎

9.2   Boundary Value Problems (BVPs)

Now we introduce a finite difference method for solving a BVP. A general two-point BVP has the form

$$u''(x) = f(x, u(x), u'(x)), \quad a < x < b \tag{9.97}$$

$$u(a) = \alpha, \quad u(b) = \beta \tag{9.98}$$

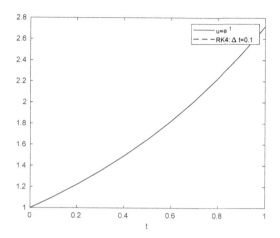

FIGURE 9.4
Comparison between the exact solution and the approximate solution computed by RK4 for an IVP.

> **Remark** For a general BVP with a second-order ODE, two boundary conditions do not necessarily guarantee that the BVP has a unique solution. For example, the BVP
>
> $$u''(x) = -u(x), \quad 0 < x < \pi$$
> $$u(0) = 1, \quad u(\pi) = 2$$
>
> has no solution.

Below we consider the simplest two-point BVP

$$u''(x) = f(x), \quad x \in [0, 1] \tag{9.99}$$
$$u(0) = \alpha, \quad u(1) = \beta \tag{9.100}$$

9.2.1 Finite Difference Methods

We first use a grid with the $n+1$ grid points $0 = x_0 < x_1 < \cdots < x_{n-1} < x_n = 1$ to cover the domain $[0, 1]$ of the equation. The grid is fine enough such that the length of each subinterval $h_i = x_i - x_{i-1}$ $(i = 1, 2, \ldots, n)$ is small enough. We then seek a discrete solution $u_i \approx u(x_i)$ $(i = 0, 1, \ldots, n)$ of the problem by solving algebraic equations for u_i. The algebraic equations for u_i can be derived by numerical differentiation that approximates the derivatives in the differential equation (and the boundary conditions if needed). On a uniform grid with the space step size $h_i = h = 1/n$, one such numerical differentiation scheme is the centered finite difference approximation

$$\frac{d^2u(x_i)}{dx^2} \approx \frac{u_{i+1} - 2u_i + u_{i-1}}{h^2} \tag{9.101}$$

Accordingly the differential equation is discretized as

$$\frac{u_{i+1} - 2u_i + u_{i-1}}{h^2} = f(x_i), \quad i = 1, 2, \ldots, n-1 \tag{9.102}$$

which together with the boundary conditions $u_0 = \alpha$ and $u_n = \beta$ gives a square linear system to solve for u_i ($i = 1, \ldots, n-1$):

$$\frac{1}{h^2} \begin{bmatrix} -2 & 1 & & & \\ 1 & -2 & 1 & & \\ & \ddots & \ddots & \ddots & \\ & & 1 & -2 & 1 \\ & & & 1 & -2 \end{bmatrix} \begin{bmatrix} u_1 \\ u_2 \\ \vdots \\ u_{n-2} \\ u_{n-1} \end{bmatrix} = \begin{bmatrix} f(x_1) - \alpha/h^2 \\ f(x_2) \\ \vdots \\ f(x_{n-2}) \\ f(x_{n-1}) - \beta/h^2 \end{bmatrix} \tag{9.103}$$

We may denote the linear system as

$$A\hat{\mathbf{u}} = \mathbf{f} \tag{9.104}$$

where

$$A = \frac{1}{h^2} \begin{bmatrix} -2 & 1 & & & \\ 1 & -2 & 1 & & \\ & \ddots & \ddots & \ddots & \\ & & 1 & -2 & 1 \\ & & & 1 & -2 \end{bmatrix} \tag{9.105}$$

$$\hat{\mathbf{u}} = \begin{bmatrix} u_1 \\ u_2 \\ \vdots \\ u_{n-2} \\ u_{n-1} \end{bmatrix}, \quad \mathbf{f} = \begin{bmatrix} f(x_1) - \alpha/h^2 \\ f(x_2) \\ \vdots \\ f(x_{n-2}) \\ f(x_{n-1}) - \beta/h^2 \end{bmatrix} \tag{9.106}$$

This is a tridiagonal linear system and can be solved using the technique given in Chapter 4.

9.2.1.1 Local Truncation Error and Global Error

The approximate discrete solutions u_i ($i = 1, 2, \ldots, n-1$) satisfies the discretized equation

$$\frac{u_{i+1} - 2u_i + u_{i-1}}{h^2} = f(x_i), \quad i = 1, 2, \ldots, n-1 \tag{9.107}$$

The true solution $u(x_i)$ $(i = 1, 2, \ldots, n-1)$ satisfies the ODE in the BVP, but generally does not satisfy the discretized equation. If we replace u_i in the discretized equation by $u(x_i)$, the discrepancy between the left- and right-hand sides is the local truncation error (LTE):

$$\tau_i = \frac{u(x_{i+1}) - 2u(x_i) + u(x_{i-1})}{h^2} - f(x_i), \quad i = 1, 2, \ldots, n-1 \tag{9.108}$$

Let's define

$$\mathbf{u} = \begin{bmatrix} u(x_1) \\ u(x_2) \\ \vdots \\ u(x_{n-2}) \\ u(x_{n-1}) \end{bmatrix}, \quad \tau = \begin{bmatrix} \tau_1 \\ \tau_2 \\ \vdots \\ \tau_{n-2} \\ \tau_{n-1} \end{bmatrix} \tag{9.109}$$

where $\mathbf{u}$ is the vector of the true solution and τ the vector of local truncation error. Then we have $\tau = A\mathbf{u} - \mathbf{f}$, i.e.

$$A\mathbf{u} = \mathbf{f} + \tau \tag{9.110}$$

In Chapter 8, we know that the centered finite difference approximation of a second derivative is of second order. So we have

$$\frac{u(x_{i+1}) - 2u(x_i) + u(x_{i-1})}{h^2} = u''(x_i) + O(h^2) \tag{9.111}$$

The ODE gives $u''(x_i) - f(x_i) = 0$. So we obtain $\tau_i = O(h^2)$. Clearly as $h \to 0$, $\tau_i \to 0$ and the discretized equation is consistent with the ODE.

Define the vector of global error $\mathbf{e}$ as $\mathbf{e} = \mathbf{u} - \hat{\mathbf{u}}$. Subtracting $A\hat{\mathbf{u}} = \mathbf{f}$ from $A\mathbf{u} = \mathbf{f} + \tau$, we obtain

$$A\mathbf{e} = \tau, \quad \mathbf{e} = A^{-1}\tau \tag{9.112}$$

9.2.1.2 Consistency, Stability and Convergence

We have derived that

$$\mathbf{e} = A^{-1}\tau \tag{9.113}$$

So we have

$$\|\mathbf{e}\|_p = \|A^{-1}\tau\|_p \le \|A^{-1}\|_p \|\tau\|_p \tag{9.114}$$

where $\|\cdot\|_p$ denotes the p-norm of a vector or matrix (see Chapter 4). We say the finite difference method for the BVP is stable if, for sufficiently small h, A^{-1} exists and satisfies

$$\|A^{-1}\|_p \le C \tag{9.115}$$

where C is a constant independent of h.

Obviously, if a finite difference method is consistent, i.e.

$$\|\tau\|_p \to 0 \tag{9.116}$$

as $h \to 0$ (which is the case for the above finite difference method in which $\|\tau\|_p = O(h^2)$), and is stable, i.e.

$$\|A^{-1}\|_p \le C \tag{9.117}$$

for sufficiently small h, then

$$\|e\|_p \le \|A^{-1}\|_p \|\tau\|_p \to 0 \tag{9.118}$$

i.e. $\hat{u} \to u$ as $h \to 0$. We say the method is convergent. To conclude, we have the following fundamental theorem of finite difference methods.

Theorem 9.2 (Lax Equivalence Theorem) *A consistent finite difference method for a well-posed linear boundary value problem is convergent if and only if it is stable.*

Remark The theorem can be concisely stated as

$$\text{consistency} + \text{stability} \Rightarrow \text{convergence}$$

9.3 Exercises

Exercise 9.1 Write the IVP

$$\theta''(t) = -\theta(t), \quad \theta(0) = 0, \quad \theta'(0) = 1$$

as an equivalent system of first-order differential equations.

Exercise 9.2 The true solution of the IVP

$$u'(t) = -u(t), \quad u(0) = 1$$

is $u(t) = e^{-t}$. Apply Euler's method to compute u_1 and u_2 with $\Delta t = 0.1$, where $u_1 \approx u(\Delta t)$ and $u_2 \approx u(2\Delta t)$. Compute the error in each approximation.

Exercise 9.3 The true solution of the IVP

$$u'(t) = -u(t), \quad u(0) = 1$$

is $u(t) = e^{-t}$. Apply implicit Euler's method to compute u_1 and u_2 with $\Delta t = 0.1$, where $u_1 \approx u(\Delta t)$ and $u_2 \approx u(2\Delta t)$. Compute the error in each approximation.

Exercise 9.4 The true solution of the IVP

$$u'(t) = \lambda u(t), \quad u(0) = u_0$$

is $u(t) = u_0 e^{\lambda t}$. If $\lambda < 0$, the solution decays with t. Show that the implicit Euler's method is absolutely stable for any value of $\Delta t > 0$, that is the approximate solution decays with t_n if $\lambda < 0$.

Exercise 9.5 For the IVP

$$u'(t) = f(u(t), t), \quad u(t_0) = u_0$$

We have the following Taylor series method

$$P_1 = f(u^{(n)}, t_n)$$

$$P_2 = \frac{\partial f(u^{(n)}, t_n)}{\partial u} P_1 + \frac{\partial f(u^{(n)}, t_n)}{\partial t}$$

$$u^{(n+1)} = u^{(n)} + P_1 \Delta t + \frac{P_2}{2} (\Delta t)^2$$

Consider the IVP

$$u'(t) = -u(t), \quad u(0) = 1$$

whose true solution is $u(t) = e^{-t}$. Apply the above Taylor series method to compute u_1 and u_2 with $\Delta t = 0.1$, where $u_1 \approx u(\Delta t)$ and $u_2 \approx u(2\Delta t)$. Compute the error in each approximation.

Exercise 9.6 Verify that the IVP

$$u'(t) = \sqrt[3]{u(t)}, \quad u(0) = 0$$

has two solutions $u(t) = 0$ and $u(t) = \sqrt{8t^3/27}$ for $t \geq 0$. What happens if Taylor series methods are applied?

Exercise 9.7 Consider the IVP

$$u'(t) = u(t), \quad u(0) = u_0$$

whose true solution is $u(t) = e^t$. If a roundoff error ε is introduced in u_0 such that $u(0) = u_0 + \varepsilon$, what is the error and relative error in the approximation at $t = 10$ computed by Euler's method with $\Delta t = 0.1$. What if the ODE is $u'(t) = -u(t)$?

Exercise 9.8 Apply the classical RK method of order 4 to solve the IVP

$$u'(t) = -t(u(t))^2, \quad u(0) = 2$$

for two steps with $\Delta t = 0.1$. Compute the error after each step. The true solution of the IVP is $u(t) = 2/(t^2 + 1)$.

Exercise 9.9 Show that the classical RK method of order 4 is reduced to a simple form if it is applied to and IVP with the ODE $u'(t) = f(t)$.

Exercise 9.10 The true solution of the IVP

$$u'(t) = \lambda u(t), \quad u(0) = u_0$$

is $u(t) = u_0 e^{\lambda t}$. If $\lambda < 0$, the solution decays with t. Let $z = \lambda \Delta t$, where $\lambda < 0$. Find the condition for z under which the classical RK method of order 4 is absolutely stable.

Exercise 9.11 Consider the BVP

$$u''(x) = 0, \quad u'(0) = \alpha, \quad u'(1) = \beta$$

The boundary conditions in this BVP give the values of the derivatives of the solution at the two boundaries. Such boundary conditions are called Neumann boundary conditions. (1) Show that the BVP has no solution if $\alpha \neq \beta$. (2) Show that the BVP has infinitely many solutions if $\alpha = \beta$.

Exercise 9.12 Consider the BVP

$$u''(x) = -\sin x, \quad u(0) = 0, \quad u(1) = 1$$

(1) What is the true solution of the BVP? (2) If the second-order centered finite difference approximation is used to solve the BVP on a uniform grid with space step h, what is the linear system $A\hat{u} = f$? (3) Solve the linear system when $h = 1/4$ to find the approximate discrete solution at the grid, and then compute the error at each grid point.

Exercise 9.13 Consider the BVP

$$u''(x) = (1+x)u(x), \quad u(0) = 0, \quad u(1) = 1$$

If the second-order centered finite difference approximation is used to solve the BVP on a uniform grid with space step h, what is the linear system $A\hat{u} = f$?

Exercise 9.14 Consider the BVP

$$u''(x) = -u(x), \quad u(0) = 0, \quad u(1) = 1$$

with the true solution $u(x) = \sin x / \sin 1$. (1) If the second-order centered finite difference approximation is used to solve the BVP on a uniform grid with space step $h = 1/4$, what is the linear system $A\hat{u} = f$? (2) Solve the linear system to find the approximate discrete solution on the grid and find the error at each grid point.

9.4 Programming Problems

Problem 9.1 The logistic differential equation

$$\frac{dP}{dt} = rP\left(1 - \frac{P}{K}\right) = rP - r\frac{P^2}{K}$$

is commonly used to model population growth, where $P(t)$ represents the population size at time t, the constant r defines the growth rate and the constant K is called the carrying capacity (the maximum population that can be sustained by a specific environment, as it can be shown below that the solution of the equation satisfies $\lim_{t\to\infty} P(t) = K$). In this equation, the early unimpeded growth rate is modeled by the term $+rP$. Later, as the population grows, the second term $-rP^2/K$ models the antagonistic effect due to the competition among the population for some critical resource such as food or living space. The competition diminishes the combined growth rate, until the population ceases to grow (this is called maturity of the population). The solution of the equation with the initial condition $P(0) = P_0$ is

$$P(t) = \frac{KP_0 e^{rt}}{K + P_0(e^{rt} - 1)} = \frac{K}{1 + \left(\frac{K - P_0}{P_0}\right)e^{-rt}}$$

Note that $\lim_{t\to\infty} P(t) = K$.

Consider the case in which $r = 1$, $K = 1$ and $P_0 = 1/2$. We have the following IVP

$$\frac{dP}{dt} = P(1 - P), \quad P(0) = 1/2$$

The true solution is the standard logistic function

$$P(t) = \frac{1}{1 + e^{-t}} = \frac{1}{2} + \frac{1}{2}\tanh\left(\frac{t}{2}\right)$$

Plot the true solution for $t \in [0, 10]$.

Apply the classical RK method of order 4 to solve the IVP for $0 \leq t \leq 10$ with $\Delta t = 0.1, 0.05, 0.025$ and 0.0125, respectively. Compute the error of the numerical solution at $t = 10$ for each value of Δt and denote the error as $E_{\Delta t}$. Plot $E_{\Delta t}$ versus Δt in log-log scale. Compute the ratio $E_{2\Delta t}/E_{\Delta t}$ for $\Delta t = 0.05, 0.025$ and 0.0125, respectively. Comment on your results.

Now change the method to Taylor series method of order 2 and redo the simulation.

Problem 9.2 Under certain conditions, the temperature of a fluid flowing through a straight pipe can be obtained by solving a two-point boundary value problem (BVP)

$$\begin{aligned}
u''(x) + \omega u'(x) + \gamma u(x) &= f(x), \quad x \in [0, 1] \\
u(0) = u(1) &= 0,
\end{aligned}$$

where $u(x)$ is the temperature distribution along the pipe, ω and γ are given constants and $f(x)$ is a given function.

We can solve the BVP numerically using the finite difference method on a uniform grid with the space step size $h = 1/n$ ($n + 1$ is the number of grid points) and the centered finite difference approximations,

$$u''(x_i) \approx \frac{u_{i+1} - 2u_i + u_{i-1}}{h^2}, \quad \text{and} \quad u'(x_i) \approx \frac{u_{i+1} - u_{i-1}}{2h},$$

Form the linear system $A\hat{u} = f$ for the BVP.

Verify that $u(x) = \sin(\pi x)$ is the true solution in the case that $\omega = \pi$, $\gamma = \pi^2$ and $f(x) = \pi^2 \cos(\pi x)$. Solve the linear system for this case to find the approximate solution of the BVP with $n = 10, 20, 40, 80$ and 160. Compute the error vector for each value of h and let E_h denote its maximum absolute entry. Plot E_h versus h in log-log scale. Compute the ratio E_{2h}/E_h for $n = 20, 40, 80$ and 160, respectively. Comment on your results.

10

Basic Iterative Methods for Linear Systems

In this chapter, we present some basic iterative methods for linear systems. The matrix and vector norms introduced for conditioning of linear systems are used here to discuss convergence of iterative methods.

Let's consider the BVP for the steady temperature distribution in a uniform thin square plate. The dimensions of the plate are described by $\{(x,y)|0 \le x \le 1, 0 \le y \le 1\}$, as shown in Fig. 10.1. The plate is heated by embedded heat source. The four edges of the plate are fixed to the constant temperature 0. At equilibrium, the steady temperature distribution $u(x,y)$ in the plate satisfies the following Poisson equation and homogeneous Dirichlet boundary conditions

$$\frac{\partial^2 u}{\partial x^2} + \frac{\partial^2 u}{\partial y^2} = -g(x,y); \quad 0 \le x \le 1, 0 \le y \le 1 \tag{10.1}$$

$$u(0,y) = u(1,y) = u(x,0) = u(x,1) = 0 \tag{10.2}$$

where $g(x,y)$ is proportional to the heat source intensity.

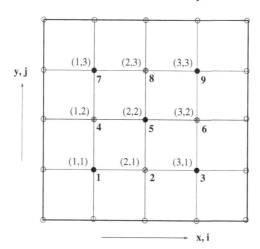

FIGURE 10.1

A square plate represented by discrete grid points. The boundary points are open circles. The interior points are solid or shaded circles.

DOI: 10.1201/9781003201694-10

We first cover the domain $[0,1] \times [0,1]$ of the equation using a grid with the $(m+1) \times (n+1)$ grid points (x_i, y_j) $(i = 0, 1, \ldots, m$ and $j = 0, 1, \ldots, n)$, where $0 = x_0 < x_1 < \cdots < x_{m-1} < x_m = 1$ and $0 = y_0 < y_1 < \cdots < y_{n-1} < y_n = 1$. We then seek a discrete solution $u_{i,j}$ of the problem by solving algebraic equations for $u_{i,j}$, where

$$u_{i,j} \approx u(x_i, y_j) \tag{10.3}$$

The algebraic equations for $u_{i,j}$ can be derived by numerical differentiation. For simplicity, below we consider a uniform grid $h = x_i - x_{i-1} = y_j - y_{j-1}$ with $m = n = 4$. Because the temperature at the four boundaries $(i = 0, 4$ or $j = 0, 4)$ is zero (the boundary conditions), we only need to find the approximate temperature at the interior grid points (i, j) $(i = 1, 2, 3$ and $j = 1, 2, 3)$. We can use centered finite difference approximations

$$\frac{\partial^2 u(x_i, y_j)}{\partial x^2} \approx \frac{u_{i-1,j} - 2u_{i,j} + u_{i+1,j}}{h^2} \tag{10.4}$$

$$\frac{\partial^2 u(x_i, y_j)}{\partial y^2} \approx \frac{u_{i,j-1} - 2u_{i,j} + u_{i,j+1}}{h^2} \tag{10.5}$$

each with the truncation errors of $O(h^2)$. Accordingly the Poisson equation is discretized as

$$-u_{i,j-1} - u_{i-1,j} + 4u_{i,j} - u_{i+1,j} - u_{i,j+1} = h^2 g(x_i, y_j), \quad i = 1, 2, 3; \ j = 1, 2, 3$$

$$\tag{10.6}$$

Note that $u_{i,j}$ at (i, j) is directly related with only the four neighboring (down, left, right and up) grid points.

If we order the 9 unknowns $u_{i,j}$ from 1 to 9 as shown in Fig. 10.1 and notice the zero temperature at the boundaries, we end up with the following linear system

$$
\begin{bmatrix}
4 & -1 &   & -1 &   &   &   &   &   \\
-1 & 4 & -1 &   & -1 &   &   &   &   \\
   & -1 & 4 &   &   & -1 &   &   &   \\
-1 &   &   & 4 & -1 &   & -1 &   &   \\
   & -1 &   & -1 & 4 & -1 &   & -1 &   \\
   &   & -1 &   & -1 & 4 &   &   & -1 \\
   &   &   & -1 &   &   & 4 & -1 &   \\
   &   &   &   & -1 &   & -1 & 4 & -1 \\
   &   &   &   &   & -1 &   & -1 & 4
\end{bmatrix}
\begin{bmatrix}
u_{1,1} \\ u_{2,1} \\ u_{3,1} \\ u_{1,2} \\ \mathbf{u_{2,2}} \\ u_{3,2} \\ u_{1,3} \\ u_{2,3} \\ u_{3,3}
\end{bmatrix}
= h^2
\begin{bmatrix}
g(x_1, y_1) \\ g(x_2, y_1) \\ g(x_3, y_1) \\ g(x_1, y_2) \\ g(x_2, y_2) \\ g(x_3, y_2) \\ g(x_1, y_3) \\ g(x_2, y_3) \\ g(x_3, y_3)
\end{bmatrix}
\tag{10.7}
$$

where zero entries in the coefficient matrix are left blank. In the middle row, the five nonzero entries are associated with five neighboring grid points, and the diagonal entry is associated with the center grid point. With the current ordering of unknowns, the nonzero entries associated with the down and left grid points are below the diagonal while the nonzero entries associated with the right and up grid points are above the diagonal.

If we use more and more grid points, the coefficient matrix has the similar structure (with a large order) as above but becomes sparser and sparser (fewer and fewer nonzero entries). For example, if $m = n = 100$, the $10^4 \times 10^4$ matrix has 10^8 entries, of which only about 5×10^4 entries or 0.05% are nonzero. When such sparse linear systems are solved by a direct method such as the PA=LU decomposition (see Chapter 4), there may be many fill-ins that fill the nonzero entries in L and U at the locations where a coefficient matrix has zeros. The large number of fill-ins make a direct method costly in both memory storage and computational time. So we need iterative methods to solve such linear systems. An iterative method generates a sequence of approximations of the solution of a linear system.

10.1 Jacobi and Gauss-Seidel Methods

When solving a univariate nonlinear equation $f(x) = 0$ (see Chapter 5), we may convert the equation $f(x) = 0$ to the form $x = g(x)$ to construct a fixed point iteration

$$x_{k+1} = g(x_k), \quad k = 0, 1, 2, \ldots \tag{10.8}$$

Similarly, we may convert a linear system $A\mathbf{x} - \mathbf{b} = \mathbf{0}$ to the form $\mathbf{x} = B\mathbf{x} + \mathbf{d}$ to construct a **stationary iteration**

$$\mathbf{x}_{k+1} = B\mathbf{x}_k + \mathbf{d}, \quad k = 0, 1, 2, \ldots \tag{10.9}$$

where B is called the **iteration matrix**. If the iteration converges, that is

$$\lim_{k \to \infty} x_k = \mathbf{x}^* \tag{10.10}$$

for any given initial $\mathbf{x}_0$, then $\mathbf{x}^*$ satisfies $\mathbf{x}^* = B\mathbf{x}^* + \mathbf{d}$ and is the solution of $A\mathbf{x} = \mathbf{b}$. We can split the coefficient matrix A as $A = M - N$, then $A\mathbf{x} = \mathbf{b}$ is equivalent to

$$\mathbf{x} = M^{-1}N\mathbf{x} + M^{-1}\mathbf{b} \tag{10.11}$$

which is of the form $\mathbf{x} = B\mathbf{x} + \mathbf{d}$ with $B = M^{-1}N$ and $\mathbf{d} = M^{-1}\mathbf{b}$. The matrix M should be chosen such that M^{-1} mimics A^{-1} but can be found easily, which seems two contradictory requirements and leaves lots of room to play.

Below we introduce some simple iterative methods for $A\mathbf{x} = \mathbf{b}$ due to simple splittings of A, where

$$A = \begin{bmatrix} a_{11} & a_{12} & a_{13} & \cdots & a_{1n} \\ a_{21} & a_{22} & a_{23} & \cdots & a_{2n} \\ a_{31} & a_{32} & a_{33} & \cdots & a_{3n} \\ \vdots & \vdots & \vdots & & \vdots \\ a_{n1} & a_{n2} & a_{n3} & \cdots & a_{nn} \end{bmatrix}, \quad \mathbf{x} = \begin{bmatrix} x_1 \\ x_2 \\ x_3 \\ \vdots \\ x_n \end{bmatrix}, \quad \mathbf{b} = \begin{bmatrix} b_1 \\ b_2 \\ b_3 \\ \vdots \\ b_n \end{bmatrix} \quad (10.12)$$

We can write A as

$$A = D - L - U \tag{10.13}$$

where

$$D = \begin{bmatrix} a_{11} & & & & \\ & a_{22} & & & \\ & & a_{33} & & \\ & & & \ddots & \\ & & & & a_{nn} \end{bmatrix} \tag{10.14}$$

$$L = \begin{bmatrix} 0 & & & & \\ -a_{21} & 0 & & & \\ -a_{31} & -a_{32} & 0 & & \\ \vdots & \vdots & & \ddots & \\ -a_{n1} & -a_{n2} & -a_{n3} & \cdots & 0 \end{bmatrix} \tag{10.15}$$

$$U = \begin{bmatrix} 0 & -a_{12} & -a_{13} & \cdots & -a_{1n} \\ & 0 & -a_{23} & \cdots & -a_{2n} \\ & & 0 & \cdots & -a_{3n} \\ & & & \ddots & \vdots \\ & & & & 0 \end{bmatrix} \tag{10.16}$$

i.e. D is from the diagonal part of A, $-L$ from the lower triangular part of A and $-U$ from the upper triangular part of A.

10.1.1 Jacobi Method

We have $A = D - L - U$. If we split A as $A = M - N$ with $M = D$ and $N = L + U$, then $A\mathbf{x} = \mathbf{b}$ can be written as

$$\mathbf{x} = D^{-1}(L + U)\mathbf{x} + D^{-1}\mathbf{b} \tag{10.17}$$

We obtain Jacobi iteration (simultaneous relaxation)

$$\mathbf{x}_{k+1} = D^{-1}(L + U)\mathbf{x}_k + D^{-1}\mathbf{b} \tag{10.18}$$

which can be written as

$$\mathbf{x}_{k+1} = D^{-1}[\mathbf{b} + (L+U)\mathbf{x}_k] \tag{10.19}$$

that is

$$
\begin{bmatrix} x_1^{(k+1)} \\ x_2^{(k+1)} \\ x_3^{(k+1)} \\ \vdots \\ x_n^{(k+1)} \end{bmatrix} =
\begin{bmatrix} a_{11}^{-1} & & & & \\ & a_{22}^{-1} & & & \\ & & a_{33}^{-1} & & \\ & & & \ddots & \\ & & & & a_{nn}^{-1} \end{bmatrix}
$$

$$
\left(\begin{bmatrix} b_1 \\ b_2 \\ b_3 \\ \vdots \\ b_n \end{bmatrix} -
\begin{bmatrix} 0 & a_{12} & a_{13} & \cdots & a_{1n} \\ a_{21} & 0 & a_{23} & \cdots & a_{2n} \\ a_{31} & a_{32} & 0 & \cdots & a_{3n} \\ \vdots & \vdots & \vdots & & \vdots \\ a_{n1} & a_{n2} & a_{n3} & \cdots & 0 \end{bmatrix}
\begin{bmatrix} x_1^{(k)} \\ x_2^{(k)} \\ x_3^{(k)} \\ \vdots \\ x_n^{(k)} \end{bmatrix} \right) \tag{10.20}
$$

In the component form, the iteration is

$$x_i^{(k+1)} = \frac{1}{a_{ii}} \left(b_i - \sum_{j \neq i} a_{ij} x_j^{(k)} \right), \quad i = 1, 2, \ldots, n \tag{10.21}$$

Note that the iteration formulas for all components at each iteration step are not coupled. They can be applied in any order or even simultaneously (in parallel). The Jacobi method is therefore also called **simultaneous relaxation**. The Gauss-Seidel method (introduced soon) is different, in which the order to sweep the values of i matters.

Example 10.1 Let's consider the linear system

$$
\begin{aligned}
10x_1 + 2x_2 + 3x_3 &= 10 \\
4x_1 + 20x_2 + 8x_3 &= 30 \\
8x_1 + 6x_2 + 30x_3 &= 40
\end{aligned}
$$

Even though we never solve a 3×3 dense linear system using an iterative method in practice, we apply the Jacobi method/iteration here to solve the above system for demonstration. The iteration reads

$$x_1^{(k+1)} = \frac{1}{10}(10 - 2x_2^{(k)} - 3x_3^{(k)})$$

$$x_2^{(k+1)} = \frac{1}{20}(30 - 4x_1^{(k)} - 8x_3^{(k)})$$

$$x_3^{(k+1)} = \frac{1}{30}(40 - 8x_1^{(k)} - 6x_2^{(k)})$$

where $k = 0, 1, 2, \ldots$. Let $\mathbf{x}^{(k)} = [x_1^{(k)}, x_2^{(k)}, x_3^{(k)}]^T$. If we start with $\mathbf{x}^{(0)} = [0, 0, 0]^T$, we have

$$\mathbf{x}^{(1)} = [1, 3/2, 4/3]^T$$

$$\mathbf{x}^{(2)} \approx [0.3, 0.76667, 0.76667]^T$$

$$\mathbf{x}^{(3)} \approx [0.61667, 1.13333, 1.10000]^T$$

$$\mathbf{x}^{(4)} \approx [0.44333, 0.93667, 0.94222]^T$$

$$\mathbf{x}^{(5)} \approx [0.53000, 1.03444, 1.02778]^T$$

$$\vdots$$

The iteration seems to converge to the true solution $\mathbf{x}^* = [0.5, 1, 1]^T$. ∎

Remark The discretized Poisson equation in the motivating example at the beginning of this chapter reads

$$-u_{i,j-1} - u_{i-1,j} + 4u_{i,j} - u_{i+1,j} - u_{i,j+1} = h^2 g(x_i, y_j) \qquad (10.22)$$

which forms a linear system of order n^2 (with n^2 unknowns) as $i = 1, 2, \ldots, n$ and $j = 1, 2, \ldots, n$. Note that the diagonal entry associated with the unknown $u_{i,j}$ is 4 in the coefficient matrix. The Jacobi iteration for the linear system in the component form is simply

$$u_{i,j}^{(k+1)} = \frac{1}{4}[h^2 g(x_i, y_j) + u_{i,j-1}^{(k)} + u_{i-1,j}^{(k)} + u_{i+1,j}^{(k)} + u_{i,j+1}^{(k)}] \qquad (10.23)$$

where i and j can take $1, 2, \ldots, n$ in any order or in parallel to sweep through the entire grid.

The Jacobi iteration

$$\mathbf{x}_{k+1} = D^{-1}[\mathbf{b} + (L+U)\mathbf{x}_k] \qquad (10.24)$$

can be written in the form of (by $L + U = D - A$)

$$\mathbf{x}_{k+1} = \mathbf{x}_k + D^{-1}\mathbf{r}_k \qquad (10.25)$$

where $\mathbf{r}_k = \mathbf{b} - A\mathbf{x}_k$ is the residual vector for the approximation $\mathbf{x}_k$. So the Jacobi method is a **residual correction method** (to correct $\mathbf{x}_k$ using the residual $\mathbf{r}_k$). To apply this iteration formula, we need to compute the matrix-vector product $A\mathbf{x}_k$ to find $\mathbf{r}_k$. When A is a sparse matrix with few nonzero entries, $A\mathbf{x}_k$ can be computed very economically.

10.1.2 Gauss-Seidel (G-S) Method

We have $A = D - L - U$. If we split A as $A = M - N$ with $M = D - L$ and $N = U$, then $A\mathbf{x} = \mathbf{b}$ can be written as

$$\mathbf{x} = (D - L)^{-1}U\mathbf{x} + (D - L)^{-1}\mathbf{b} \tag{10.26}$$

We obtain G-S iteration (relaxation)

$$\mathbf{x}_{k+1} = (D - L)^{-1}U\mathbf{x}_k + (D - L)^{-1}\mathbf{b} \tag{10.27}$$

which can be written as

$$D\mathbf{x}_{k+1} = \mathbf{b} + L\mathbf{x}_{k+1} + U\mathbf{x}_k \tag{10.28}$$

or

$$
\begin{bmatrix} x_1^{(k+1)} \\ x_2^{(k+1)} \\ x_3^{(k+1)} \\ \vdots \\ x_n^{(k+1)} \end{bmatrix} =
\begin{bmatrix} a_{11}^{-1} & & & \\ & a_{22}^{-1} & & \\ & & a_{33}^{-1} & \\ & & & \ddots & \\ & & & & a_{nn}^{-1} \end{bmatrix}
\left(\begin{bmatrix} b_1 \\ b_2 \\ b_3 \\ \vdots \\ b_n \end{bmatrix} \right.
$$

$$
- \begin{bmatrix} 0 & & & & \\ a_{21} & 0 & & & \\ a_{31} & a_{32} & 0 & & \\ \vdots & \vdots & \vdots & \ddots & \\ a_{n1} & a_{n2} & a_{n3} & \cdots & 0 \end{bmatrix}
\begin{bmatrix} x_1^{(k+1)} \\ x_2^{(k+1)} \\ x_3^{(k+1)} \\ \vdots \\ x_n^{(k+1)} \end{bmatrix}
- \begin{bmatrix} 0 & a_{12} & a_{13} & \cdots & a_{1n} \\ & 0 & a_{23} & \cdots & a_{2n} \\ & & 0 & \cdots & a_{3n} \\ & & & \ddots & \vdots \\ & & & & 0 \end{bmatrix}
\begin{bmatrix} x_1^{(k)} \\ x_2^{(k)} \\ x_3^{(k)} \\ \vdots \\ x_n^{(k)} \end{bmatrix} \left. \right)
$$

$$\tag{10.29}$$

In the component form, the iteration is

$$x_i^{(k+1)} = \frac{1}{a_{ii}} \left(b_i - \underbrace{\sum_{j<i} a_{ij}x_j^{(k+1)}}_{\text{from lower part of A}} - \underbrace{\sum_{j>i} a_{ij}x_j^{(k)}}_{\text{from upper part of A}} \right), \quad i = 1, 2, \ldots, n \tag{10.30}$$

Note that the order to sweep the values of i matters as the right-hand side of the iteration formula involves the newly updated values $x_j^{(k+1)}$ (where $j < i$), which needs to be available for computing the left-hand side $x_i^{(k+1)}$. So the value of i must sweep from the smallest 1 toward the largest n.

Example 10.2 Let's consider the linear system

$$10x_1 + 2x_2 + 3x_3 = 10$$
$$4x_1 + 20x_2 + 8x_3 = 30$$
$$8x_1 + 6x_2 + 30x_3 = 40$$

We apply the G-S method/iteration here to solve the above system for demonstration. The iteration in the correct order reads

$$x_1^{(k+1)} = \frac{1}{10}(10 - 2x_2^{(k)} - 3x_3^{(k)})$$

$$x_2^{(k+1)} = \frac{1}{20}(30 - 4x_1^{(k+1)} - 8x_3^{(k)})$$

$$x_3^{(k+1)} = \frac{1}{30}(40 - 8x_1^{(k+1)} - 6x_2^{(k+1)})$$

where $k = 0, 1, 2, \ldots$. Let $\mathbf{x}^{(k)} = [x_1^{(k)}, x_2^{(k)}, x_3^{(k)}]^T$. If we start with $\mathbf{x}^{(0)} = [0, 0, 0]^T$, we have

$$\mathbf{x}^{(1)} \approx [1.00000, 1.30000, 0.80667]^T$$

$$\mathbf{x}^{(2)} \approx [0.49800, 0.88040, 1.02445]^T$$

$$\mathbf{x}^{(3)} \approx [0.51658, 1.04452, 0.98667]^T$$

$$\mathbf{x}^{(4)} \approx [0.49509, 0.98317, 1.00467]^T$$

$$\mathbf{x}^{(5)} \approx [0.50196, 1.00634, 0.99821]^T$$

$$\vdots$$

The iteration seems to converge to the true solution $\mathbf{x}^* = [0.5, 1, 1]^T$. ∎

Example 10.3 Let's apply the G-S method/iteration the following linear system

$$x_1 + x_2 + x_3 = 3$$
$$x_1 + x_2 - x_3 = 1$$
$$x_1 - x_2 + x_3 = 1$$

The iteration in the correct order reads

$$x_1^{(k+1)} = 3 - x_2^{(k)} - x_3^{(k)}$$

$$x_2^{(k+1)} = 1 - x_1^{(k+1)} + x_3^{(k)}$$

$$x_3^{(k+1)} = 1 - x_1^{(k+1)} + x_2^{(k+1)}$$

where $k = 0, 1, 2, \ldots$. Let $\mathbf{x}^{(k)} = [x_1^{(k)}, x_2^{(k)}, x_3^{(k)}]^T$. If we start with $\mathbf{x}^{(0)} = [0, 0, 0]^T$, we have

$$\mathbf{x}^{(1)} = [3, -2, -4]^T$$

$$\mathbf{x}^{(2)} = [2, -5, -6]^T$$

$$\mathbf{x}^{(3)} = [6, -11, -16]^T$$

$$\mathbf{x}^{(4)} = [8, -23, -30]^T$$

$$\mathbf{x}^{(5)} = [18, -47, -64]^T$$

$$\vdots$$

The iteration does not seem to converge to the true solution $\mathbf{x}^* = [1, 1, 1]^T$. ∎

Remark The discretized Poisson equation in the motivating example at the beginning of this chapter reads

$$-u_{i,j-1} - u_{i-1,j} + 4u_{i,j} - u_{i+1,j} - u_{i,j+1} = h^2 g(x_i, y_j) \qquad (10.31)$$

which forms a linear system of order n^2 (with n^2 unknowns) when $i = 1, 2, \ldots, n$ and $j = 1, 2, \ldots, n$. Note that the diagonal entry associated with the unknown $u_{i,j}$ is 4 in the coefficient matrix. The G-S iteration for the linear system in the component form is simply

$$u_{i,j}^{(k+1)} = \frac{1}{4} \left[h^2 g(x_i, y_j) + \underbrace{u_{i,j-1}^{(k+1)} + u_{i-1,j}^{(k+1)}}_{\text{down and left: below diagonal}} + \underbrace{u_{i+1,j}^{(k)} + u_{i,j+1}^{(k)}}_{\text{right and up: above diagonal}} \right]$$

$$(10.32)$$

where j increases from 1 toward n, and for each value of j, i increases from 1 toward n to sweep through the entire grid.

Below is the pseudocode of the G-S iteration with zero initial guess for the discretized Poisson equation in the motivating example at the beginning of this chapter.

The G-S iteration

$$\mathbf{x}_{k+1} = (D - L)^{-1} U \mathbf{x}_k + (D - L)^{-1} \mathbf{b} \qquad (10.33)$$

can be written as (by $U = (D - L) - A$)

$$\mathbf{x}_{k+1} = \mathbf{x}_k + (D - L)^{-1} \mathbf{r}_k \qquad (10.34)$$

where $\mathbf{r}_k = \mathbf{b} - A\mathbf{x}_k$ is the residual vector for the approximation $\mathbf{x}_k$. So the G-S method is also a residual correction method.

Algorithm 19 The G-S method for a Poisson equation

---

for j from 0 to $n+1$ **do**
  for i from 0 to $n+1$ **do**
    $u_{ij} \leftarrow 0$
  end for
end for
for j from 1 to n **do**
  for i from 1 to n **do**
    $u_{ij} \leftarrow (1/4)(h^2 g_{ij} + u_{i,j-1} + u_{i-1,j} + u_{i+1,j} + u_{i,j+1})$
  end for
end for

10.2 Convergence Analysis

An iterative method may or may not converge to the desired true solution, as exampled in the previous section. Consider a stationary iteration such as the Jacobi iteration and the G-S iteration

$$\mathbf{x}_{k+1} = B\mathbf{x}_k + \mathbf{d}, \quad k = 0, 1, 2, \ldots \tag{10.35}$$

Note that the iteration matrix $B = D^{-1}(L+U)$ for the Jacobi iteration and $B = (D-L)^{-1}U$ for the G-S iteration. Below we establish the condition under which the stationary iteration converges to the true solution $\mathbf{x}^*$ of the linear system $\mathbf{x} = B\mathbf{x} + \mathbf{d}$ (which is equivalent to $A\mathbf{x} = \mathbf{b}$). As usual, the convergence analysis establishes the relation between the error of the next approximation with the errors of the current and previous approximations to find out the conditions under which the errors eventually shrink to zero.

The stationary iteration

$$\mathbf{x}_{k+1} = B\mathbf{x}_k + \mathbf{d} \tag{10.36}$$

relates two consecutive approximations of the true solution $\mathbf{x}^*$. The true solution $\mathbf{x}^*$ satisfies

$$\mathbf{x}^* = B\mathbf{x}^* + \mathbf{d} \tag{10.37}$$

Subtracting the above two equations gives the error relation

$$\mathbf{e}_{k+1} = B\mathbf{e}_k, \quad k = 0, 1, 2, \ldots \tag{10.38}$$

where $\mathbf{e}_{k+1} = \mathbf{x}^* - \mathbf{x}_{k+1}$ and $\mathbf{e}_k = \mathbf{x}^* - \mathbf{x}_k$ are the error vectors. We therefore have

$$\mathbf{e}_{k+1} = B\mathbf{e}_k = B(B\mathbf{e}_{k-1}) = \cdots = B^{k+1}\mathbf{e}_0 \tag{10.39}$$

where $\mathbf{e}_0 = \mathbf{x}^* - \mathbf{x}_0$ is the error vector of the initial guess $\mathbf{x}_0$. The convergence of the iteration is determined by the iteration matrix B.

Theorem 10.1 (Stationary Iteration Theorem) *The stationary iteration*

$$\mathbf{x}_{k+1} = B\mathbf{x}_k + \mathbf{d}, \quad k = 0, 1, 2, \ldots$$

converges to the solution $\mathbf{x}^*$ *of the linear system* $\mathbf{x} = B\mathbf{x} + \mathbf{d}$ *for any initial guess* $\mathbf{x}_0$ *if*

$$\|B\|_p < 1$$

where $\|B\|_p$ *is a p-norm of the matrix B.*

Proof *The proof of the theorem is based on the error relation*

$$\mathbf{e}_{k+1} = B^{k+1}\mathbf{e}_0 \tag{10.40}$$

We have

$$\|\mathbf{e}_{k+1}\|_p = \|B^{k+1}\mathbf{e}_0\|_p \leq \|B\|_p \|B^k\mathbf{e}_0\|_p \leq \cdots \leq \|B_p\|^{k+1}\|\mathbf{e}_0\|_p \tag{10.41}$$

If $\|B\|_p < 1$, *as* $k \to \infty$, $\|B_p\|^{k+1}\|\mathbf{e}_0\|_p \to 0$ *and therefore* $\|\mathbf{e}_{k+1}\|_p \to 0$. ∎

Remark The condition $\|B\|_p < 1$ in the stationary iteration theorem is a sufficient condition for the convergence, but it is not a necessary condition. A sufficient and necessary condition for the convergence is $\rho(B) < 1$, where $\rho(B)$ is called the **spectral radius** of B and is defined as the largest magnitude of all the eigenvalues of B, i.e.

$$\rho(B) = \max_{1 \leq i \leq n} |\lambda_i(B)| \tag{10.42}$$

where $\lambda_i(B)$ is the i-th eigenvalue of the $n \times n$ matrix B. If $B = P\Lambda P^{-1}$ is diagonalizable with $\rho(B) < 1$, then $B^{k+1} = P\Lambda^{k+1}P^{-1} \to 0$ as $k \to \infty$ and the convergence is obvious. If B is not diagonalizable, we may use the Jordan canonical form of B to prove the result.

By definition of $\|B\|_p$, $\|B\|_p \geq \|B\mathbf{v}\|_p$ for any unit vector $\mathbf{v}$. If we choose $\mathbf{v}$ as a unit eigenvector $\mathbf{v}_i$ of B, we have $\|B\|_p \geq \|B\mathbf{v}_i\|_p = \|\lambda_i(B)\mathbf{v}_i\|_p = |\lambda_i(B)|$. We thus have

$$\|B\|_p \geq \rho(B) \tag{10.43}$$

So the sufficient and necessary condition $\rho(B) < 1$ for convergence is weaker than the condition $\|B\|_p < 1$.

Remark The $n \times n$ matrix A (with entries a_{ij}) is called a strictly **diagonally dominant** matrix if

$$|a_{ii}| > \sum_{j \neq i} |a_{ij}|, \quad i = 1, 2, \ldots, n \tag{10.44}$$

If A has strict diagonal dominance, then A is nonsingular (as $Ax = 0$ cannot have a nonzero solution), and the Jacobi and G-S methods converge (as the iteration matrix B of each method satisfies $\rho(B) < 1$) for the linear system $Ax = b$.

Since $\|e_{k+1}\|_p \leq \|B_p\|^{k+1} \|e_0\|_p$, the speed of convergence depends on $\|B\|_p$. The smaller $\|B\|_p$ is, the faster the convergence is.

Remark The smaller the spectral radius $\rho(B)$ is, the faster the convergence is.

Lastly, we need to know when to terminate the Jacobi or G-S iteration. By $e_{k+1} = Be_k$, we have

$$\|x^* - x_{k+1}\|_p \leq \|B\|_p \|x^* - x_k\|_p \leq \|B\|_p \|x^* - x_{k+1} + x_{k+1} - x_k\|_p$$

$$\leq \|B\|_p (\|x^* - x_{k+1}\|_p + \|x_{k+1} - x_k\|_p) \Rightarrow$$

$$\|x^* - x_{k+1}\|_p \leq \frac{\|B\|_p}{1 - \|B\|_p} \|x_{k+1} - x_k\|_p \tag{10.45}$$

Given the error tolerance τ, we can terminate the iteration when the upper bound for the error $e_{k+1} = x^* - x_{k+1}$ satisfies

$$\frac{\|B\|_p}{1 - \|B\|_p} \|x_{k+1} - x_k\|_p < \tau \tag{10.46}$$

For simplicity, we may terminate the iteration when

$$\|x_{k+1} - x_k\|_p < \tau \tag{10.47}$$

10.3   Other Iterative Methods

There are many other advanced iterative methods for solving sparse linear systems. Here is a list of a few names: Multigrid Method (MG), Conjugate Gradient (CG) and Generalized Minimal Residual Methods (GMRES). These methods are widely used in practical applications.

10.4   Exercises

Exercise 10.1 Consider the linear system

$$5x_1 + x_2 = 0$$
$$x_1 + 5x_2 = 1$$

The true solution is $\mathbf{x}^* = [x_1^*, x_2^*]^T = [-1/24, 5/24]^T$. Apply the Jacobi method and the G-S method to the system with $\mathbf{x}^{(0)} = [0,0]^T$ until $\|\mathbf{e}_k\|_\infty = \|\mathbf{x}^* - \mathbf{x}^{(k)}\|_\infty \le 0.05$. How many iterations are used in each method? Which method converges faster?

Now we interchange the two equations and consider the following equivalent linear system

$$x_1 + 5x_2 = 1$$
$$5x_1 + x_2 = 0$$

Apply the Jacobi method and the G-S method with $\mathbf{x}^{(0)} = [0,0]^T$ to the system for 5 iteration steps. Which method diverges more rapidly?

$$\min_{\mathbf{x}} \|\mathbf{b} - A\mathbf{x}\|_2^2$$

Exercise 10.2 Consider the linear system $A\mathbf{x} = \mathbf{b}$

$$5x_1 + x_2 = 0$$
$$x_1 + 5x_2 = 1$$

(1) What are the matrices D, L and U in the splitting of A as $A = D - L - U$? What is the iteration matrix $B = D^{-1}(L+U)$ for the Jacobi iteration? Find $\|B\|_p$ using the definition of the norm

$$\|B\|_p = \max_{\mathbf{v} \neq 0} \frac{\|B\mathbf{v}\|_p}{\|\mathbf{v}\|_p}$$

Is the Jacobi method convergent?

(2) Answer the above questions if the linear system is changed to

$$x_1 + 5x_2 = 1$$
$$5x_1 + x_2 = 0$$

Exercise 10.3 Conduct two steps of the Jacobi method and G-S method with a zero initial vector for each of the linear system $A\mathbf{x} = \mathbf{b}$.

(1)

$$A = \begin{bmatrix} 2 & -1 & 0 \\ -1 & 3 & -1 \\ 0 & -1 & 2 \end{bmatrix}, \quad \mathbf{b} = \begin{bmatrix} 1 \\ 8 \\ -5 \end{bmatrix}$$

The true solution of the system is $\mathbf{x}^* = [x_1^*, x_2^*, x_3^*]^T = [2, 3, -1]^T$.

(2)

$$A = \begin{bmatrix} 2 & 2 & -4 \\ 1 & 1 & 5 \\ 1 & 3 & 6 \end{bmatrix}, \quad \mathbf{b} = \begin{bmatrix} 10 \\ -2 \\ -5 \end{bmatrix}$$

The true solution of the system is $\mathbf{x}^* = [x_1^*, x_2^*, x_3^*]^T = [4, -1, -1]^T$.

Exercise 10.4 Consider the linear system $A\mathbf{x} = \mathbf{b}$, where

$$A = \begin{bmatrix} 2 & -1 & 0 \\ -1 & 3 & -1 \\ 0 & -1 & 2 \end{bmatrix}, \quad \mathbf{b} = \begin{bmatrix} 1 \\ 8 \\ -5 \end{bmatrix}$$

The true solution of the system is $\mathbf{x}^* = [x_1^*, x_2^*, x_3^*]^T = [2, 3, -1]^T$. Is the coefficient matrix A strictly diagonally dominant? Are the Jacobi method and G-S method convergent for the linear system?

Exercise 10.5 Suppose the coefficient matrix A of the linear system $A\mathbf{x} = \mathbf{b}$ is split as $A = M - N$ to obtain an iterative method

$$M\mathbf{x}_{k+1} = N\mathbf{x}_k + \mathbf{b}, \quad k = 0, 1, 2, \ldots$$

Show that the iteration matrix B of the method is $B = I - M^{-1}A$. What is the condition to guarantee the convergence of the method for any starting $\mathbf{x}_0$?

10.5 Programming Problems

Problem 10.1 Consider heat conduction in a square plate of size 1×1. Suppose the four boundaries of the plate are fixed to zero temperature, and the plate is subject to distributed heating and cooling. In a rectangular coordinate system, the steady temperature distribution $T(x, y)$ on the plate is governed by the following Poisson equation

$$\frac{\partial^2 T}{\partial x^2} + \frac{\partial^2 T}{\partial y^2} = -f(x, y)$$

where $f(x, y)$ is proportional to the intensity of the distributed heating and cooling.

As shown in the motivation example at the beginning of this chapter, we can solve the above Poisson equation numerically. We use a uniform grid of $(n+1) \times (n+1)$ grid points with the space step size $h = 1/n$.

Since the temperature at the boundary is known to be zero, we solve for the temperature at interior grid points (i, j) $(i, j = 1, 2, \ldots, n-1)$. Using the finite difference method, we can obtain the following linear equations for the approximate temperature T_{ij} $(i, j = 1, 2, \ldots, n-1)$

$$-T_{i-1,j} - T_{i,j-1} + 4T_{i,j} - T_{i+1,j} - T_{i,j+1} = h^2 F_{i,j}$$

where $F_{i,j}$ is the value of the heating/cooling function $f(x,y)$ at the grid point (i, j).
Let the heating/cooling function be

$$f(x,y) = 8\pi^2 \sin(2\pi x) \sin(2\pi y)$$

The exact solution corresponding to this $f(x,y)$ is

$$T(x,y) = \sin(2\pi x) \sin(2\pi y)$$

Now use the G-S method shown in Algorithm 19 to solve all the linear equations for $n = 10, 20, 40, 80$. Stop the iteration when the 2-norm of the error in the computed solution is less than 10^{-6}. Print the number of iterations for each value of n and discuss your results.

11

Discrete Least Squares Problems

In this chapter, we start with the definition of a linear discrete least squares (LS) problem to set up its matrix-vector form. We then derive the normal equation of the problem using both calculus and linear algebra approaches. We then present how to solve a linear discrete LS problem using the QR decomposition. Finally, we describe artificial neural networks from the perspective of nonlinear discrete LS problems.

A high-speed flow over a solid wall (for example, the air flow over a wing of a high-speed aircraft) establishes a very thin turbulent flow layer attached to the wall, called the turbulent boundary layer, in which the mean velocity of the flow undergoes a sharp change within a short distance from the wall. Within the turbulent boundary, there is a logarithmic region in which the mean flow velocity u^+ at the distance y^+ from the surface satisfies the log law

$$u^+ = \frac{1}{\kappa}\ln y^+ + C^+ \tag{11.1}$$

where both u^+ and y^+ are properly nondimensionalized by local flow parameters, and κ is called von Karman constant.

Suppose a fluid dynamicist wants to determine the values of κ and C^+ by experiments. He/she measures u^+ at different y^+ and obtains the data shown in Fig. 11.1. Ideally the n data points $(\ln y_i^+, u_i^+)$, $i = 1, 2, \ldots, n$, should fall on a straight line to satisfy the log law such that

$$\begin{bmatrix} 1 & \ln y_1^+ \\ 1 & \ln y_2^+ \\ \vdots & \vdots \\ 1 & \ln y_n^+ \end{bmatrix} \begin{bmatrix} 1/\kappa \\ C^+ \end{bmatrix} = \begin{bmatrix} u_1^+ \\ u_2^+ \\ \vdots \\ u_n^+ \end{bmatrix} \tag{11.2}$$

which is an over-determined linear system $A\mathbf{x} = \mathbf{b}$ with a very thin coefficient matrix (very thin if $n \gg 2$, the number of the data is much larger than the number of the unknown parameters). However, because of experimental errors (and the modeling error in the log law), the n data points $(\ln y_i^+, u_i^+)$, $i = 1, 2, \ldots, n$, do not fall on a line, and the over-determined linear system is not consistent. So he/she needs to determine a line that can fit the data. By eyeball, the dashed lines in Fig. 11.1 are clearly bad choices while the solid lines seem reasonable.

To narrow down to a particular line, he/she requires that the line should fit the data best in some criterion. One criterion is that the sum of the squares of the vertical

DOI: 10.1201/9781003201694-11

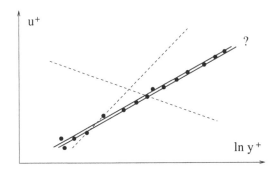

FIGURE 11.1
The experimental data shown as black dots for a turbulent boundary layer.

distances from all data points to the line is the least. This curve fitting is therefore called the **least squares (LS) fit**, and the LS fit minimizes

$$\sum_{i=1}^{n} |u_i^+ - u^+(y_i^+)|^2 \tag{11.3}$$

where $u^+(y_i^+) = \frac{1}{\kappa}\ln y_i^+ + C^+$ is the predicted value at y_i^+ from the curve/law/function, which is the i-th row of the left-hand side vector $A\mathbf{x}$ of the linear system above. Note that $u_i^+ - u^+(y_i^+)$ is the i-th row of the residual vector $\mathbf{r} = \mathbf{b} - A\mathbf{x}$. So the LS fit finds the solution $\mathbf{x} = [1/\kappa, C^+]^T$ to minimize the 2-norm of the residual vector $\mathbf{r}$:

$$\min_{\mathbf{x}} ||\mathbf{b} - A\mathbf{x}||_2^2$$

The solution of the minimization problem (the LS problem) is called the **LS solution**.

11.1   The Discrete LS Problems

Given n discrete data points (t_i, y_i), $i = 1, 2, \ldots, n$

t_i	t_1	t_2	$\cdots$	t_n
y_i	y_1	y_2	$\cdots$	y_n

TABLE 11.1
Data for the LS fit.

we seek a function $y = f(t)$ to fit the data in the LS sense such that the sum of the squares $\sum_{i=1}^{n} |y_i - f(t_i)|^2$ is the least (Table 11.1). The problem is called a **discrete**

LS problem. Note that the index i starts from 1 (instead of 0), and the independent variable is denoted as t (instead of x) for the convenience of the matrix-vector set-up of the problem below. In addition, the values t_i $(i = 1, 2, \ldots, n)$ may not be distinct (for example two different measurements at the same value of t in an experiment), which is in contrast with interpolation.

If the function $f(t)$ is constructed as a linear combination of m given basis functions $\phi_1(t), \phi_2(t), \ldots, \phi_m(t)$ as

$$f(t) = x_1\phi_1(t) + x_2\phi_2(t) + \cdots + x_m\phi_m(t) \tag{11.4}$$

we then find $f(t)$ by determining the unknown weights $x_1, x_2, \ldots, x_m$ which appear linearly in $f(t)$, and the LS problem is a **linear LS problem**. Otherwise, if the to-be-determined unknown parameters in $f(t)$ appear nonlinearly, the LS problem is a **nonlinear LS problem**. We will formulate and solve the linear LS problem first. Later we use artificial neural networks as examples of nonlinear LS problems.

Ideally the function $f(t)$ in the above linear LS problem generates (interpolates) the data such that $\sum_{i=1}^{n} |y_i - f(t_i)|^2 = 0$ and

$$\begin{bmatrix} f(t_1) \\ f(t_2) \\ \vdots \\ f(t_n) \end{bmatrix} = \begin{bmatrix} y_1 \\ y_2 \\ \vdots \\ y_n \end{bmatrix} \Leftrightarrow \begin{bmatrix} x_1\phi_1(t_1) + x_2\phi_2(t_1) + \cdots + x_m\phi_m(t_1) \\ x_1\phi_1(t_2) + x_2\phi_2(t_2) + \cdots + x_m\phi_m(t_2) \\ \vdots \\ x_1\phi_1(t_n) + x_2\phi_2(t_n) + \cdots + x_m\phi_m(t_n) \end{bmatrix} = \begin{bmatrix} y_1 \\ y_2 \\ \vdots \\ y_n \end{bmatrix}$$

$$\Leftrightarrow \begin{bmatrix} \phi_1(t_1) & \phi_2(t_1) & \cdots & \phi_m(t_1) \\ \phi_1(t_2) & \phi_2(t_2) & \cdots & \phi_m(t_2) \\ \vdots & \vdots & & \vdots \\ \phi_1(t_n) & \phi_2(t_n) & \cdots & \phi_m(t_n) \end{bmatrix} \begin{bmatrix} x_1 \\ x_2 \\ \vdots \\ x_m \end{bmatrix} = \begin{bmatrix} y_1 \\ y_2 \\ \vdots \\ y_n \end{bmatrix} \tag{11.5}$$

which leads to a linear system $A\mathbf{x} = \mathbf{b}$, where the coefficient matrix A is an $n \times m$ matrix whose entry $a_{ij} = \phi_j(t_i)$ is the value of the basis function $\phi_j(t)$ at t_i. However, the data in real applications contain errors. Instead of interpolating the data, it is desirable to filter the errors out when representing the data by a function (a law). In general, the number of weights, m, is much less than the number of the data points, n, and the linear system is over-determined and inconsistent. So we seek a LS solution to minimize $\sum_{i=1}^{n} |y_i - f(t_i)|^2$, the 2-norm of the residual vector $\mathbf{r} = \mathbf{b} - A\mathbf{x}$, and the LS problem is formulated as

$$\min_{\mathbf{x}} ||\mathbf{b} - A\mathbf{x}||_2^2$$

Remark The basis functions $\phi_j(t)$, $j = 1, 2, \ldots, m$, are chosen such that
- they are appropriate to the physics behind the problem
- they are linearly independent, meaning that $f(t) = x_1\phi_1(t) + x_2\phi_2(t) + \cdots + x_m\phi_m(t)$ is a zero function only when all the weights $x_1 = x_2 = \cdots = x_m = 0$
- they are chosen such that $A\mathbf{x} = \mathbf{b}$ is well conditioned

Example 11.1 In linear regression, we find a LS fit of the following n data points by a straight line $f(t) = x_1 + x_2 t$. Write down the matrix A and the vectors $\mathbf{x}$ and $\mathbf{b}$ in the corresponding LS problem $\min_{\mathbf{x}} ||\mathbf{b} - A\mathbf{x}||_2^2$.

t_i	t_1	t_2	$\cdots$	t_n
y_i	y_1	y_2	$\cdots$	y_n

[Solution:] In this case, the basis functions are $\phi_1(t) = 1$ and $\phi_2(t) = t$, and the matrix A and the vectors $\mathbf{x}$ and $\mathbf{b}$ in the LS problem $\min_{\mathbf{x}} ||\mathbf{b} - A\mathbf{x}||_2^2$ are

$$A = \begin{bmatrix} \phi_1(t_1) & \phi_2(t_1) \\ \phi_1(t_2) & \phi_2(t_2) \\ \vdots & \vdots \\ \phi_1(t_n) & \phi_2(t_n) \end{bmatrix} = \begin{bmatrix} 1 & t_1 \\ 1 & t_2 \\ \vdots & \vdots \\ 1 & t_n \end{bmatrix}, \quad \mathbf{x} = \begin{bmatrix} x_1 \\ x_2 \end{bmatrix}, \quad \mathbf{b} = \begin{bmatrix} y_1 \\ y_2 \\ \vdots \\ y_n \end{bmatrix} \quad (11.6)$$

∎

Remark Linear regression is an example of polynomial data fitting. In polynomial LS data fitting, we seek a LS fit of the data $\{(t_i, y_i)\}_{i=1}^n$ by a polynomial of degree $\leq m - 1$

$$p(t) = x_1 + x_2 t + \cdots + x_m t^{m-1}$$

The matrix A in this LS problem $\min_{\mathbf{x}} ||\mathbf{b} - A\mathbf{x}||_2^2$ is a Vandermonde matrix:

$$A = \begin{bmatrix} \phi_1(t_1) & \phi_2(t_1) & \cdots & \phi_m(t_1) \\ \phi_1(t_2) & \phi_2(t_2) & \cdots & \phi_m(t_2) \\ \vdots & \vdots & & \vdots \\ \phi_1(t_n) & \phi_2(t_n) & \cdots & \phi_m(t_n) \end{bmatrix} = \begin{bmatrix} 1 & t_1 & \cdots & t_1^{m-1} \\ 1 & t_2 & \cdots & t_2^{m-1} \\ \vdots & \vdots & & \vdots \\ 1 & t_n & \cdots & t_n^{m-1} \end{bmatrix} \quad (11.7)$$

If we want to fit the data $\{(t_i, y_i)\}_{i=1}^n$ by a function in which the unknown parameters appear in a nonlinear manner, we generally need to solve a nonlinear LS problem to determine the parameters. However, sometimes we can transform the fitting function and the data such that the new fitting function for the transformed data can be found by solving a linear LS problem. The example below is a demonstration of a transformation.

Example 11.2 We want to fit the data $\{(t_i, y_i)\}_{i=1}^n$ by a power function of t as $f(t) = \alpha t^\beta$. Suppose $t_i > 0$ and $y_i > 0$, $i = 1, 2, \ldots, n$. We can apply the natural logarithm to the power function $y = \alpha t^\beta$ to obtain

$$\ln y = \ln \alpha + \beta \ln t$$

Let $u = \ln y$, $s = \ln t$, $a = \ln \alpha$ and $b = \beta$. We have

$$u = a + bs$$

which is a linear function of s. We can then fit the transformed data $\{(s_i, u_i)\}_{i=1}^n = \{(\ln t_i, \ln y_i)\}_{i=1}^n$ by this linear function $g(s) = a + bs$ in the LS sense to determine a and b. Finally we obtain $\alpha = e^a$, $\beta = b$, and the fitting power function $f(t) = \alpha t^\beta$. ∎

11.2   The Normal Equation by Calculus

In the linear LS problem, we minimize $\sum_{i=1}^n |y_i - f(t_i)|^2$, where $f(t_i) = x_1 \phi_1(t_i) + x_2 \phi_2(t_i) + \cdots + x_m \phi_m(t_i) = \sum_{j=1}^m x_j \phi_j(t_i)$. Define

$$F(x_1, x_2, \ldots, x_m) = \sum_{i=1}^n \left[y_i - \sum_{j=1}^m x_j \phi_j(t_i) \right]^2 \tag{11.8}$$

A calculus formulation of the LS problem is to find $x_1, x_2, \ldots, x_m$ to minimize the multivariate function $F(x_1, x_2, \ldots, x_m)$.

Example 11.3 Given the data $\{(t_i, y_i)\}_{i=1}^2 = \{(2, 1), (2, 3)\}$ (note that unlike in interpolation, the nodes t_i in a LS problem may not be all distinct), we seek a constant function $f(t) = x_1$ (i.e. $f(t) = x_1 \phi_1(t)$ with $\phi_1(t) = 1$) to fit the data in the LS sense. What is your guess of the value for x_1? Write down the function $F(x_1)$ to be minimized in the LS problem and then find the LS solution x_1.
[Solution:] We guess x_1 to be the average of y_1 and y_2, that is $x_1 = (1+3)/2 = 2$. In the LS problem, the function $F(x_1)$ to be minimized is

$$F(x_1) = \sum_{i=1}^n \left[y_i - \sum_{j=1}^m x_j \phi_j(t_i) \right]^2 = \sum_{i=1}^2 [y_i - x_1]^2 = (1 - x_1)^2 + (3 - x_1)^2$$

whose graph is a parabola opening upward. Note that

$$F'(x_1) = 2(x_1 - 1) + 2(x_1 - 3) = 4x_1 - 8, \quad F''(x_1) = 4 > 0$$

The function $F(x_1)$ is called a convex function as its graph is concave up. It has a global minimum at the critical point $x_1 = 2$ solved from $F'(x_1) = 0$. So the LS solution of the problem is indeed $f(t) = 2$. ∎

Similar as the above example, the function $F(x_1, x_2, \ldots, x_m)$ to be minimized in a LS problem is a **convex function** that has a global minimum at the critical point where the gradient of the function F is zero, i.e.

$$\nabla F = \left\langle \frac{\partial F}{\partial x_1}, \frac{\partial F}{\partial x_2}, \cdots, \frac{\partial F}{\partial x_m} \right\rangle = 0 \tag{11.9}$$

By solving $\nabla F = 0$, i.e. $\partial F / \partial x_k = 0$ for $k = 1, 2, \ldots, m$, we can find the LS solution for $x_1, x_2, \ldots, x_m$.

Note that $\phi_j(t_i)$ is the entry a_{ij} at the i-th row and j-th column of the matrix A, we have $F(x_1, x_2, \ldots, x_m) = \sum_{i=1}^{n} \left[y_i - \sum_{j=1}^{m} a_{ij} x_j \right]^2$ and

$$
\begin{aligned}
\frac{\partial F}{\partial x_k} &= 2 \sum_{i=1}^{n} \left[y_i - \sum_{j=1}^{m} a_{ij} x_j \right] \frac{\partial}{\partial x_k} \left[y_i - \sum_{j=1}^{m} a_{ij} x_j \right] \\
&= -2 \sum_{i=1}^{n} \left[y_i - \sum_{j=1}^{m} a_{ij} x_j \right] a_{ik} = -2 \sum_{i=1}^{n} \left[a_{ik} y_i - a_{ik} \sum_{j=1}^{m} a_{ij} x_j \right]
\end{aligned} \tag{11.10}
$$

Note that $\sum_{j=1}^{m} a_{ij} x_j$ is the i-th entry $A\mathbf{x}$. Let's call it $\hat{y}_i$. Then $\partial F / \partial x_k = 0$ gives

$$
\sum_{i=1}^{n} a_{ik} \hat{y}_i = \sum_{i=1}^{n} a_{ik} y_i \tag{11.11}
$$

which can be written in the inner product form as

$$
\begin{bmatrix} a_{1k} & a_{2k} & \cdots & a_{nk} \end{bmatrix} \begin{bmatrix} \hat{y}_1 \\ \hat{y}_2 \\ \vdots \\ \hat{y}_n \end{bmatrix} = \begin{bmatrix} a_{1k} & a_{2k} & \cdots & a_{nk} \end{bmatrix} \begin{bmatrix} y_1 \\ y_2 \\ \vdots \\ y_n \end{bmatrix} \tag{11.12}
$$

where

$$
\begin{bmatrix} \hat{y}_1 \\ \hat{y}_2 \\ \vdots \\ \hat{y}_n \end{bmatrix} = A\mathbf{x}, \qquad \begin{bmatrix} y_1 \\ y_2 \\ \vdots \\ y_n \end{bmatrix} = \mathbf{b} \tag{11.13}
$$

Note that $\begin{bmatrix} a_{1k} & a_{2k} & \cdots & a_{nk} \end{bmatrix}$ is the transpose of the k-th column of A and thus the k-th row of A^T. Let's denote is as $\mathbf{a}_k^T$. Then we have

$$
\mathbf{a}_k^T (A\mathbf{x}) = \mathbf{a}_k^T \mathbf{b} \tag{11.14}
$$

which holds for $k = 1, 2, \ldots, m$, i.e.

$$
\begin{bmatrix} \mathbf{a}_1^T \\ \mathbf{a}_2^T \\ \vdots \\ \mathbf{a}_m^T \end{bmatrix} A\mathbf{x} = \begin{bmatrix} \mathbf{a}_1^T \\ \mathbf{a}_2^T \\ \vdots \\ \mathbf{a}_m^T \end{bmatrix} \mathbf{b} \tag{11.15}
$$

So finally we obtain

$$
A^T A \mathbf{x} = A^T \mathbf{b} \tag{11.16}
$$

which is called the **normal equation** for the LS problem. By solving the normal equation for $\mathbf{x}$, we have the LS solution for $x_1, x_2, \ldots, x_m$ and therefore the LS fit $f(t) = x_1 \phi_1(t) + x_2 \phi_2(t) + \cdots + x_m \phi_m(t)$.

t_i	t_1	t_2	$\cdots$	t_n
y_i	y_1	y_2	$\cdots$	y_n

Example 11.4 In linear regression, we find a LS fit of the n data points by a straight line $f(t) = x_1 + x_2 t$, that is $f(t) = x_1\phi_1(t) + x_2\phi_2(t)$ with $\phi_1(t) = 1$ and $\phi_2(t) = t$. Write down the normal equation for solving for the weights x_1 and x_2 in the LS problem.

[Solution:] With $\phi_1(t) = 1$ and $\phi_2(t) = t$, we have

$$
\begin{bmatrix} \phi_1(t_1) & \phi_2(t_1) \\ \phi_1(t_2) & \phi_2(t_2) \\ \vdots & \vdots \\ \phi_1(t_n) & \phi_2(t_n) \end{bmatrix} = \begin{bmatrix} 1 & t_1 \\ 1 & t_2 \\ \vdots & \vdots \\ 1 & t_n \end{bmatrix}, \quad \mathbf{x} = \begin{bmatrix} x_1 \\ x_2 \end{bmatrix}, \quad \mathbf{b} = \begin{bmatrix} y_1 \\ y_2 \\ \vdots \\ y_n \end{bmatrix} \tag{11.17}
$$

The normal equation for the LS problem is $A^T A \mathbf{x} = A^T \mathbf{b}$, where

$$
A^T A = \begin{bmatrix} 1 & 1 & \cdots & 1 \\ t_1 & t_2 & \cdots & t_n \end{bmatrix} \begin{bmatrix} 1 & t_1 \\ 1 & t_2 \\ \vdots & \vdots \\ 1 & t_n \end{bmatrix} = \begin{bmatrix} n & \sum_{i=1}^n t_i \\ \sum_{i=1}^n t_i & \sum_{i=1}^n t_i^2 \end{bmatrix} \tag{11.18}
$$

$$
A^T \mathbf{b} = \begin{bmatrix} 1 & 1 & \cdots & 1 \\ t_1 & t_2 & \cdots & t_n \end{bmatrix} \begin{bmatrix} y_1 \\ y_2 \\ \vdots \\ y_n \end{bmatrix} = \begin{bmatrix} \sum_{i=1}^n y_i \\ \sum_{i=1}^n t_i y_i \end{bmatrix} \tag{11.19}
$$

So the normal equation becomes

$$
\begin{bmatrix} n & \sum_{i=1}^n t_i \\ \sum_{i=1}^n t_i & \sum_{i=1}^n t_i^2 \end{bmatrix} \begin{bmatrix} x_1 \\ x_2 \end{bmatrix} = \begin{bmatrix} \sum_{i=1}^n y_i \\ \sum_{i=1}^n t_i y_i \end{bmatrix} \tag{11.20}
$$

∎

Example 11.5 Find a LS fit of the 3 data points by a straight line $f(t) = x_1 + x_2 t$.

t_i	1	2	3
y_i	2	1	3

[Solution:] In this linear regression by a LS fit, x_1 and x_2 satisfy

$$
\begin{bmatrix} n & \sum_{i=1}^n t_i \\ \sum_{i=1}^n t_i & \sum_{i=1}^n t_i^2 \end{bmatrix} \begin{bmatrix} x_1 \\ x_2 \end{bmatrix} = \begin{bmatrix} \sum_{i=1}^n y_i \\ \sum_{i=1}^n t_i y_i \end{bmatrix} \tag{11.21}
$$

which is

$$
\begin{bmatrix} 3 & 1+2+3 \\ 1+2+3 & 1^2+2^2+3^2 \end{bmatrix} \begin{bmatrix} x_1 \\ x_2 \end{bmatrix} = \begin{bmatrix} 2+1+3 \\ 1\times2+2\times1+3\times3 \end{bmatrix}
$$

$$
\begin{bmatrix} 3 & 6 \\ 6 & 14 \end{bmatrix} \begin{bmatrix} x_1 \\ x_2 \end{bmatrix} = \begin{bmatrix} 6 \\ 13 \end{bmatrix}
$$

Solving this linear system gives $x_1 = -3/4$ and $x_2 = 5/4$. So the LS fit is $f(t) = -3/4 + 5t/4$ ∎

11.3   The Normal Equation by Linear Algebra

The linear discrete LS problem ends up solving the following minimization problem

$$
\min_{\mathbf{x}} ||\mathbf{b} - A\mathbf{x}||_2^2 \tag{11.22}
$$

i.e. finding a vector $\mathbf{x}$ such that the distance between $\mathbf{b}$ and $A\mathbf{x}$ is shortest.

From linear algebra, we know that $A\mathbf{x}$ is a linear combination of the columns of A using entries in the vector $\mathbf{x}$ as weights, and $A\mathbf{x}$ is in Col A, the column space of A (the set of all possible linear combinations of the columns of A). So the LS problem is to find a vector $A\mathbf{x}$ in Col A that is closest to the vector $\mathbf{b}$ with the distance defined by the 2-norm. As illustrated in Fig. 11.2, the LS solution $\mathbf{x}$ thus must satisfy $A\mathbf{x} = \text{proj}_{\text{Col }A}\mathbf{b}$ ($A\mathbf{x}$ is the orthogonal projection of $\mathbf{b}$ onto Col A), and the corresponding residual vector $\mathbf{r} = \mathbf{b} - A\mathbf{x}$ must be perpendicular to Col A ($\mathbf{r} = \mathbf{b} - A\mathbf{x}$ is in $(\text{Col }A)^{\perp}$). The **fundamental theorem of linear algebra** says

$$
(\text{Col }A)^{\perp} = \text{Nul }A^T \tag{11.23}
$$

As $\mathbf{r} = \mathbf{b} - A\mathbf{x}$ is in $(\text{Col }A)^{\perp} = \text{Nul }A^T$, we have

$$
A^T(\mathbf{r} = A^T(\mathbf{b} - A\mathbf{x}) = 0 \tag{11.24}
$$

which leads to the normal equation for the LS problem

$$
A^T A\mathbf{x} = A^T \mathbf{b} \tag{11.25}
$$

Example 11.6 Given the data (velocity $\bar{u}^+$ versus distance y^+) for a turbulent boundary layer,

y_i^+	20	40	80
$\bar{u}_i^+$	13	14	16

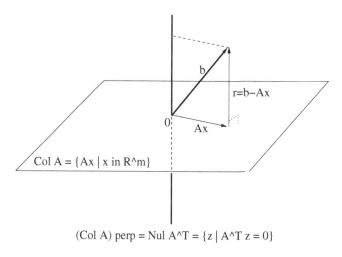

FIGURE 11.2
The LS solution via the fundamental theorem of linear algebra.

find a LS fit of the data by the law $\bar{u}^+(y^+) = \frac{1}{\kappa}\ln y^+ + C^+$.
Solution: Write $\bar{u}^+(y^+)$ as $\bar{u}^+(y^+) = x_1\phi_1(y^+) + x_2\phi_2(y^+)$, where $x_1 = 1/\kappa$, $\phi_1(y^+) = \ln y^+$, $x_2 = C^+$ and $\phi_2(y^+) = 1$. We have

$$A = \begin{bmatrix} \phi_1(y_1^+) & \phi_2(y_1^+) \\ \phi_1(y_2^+) & \phi_2(y_2^+) \\ \phi_1(y_3^+) & \phi_2(y_3^+) \end{bmatrix} = \begin{bmatrix} \ln 20 & 1 \\ \ln 40 & 1 \\ \ln 80 & 1 \end{bmatrix}, \quad \mathbf{x} = \begin{bmatrix} x_1 \\ x_2 \end{bmatrix}, \quad \mathbf{b} = \begin{bmatrix} \bar{u}_1^+ \\ \bar{u}_2^+ \\ \bar{u}_3^+ \end{bmatrix} = \begin{bmatrix} 13 \\ 14 \\ 16 \end{bmatrix}$$

The normal equation for the LS problem is $A^T A\mathbf{x} = A^T\mathbf{b}$, where

$$A^T A = \begin{bmatrix} \ln 20 & \ln 40 & \ln 80 \\ 1 & 1 & 1 \end{bmatrix} \begin{bmatrix} \ln 20 & 1 \\ \ln 40 & 1 \\ \ln 80 & 1 \end{bmatrix} \approx \begin{bmatrix} 41.78 & 11.07 \\ 11.07 & 3 \end{bmatrix}$$

$$A^T\mathbf{b} = \begin{bmatrix} \ln 20 & \ln 40 & \ln 80 \\ 1 & 1 & 1 \end{bmatrix} \begin{bmatrix} 13 \\ 14 \\ 16 \end{bmatrix} \approx \begin{bmatrix} 160.70 \\ 43 \end{bmatrix}$$

So the normal equation becomes

$$\begin{bmatrix} 41.78 & 11.07 \\ 11.07 & 3 \end{bmatrix} \begin{bmatrix} x_1 \\ x_2 \end{bmatrix} \approx \begin{bmatrix} 160.70 \\ 43 \end{bmatrix}$$

which is solved to give $x_1 \approx 2.16$ and $x_2 \approx 6.35$. So we have $\kappa = 1/x_1 \approx 0.46$, $C^+ = x_2 \approx 6.35$ and the log law $\bar{u}^+(y^+) = \frac{1}{0.46}\ln y^+ + 6.35$. ∎

11.4   LS Problems by A=QR

In practice, the LS problem

$$\min_{\mathbf{x}} ||\mathbf{b} - A\mathbf{x}||_2^2 \tag{11.26}$$

is generally solved by the QR factorization of A.

We know from linear algebra that if the $n \times m$ matrix A has m linearly independent columns (i.e. A has full column rank with rank $A = m$), then the Gram-Schmidt process can be applied to the columns of A to produce an orthonormal basis for Col A, which leads to the **QR decomposition** of A

$$A = QR \tag{11.27}$$

where Q is an $n \times m$ matrix (of the same shape as A) containing the orthonormal basis vectors as the columns, and R is an $m \times m$ nonsingular upper triangular matrix. There are other more economic ways to find the QR decomposition of A. Below we focus on how to solve the LS problem with a given QR decomposition $A = QR$.

We know that the LS solution satisfies the normal equation

$$A^T A\mathbf{x} = A^T \mathbf{b} \tag{11.28}$$

where $A^T A$ is invertible as A has linearly independent columns. We therefore can write the LS solution as

$$\mathbf{x} = (A^T A)^{-1} A^T \mathbf{b} \tag{11.29}$$

Plugging $A = QR$ in the above expression gives

$$\mathbf{x} = (R^T Q^T QR)^{-1} R^T Q^T \mathbf{b} \tag{11.30}$$

Realizing that $Q^T Q = I_m$ (as Q has orthonormal columns), we have

$$\mathbf{x} = R^{-1} Q^T \mathbf{b} \tag{11.31}$$

So finally we can find the LS solution by solving an upper triangular linear system

$$R\mathbf{x} = Q^T \mathbf{b} \tag{11.32}$$

using backward substitution.

Example 11.7 Solve the LS problem

$$\min_{\mathbf{x}} ||\mathbf{b} - A\mathbf{x}||_2^2$$

where

$$A = \begin{bmatrix} 2 & 2 \\ 1 & 2 \\ 2 & 6 \end{bmatrix}, \quad \mathbf{b} = \begin{bmatrix} -4 \\ 24 \\ 4 \end{bmatrix}$$

using the QR decomposition $A = QR$, where

$$Q = \begin{bmatrix} 2/3 & -\sqrt{2}/2 \\ 1/3 & 0 \\ 2/3 & \sqrt{2}/2 \end{bmatrix}, \quad R = \begin{bmatrix} 3 & 6 \\ 0 & 2\sqrt{2} \end{bmatrix}$$

[**Solution:**] Compute $Q^T \mathbf{b}$

$$Q^T \mathbf{b} = \begin{bmatrix} 2/3 & 1/3 & 2/3 \\ -\sqrt{2}/2 & 0 & \sqrt{2}/2 \end{bmatrix} \begin{bmatrix} -4 \\ 24 \\ 4 \end{bmatrix} = \begin{bmatrix} 8 \\ 4\sqrt{2} \end{bmatrix}$$

Solve the upper triangular system $R\mathbf{x} = Q^T \mathbf{b}$

$$\begin{bmatrix} 3 & 6 \\ 0 & 2\sqrt{2} \end{bmatrix} \begin{bmatrix} x_1 \\ x_2 \end{bmatrix} = \begin{bmatrix} 8 \\ 4\sqrt{2} \end{bmatrix}$$

by backward substitution to obtain $\mathbf{x} = [-4/3, 2]^T$. ∎

11.5 Artificial Neural Network

In previous sections, we consider discrete linear LS problems. In a nonlinear LS problem, the parameters in a function to fit given data in the LS sense appear nonlinearly. For example, the **logistic function** is given as

$$f(x) = \frac{L}{1 + e^{-k(x-x_0)}}$$

in which the parameters L, k and x_0 appear nonlinearly. If we determine the parameters in the logistic function to fit given data in the LS sense, we end up with a nonlinear LS problem. Below we briefly introduce artificial neural networks, which are widely used in machine learning. The parameters in a neural network is generally determined by solving a nonlinear LS problem.

A feedforward neural network is regarded as a nonlinear function approximator. It models the nonlinear function

$$\mathbf{u} = \mathbf{f}(\mathbf{x}) \tag{11.33}$$

where $\mathbf{x} = (x_1, x_2, \ldots, x_M)^T$ is the input vector, and $\mathbf{u} = (u_1, u_2, \ldots, u_N)^T$ is the output vector. Neural networks are loosely inspired by neuroscience. As illustrated in Fig. 11.3, a neural network is composed of inputs, hidden layers and outputs, and the layers are made of nodes where computation happens as neurons in the human brain fired with stimuli. As shown in Fig. 11.3, the computation of a node consists of a

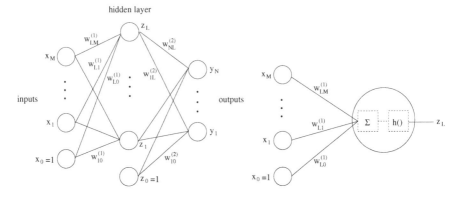

FIGURE 11.3
Left: Network diagram for a feedforward neural network with the input, hidden and output variables represented by nodes and the weight parameters represented by links; Right: A hidden node in a neural network, in which the computation consists of the weighted sum of the inputs and a transformation of the sum by an activation function.

weighted sum of the inputs to this node and a transformation by a nonlinear activation function $h(\cdot)$. The output z_l in Fig. 11.3 is given by

$$z_l = \mathbf{h}\left(\sum_{i=1}^{M} w_{li}^{(1)} x_i + w_{l0}^{(1)}\right), \quad l = 1, 2, \ldots, L \tag{11.34}$$

where the parameter $w_{li}^{(1)}$ is the weight on the link connecting the node having the output z_l and the node having the input x_i in the neural network, the parameter $w_{l0}^{(1)}$ is called a bias, and the activation function $h(\cdot)$ can be chosen as a **logistic sigmoid function**

$$\sigma(s) = \frac{1}{1 + e^{-s}} \tag{11.35}$$

There are many other choices for $h(\cdot)$, such as the hyperbolic tangent, the rectified linear unit (ReLU) and the softmax unit, etc., depending on the applications. The outputs of the neural network in Fig. 11.3 are therefore given by

$$y_j = h\left(\sum_{l=1}^{L} w_{jl}^{(2)} z_l + w_{j0}^{(2)}\right), \quad j = 1, 2, \ldots, N \tag{11.36}$$

Neural networks with hidden layers provide a universal approximation framework. They are able to represent any nonlinear functions regardless of nonlinearity provided that the neural networks have enough hidden units. The depth of a neural network is indicated by the number of hidden layers. A **deep neural network** (used in so-called **deep learning**) has more than one hidden layer.

To train a neural network is to determine the weights and biases in the neural network using the **training data set** $\{(\mathbf{x}_k, \mathbf{u}_k)\}_{k=1}^{K}$ generated by the nonlinear function $\mathbf{u} = \mathbf{f}(\mathbf{x})$ such that the neural network models the nonlinear function. An error function that measures the misfit between the network predictions and the training data is minimized in network training. One simple choice of the error function is given by the sum of the squares of the errors as

$$E(\mathbf{w}) = \sum_{k=1}^{K} \|\mathbf{u}_k - \mathbf{y}_k\|_2^2, \qquad (11.37)$$

where $\mathbf{w}$ is the vector of all the network parameters (weights and biases) to be determined, $\mathbf{y}_k = (y_1, y_2, \ldots, y_N)_k^T$ is the kth prediction, $\mathbf{u}_k = (u_1, u_2, \ldots, u_N)_k^T$ is the kth target. With the use of a smooth nonlinear activation function $h(\cdot)$ such as the sigmoid function , the error function $E(\mathbf{w})$ is differentiable with respect to the network parameters. So gradient decent algorithms can be used to minimize the error function. The gradient and the Hessian matrix can be obtained using the technique of error **backpropagation (BP)** that is based on the chain rule. After the neural network is trained, a testing data set generated by the same nonlinear function $\mathbf{f}(\mathbf{x})$ can be used to test its performance by looking at the validation error defined as the difference between the network predictions and the testing data (the **validation data set**).

The numbers of inputs and outputs in a neural network are generally determined by the dimensionality of the data, whereas the number of hidden nodes that control the number of parameters can be adjusted to give the best predictive performance. When the number of hidden nodes is too large, the network predictions may oscillate widely between data points, and the function $\mathbf{f}(\mathbf{x})$ may be poorly represented. This behavior is called **over-fitting**. To avoid over-fitting the neural network can be regularized. The regularization controls the network complexity by forcing the weights to be small, leading to a smoother network response. One of the successful is Bayesian regularization. Another approach to avoid over-fitting is the procedure of **early stopping**. The validation error in the training process often decreases first but then increases as the network starts to over-fit. Training can therefore be stopped at the point of the smallest validation error.

11.6   Exercises

Exercise 11.1 To find a LS fit of the data given in Table 11.2 by the function $p(t) = x_1 + x_2 t$, do the follows.

(1) Write down what is to be minimized (i.e. sum of squares) as a multivariate function $g(x_1, x_2)$. (2) Find ∇g. (3) find the LS solution (i.e. the values of x_1 and x_2) by solving $\nabla g = 0$. (4) Sketch the data as open circles and the LS fitting curve $y = p(t)$ in the same figure.

t_i	-1	0	1
y_i	0	1	0

TABLE 11.2
Data for LS fitting.

Exercise 11.2 To find a LS fit of the data given in Table 11.3 by the function $p(t) = x_1\phi_1(t) + x_2\phi_2(t) + x_3\phi_3(t)$, we solve the following LS problem

$$\min_{\mathbf{x}} ||\mathbf{b} - A\mathbf{x}||_2^2$$

(1) Write down the vectors $\mathbf{x}$, $\mathbf{b}$ and the matrix A if the basis functions are $\phi_1(t) = 1$, $\phi_2(t) = t$ and $\phi_3(t) = t^2$.
(2) Write down the vectors $\mathbf{x}$, $\mathbf{b}$ and the matrix A if the basis functions are $\phi_1(t) = 2$, $\phi_2(t) = t + 1$ and $\phi_3(t) = (t-1)^2$.

t_i	-2	-1	0	1	2
y_i	-2	0	-1	0	-2

TABLE 11.3
Data for LS fitting.

Exercise 11.3 If we want to fit the data $\{(t_i, y_i)\}_{i=1}^{n}$ by a function of t as $y = 1/(a + bt)^3$, we can transform the LS fitting function so that a and b can be determined by setting up and solving a linear LS problem. Show how to do so. What if $y = a/(b+t)$?

Exercise 11.4 Find a LS fit of the data given in Table 11.4 by the function $p(t) = x_1 + x_2 t$ by solving the normal equation for x_1 and x_2.

t_i	-1	0	1
y_i	0	1	0

TABLE 11.4
Data for LS fitting.

Exercise 11.5 Find the LS fit $f(t) = x_1 + x_2 t$ of the data given in Table 11.5 by following the steps below.

(1) Write down the matrix A and the vector $\mathbf{b}$ in the LS problem $\min_{\mathbf{x}} ||\mathbf{b} - A\mathbf{x}||_2^2$.
(2) Write down the normal equation for solving the LS problem.
(3) Solve the normal equation for x_1 and x_2 and sketch the data and the LS fit $f(t) = x_1 + x_2 t$ in a figure.

t	-2	0	2
y	0	2	1

TABLE 11.5
Data for LS fitting.

Exercise 11.6 Find a LS fit of the data given in Table 11.6 by the function $p(t) = x_1\phi_1(t) + x_2\phi_2(t) + x_3\phi_3(t)$, where $\phi_1(t) = 1$, $\phi_2(t) = t$ and $\phi_3(t) = t^2$, by solving the normal equation for x_1, x_2 and x_3.

t_i	-2	-1	0	1	2
y_i	-2	0	-1	0	-2

TABLE 11.6
Data for LS fitting.

Exercise 11.7 Consider the LS problem

$$\min_{\mathbf{x}} ||\mathbf{b} - A\mathbf{x}||_2^2$$

where

$$A = \begin{bmatrix} 1 & -1 \\ 1 & 4 \\ 1 & -1 \\ 1 & 4 \end{bmatrix}, \quad \mathbf{b} = \begin{bmatrix} -1 \\ 6 \\ 5 \\ 7 \end{bmatrix}$$

The QR decomposition of A is

$$A = QR = \begin{bmatrix} 0.5 & -0.5 \\ 0.5 & 0.5 \\ 0.5 & -0.5 \\ 0.5 & 0.5 \end{bmatrix} \begin{bmatrix} 2 & 3 \\ 0 & 5 \end{bmatrix}$$

(1) Use the Gram-Schmidt process to verify the QR decomposition of A.
(2) Use the QR decomposition of A to solve the LS problem.

Exercise 11.8 Consider the LS problem

$$\min_{\mathbf{x}} \|\mathbf{b} - A\mathbf{x}\|_2^2$$

where

$$A = \begin{bmatrix} 1 & 3 & 5 \\ 1 & 1 & 0 \\ 1 & 1 & 2 \\ 1 & 3 & 3 \end{bmatrix}, \quad \mathbf{b} = \begin{bmatrix} 3 \\ 5 \\ 7 \\ -3 \end{bmatrix}$$

The QR decomposition of A is

$$A = QR = \begin{bmatrix} 0.5 & 0.5 & 0.5 \\ 0.5 & -0.5 & -0.5 \\ 0.5 & -0.5 & 0.5 \\ 0.5 & 0.5 & -0.5 \end{bmatrix} \begin{bmatrix} 2 & 4 & 5 \\ 0 & 2 & 3 \\ 0 & 0 & 2 \end{bmatrix}$$

(1) Use the Gram-Schmidt process to verify the QR decomposition of A.
(2) Use the QR decomposition of A to solve the LS problem.

11.7 Programming Problems

Problem 11.1 Write a MATLAB m-function

```
function y = LinearRegression(ti,yi,t)
```

to find and evaluate a linear LS fit using linear regression. The data for the LS fitting are supplied in vectors `ti` and `yi`, respectively; the t-values at which the linear polynomial is evaluated are provided in the vector `t`; and the values of the polynomial are returned in the vector `y`. Test your m-function on the data $\{(t_i, y_i)\}_{i=1}^{n}$ generated by a linear function with a small random perturbation to each y_i. In your test, plot the data and the fitting line in one figure.

Problem 11.2 Consider the LS fitting of the data in Table 11.7 by a linear polynomial.
(1) Set up an over-determined system $A\mathbf{x} = \mathbf{b}$ (i.e. A and $\mathbf{b}$) for a LS fitting using the basis functions $\phi_1(t) = 1$ and $\phi_2(t) = t$. Find the LS solution by x=A\b in MATLAB (the QR decomposition method). Calculate the 2-norm of the residual.
(2) Obtain the normal equation $A^T A\mathbf{x} = A^T \mathbf{b}$. Solve the normal equation for the LS solution. Calculate the 2-norm of the residual.
(3) Compare your solutions from (1) and (2). Plot the two polynomials and the data points on a single graph.
(4) Now set up the over-determined system $\hat{A}\hat{\mathbf{x}} = \hat{\mathbf{b}}$ if the basis functions $\phi_1(t) = 1$ and $\phi_2(t) = 3 \times 10^7(t - 0.98765435)$ are used instead. Find the LS solution by the backslash operator in MATLAB (i.e. the QR decomposition method). Calculate the 2-norm of the residual.

(5) Obtain the normal equation $\hat{A}^T \hat{A} \hat{\mathbf{x}} = \hat{A}^T \hat{\mathbf{b}}$. Solve the normal equation for the LS solution. Calculate the 2-norm of the residual.

(6) Compare your results from (4) and (5). Plot the two polynomials and the data points on a single graph.

t_i	y_i
0.98765431	2.1
0.98765432	2.1
0.98765433	2.1
0.98765434	2.0
0.98765435	1.9
0.98765436	1.9
0.98765437	1.8
0.98765438	1.7
0.98765439	1.7

TABLE 11.7
Data for LS fitting.

	t_i	1000	1050	1060	1080	1110	1130
(1)	y_i	6010	6153	6421	6399	6726	6701
(2)	y_i	9422	9300	9220	9150	9042	8800

TABLE 11.8
Data for LS fitting.

Problem 11.3 Write a MATLAB m-function LSbyQR.m that solves least squares problems $A\mathbf{x} = \mathbf{b}$ using the QR decomposition $A = QR$. The QR decomposition can be computed by the MATLAB command qr. You can use your m-function for backward substitution in Problem 4.1 to solve the upper triangular system $Rx = Q^T b$.

Write a MATLAB m-script testLSbyQR.m to apply your solver to find the LS linear polynomials using basis functions $\phi_1(t) = 50$ and $\phi_2(t) = t - 1065$ for the two data sets in Table 11.8. Compare your solutions to the solutions computed by the MATLAB command x = A\b. Plot your solutions and data points.

12

Monte Carlo Methods and Parallel Computing

In this chapter, we introduce Monte Carlo methods and parallel computing with the example of approximating the value of π. We present basic ideas and terms in Monte Carlo methods and parallel computing.

Buffon's needle problem is a question first posed in the 18th century by the French mathematician Comte de Buffon. Suppose there is a floor made of parallel strips of wood with the same width, and a needle is dropped onto the floor. What is the probability that the needle will lie across a line between two strips?

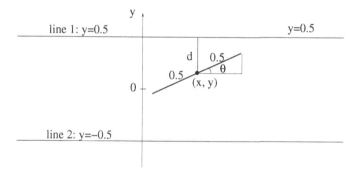

FIGURE 12.1
Buffon's needle problem.

To find the answer, one may drop a needle on the floor for many times, count the successful times for which the needle lies across a line between two strips, and estimate the probability as the ratio of the successful times to the total times. We can do such an experiment on a computer, and the computer experiment is called **Monte Carlo simulation**. We assume that the center of the needle lands at a random point between lines with equal chance. We also assume the angular orientation of the needle takes a value between 0 and $\pi/2$ randomly with equal chance. Let's consider the case shown in Fig. 12.1 in which both the width of the strip and the length of the needle are 1. We only need to consider the orientation angle $\theta \in [0, \pi/2]$ and the y-coordinate $y \in [0, 0.5]$ of the center of the needle on one strip between two lines $y = -0.5$ and $y = 0.5$ due to the symmetry of the problem. In our Monte Carlo simulation, we have a **random number generator** that generates random numbers

DOI: 10.1201/9781003201694-12

uniformly distributed in the interval $[0,1]$, and we assign the values of the random variables θ and y using independent random numbers. The needle lies across the top line if

$$d = 0.5 - y \leq 0.5 \sin \theta \qquad (12.1)$$

We can drop the needle for n times and count and denote the number of successful times as m. The probability is estimated as $p = m/n$. Below is the pseudocode, in which random() is the random number generator that generated random numbers uniformly distributed between 0 and 1.

Algorithm 20 Monte Carlo Method for Buffon's needle problem

$m \leftarrow 0$
for i from 1 to n **do**
  $y \leftarrow 0.5 \, \text{random}()$
  $\theta \leftarrow (\pi/2) \, \text{random}()$
  $d \leftarrow 0.5 - y$
  if $d \leq 0.5 \sin(\theta)$ **then**
    $m \leftarrow m + 1$
  end if
end for
$p \leftarrow m/n$

Buffon's needle problem is a geometric probability problem which can be solved using integration (see the remark below). If the needle length l is not greater than the width w of the strip, the sought probability p is

$$p = \frac{2}{\pi} \frac{l}{w} \qquad (12.2)$$

For the case $l = w = 1$ considered in the above Monte Carlo simulation, we have $\pi = 2/p$. So we can use this Monte Carlo simulation to estimate the value of π, and this kind of methods are called Monte Carlo methods.

Remark In Buffon's needle problem with $l \leq w$, the probability that the random variable y ($y \in [0, w/2]$) falls in the range $[y, y + dy]$ is

$$p_1 = \frac{dy}{w/2} \qquad (12.3)$$

and the probability that the random variable θ ($\theta \in [0, \pi/2]$) falls in the range $[\theta, \theta + d\theta]$ is

$$p_2 = \frac{d\theta}{\pi/2} \qquad (12.4)$$

Since the two random variables are independent, the joint probability is

$$p_1 p_2 = \frac{4}{w\pi} dy d\theta \qquad (12.5)$$

The needle lands across the line $y = w/2$ if

$$d = \frac{w}{2} - y \le \frac{l}{2}\sin(\theta) \Rightarrow y \ge \frac{w}{2} - \frac{l}{2}\sin(\theta) \qquad (12.6)$$

So the probability that the needle lands across the line $y = w/2$ is

$$p = \int_0^{\frac{\pi}{2}} \int_{\frac{w}{2} - \frac{l}{2}\sin\theta}^{\frac{w}{2}} \frac{4}{w\pi} dy d\theta = \frac{2}{\pi}\frac{l}{w} \qquad (12.7)$$

12.1 Monte Carlo Methods

The underlying idea of Monte Carlo methods is to use randomness (for example the random variables y and θ in Buffon's needle problem) to solve problems that might be deterministic in principle (for example the value of π). They are often used in physical and mathematical problems when it is difficult or impossible to use other approaches. The name Monte Carlo is related with the Monte Carlo Casino.

Here we introduce another Monte Carlo method to approximate the value of π using geometric probability. We draw on a canvas a square with an inner circle. We then generate a large number (denoted by n) of random points uniformly distributed within the square and count how many (denoted by m) fall in the enclosed circle. For the dimensions given in Fig. 12.2, the probability of a point fall inside the circle is $p = \pi/4$, the ratio of the area of the circle to the area of the square. So we have the approximation

$$p = \frac{\pi}{4} \approx \frac{m}{n} \Rightarrow \pi \approx \frac{4m}{n} \qquad (12.8)$$

By the symmetry of the problem, we can consider only the first quadrant. Below gives the pseudocode for the method.

In general, a Monte Carlo method includes the following steps:
- Define a domain of possible inputs
- Generate inputs randomly with given probability distributions over the domain
- Perform a deterministic computation using the inputs
- Determine the desired outputs

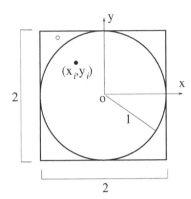

FIGURE 12.2
Monte Carlo method to approximate π.

Algorithm 21 Monte Carlo Method for Buffon's needle problem

$m \leftarrow 0$
for i from 1 to n **do**
    $x \leftarrow \text{random}()$
    $y \leftarrow \text{random}()$
    $r \leftarrow \sqrt{x^2 + y^2}$
    if $r \leq 1$ **then**
        $m \leftarrow m + 1$
    end if
end for
$\pi_A \leftarrow 4m/n$

Take the previous problem as an example. The domain of inputs is the unit square that circumscribes the quarter circle. The random inputs are scattering points generated randomly with the uniform probability distribution. The deterministic computation on each input is to test whether a point falls within the quarter circle. The desired output is determined as a ratio to approximate π. Note that if the points are not uniformly distributed, then the approximation is poor; and if the number of the points is not large, the approximation is poor too.

12.2   Parallel Computing

In this section, we use the Monte Carlo method in the previous section as an example to have a taste of parallel computing. Because of physical limits on the speed of a single processor, **high-performance parallel computers (HPC)** are developed for parallel computing to reduce the wall-clock time for solving a computational problem. An analogy to the use of parallel computers instead of a very advanced single-CPU computer is to pull a heavy wagon by many horses instead of one single gigantic horse. A parallel computer is a multiple-CPU computer system. Fig. 12.3 lists common models of parallel computers. Fig. 12.4 shows a picture of a high-performance computing cluster. Parallel computing is to use a parallel computer to solve a computational problem by computing part of the problem on each CPU in parallel and collaborating on the computation through communication (message passing). Parallel computing is now a common way to solve problems in science and engineering such as climate modeling.

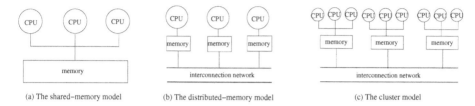

(a) The shared–memory model   (b) The distributed–memory model   (c) The cluster model

FIGURE 12.3
Different models for parallel computers.

FIGURE 12.4
The high-performance computing cluster ManeFrame II at Southern Methodist University.

Remark   The Message-Passing Interface (MPI) is a communication library developed for message passing and related operations on parallel computers and workstation networks. There are many resources on how to use MPI in different computer languages.

A computational process involves executing a sequence of instructions (the instruction stream) to manipulate a sequence of data (the data stream). For example, the instructions in the Monte Carlo method to determine π in the previous section include computing of the distance of a point from the origin and counting the number of points inside the circle, and the data are all the random points. When solving a computational problem by parallel computing, we need to split the problem into parts such each part can be run in parallel on a parallel computer. Sometimes we split only the instruction stream such that multiple instructions (MI) are executed in parallel on the same single data set (SD), and this case is called MISD. Sometimes we can split only the data such that the same single instruction stream (SI) is executed on multiple subsets of the data (MD) in parallel, and this case is called SIMD. Sometimes we can split both streams (the MIMD case).

We cannot split the instruction stream in the above Monte Carlos method because the instructions must follow a time order and cannot be executed in parallel. However, we can split the data such that each processor works on some of the points. For example, we may use p processors, and each processor processes n points. We use one **master processor** to collect the information from the other processors and compute the final approximation of π using the collected information. Below is the pseudocode to implement this SIMD Monte Carlo method. This pseudocode is for each processor.

A computational problem can be solved with parallel computing or without parallel computing, and the latter is called **sequential computing**. One reason for parallel computing is to reduce wall-clock time in solving the problem. We define the **speedup** ψ of parallel computing as the ratio of sequential execution time to parallel execution time to solve the same problem. Note that speedup ψ is generally less than the number of processors p in parallel computing for two main reasons. First, a computational problem may not be fully parallelized and have the fraction f ($0 \leq f \leq 1$) of operations that must be performed sequentially (not in parallel). Second, the communication among processors takes time that is not required in sequential computing. So the speedup ψ is bounded as

$$\psi \leq \frac{1}{f + \frac{1-f}{p}} \tag{12.9}$$

where the fraction $1 - f$ of the operations are performed in parallel on p processor. The above inequality is called **Amdahl's law**.

12.3 Exercises

Exercise 12.1 Two ways are proposed to generate random points uniformly distributed inside a unit circle $x^2 + y^2 = 1$ using a random number generator over an interval that generates random numbers uniformly distributed in the interval. In the first way, points in the square circumscribing the circle are generated by independently generating the Cartesian coordinates x and y of each point using the random number

---

Algorithm 22 Monte Carlo Method for Buffon's needle problem

$m \leftarrow 0$
if current processor is master process **then**
    for i from 1 to n **do**
        $x \leftarrow \text{random}()$
        $y \leftarrow \text{random}()$
        $r \leftarrow \sqrt{x^2 + y^2}$
        if $r \leq 1$ **then**
            $m \leftarrow m + 1$
        end if
    end for
    receive n from all processors
    receive m from all processors
else
    for i from 1 to n **do**
        $x \leftarrow \text{random}()$
        $y \leftarrow \text{random}()$
        $r \leftarrow \sqrt{x^2 + y^2}$
        if $r \leq 1$ **then**
            $m \leftarrow m + 1$
        end if
    end for
    send n to master processor
    send m to master processor
end if
if current processor is master process **then**
    N_{total} is sum of n from all processors
    M_{total} is sum of m from all processors
    $\pi_A \leftarrow 4M_{\text{total}}/N_{\text{total}}$
end if

---

generator over $[-1, 1]$, and then the points outside the circle are discarded. In the second way, points in the circled are directly generated by independently generating the polar coordinates r and θ of each point using the random number generator over $[0, 1]$ for r and over $[0, 2\pi)$ for θ. Which way is the right way?

Exercise 12.2 For Buffon's needle problem, we have considered the case in which the length l of the needle is shorter than the width w of the strip. Derive the formula for the sought probability in terms of l and w if $l > w$.

Exercise 12.3 Describe how you can use a random number generator that generates random numbers uniformly distributed in the interval $[0, 1]$ to simulate the outcomes from flipping a coin.

Exercise 12.4 Describe how you can use a random number generator that generates random numbers uniformly distributed in the interval $[0, 1]$ to simulate the outcomes from tossing a standard/fair six-sided die.

Exercise 12.5 Try with your friends to sort a deck of 52 shuffled cards into the four suits: clubs ($\clubsuit$), diamonds ($\diamondsuit$), hearts ($\heartsuit$) and spades ($\spadesuit$). Find out how long it takes for p person to sort one deck of cards with $p = 1, 2, 4$ and 13, respectively. Note that the time to distribute and collect cards must be counted in the total time. How about $p = 52$?

Exercise 12.6 If each operation in a parallel program takes about the same time, and 10% of the operations in the program must be performed sequentially, what is the maximum speedup ψ if 8 processors are used?

12.4 Programming Problems

Problem 12.1 Write a program to implement the Monte Carlo method to determine the value of π as the ratio of the area of the circle circumscribed by a square to the area of the square.

Problem 12.2 Write a program to implement the Monte Carlo method to determine the value of π from the probability of Buffon's needle problem.

Problem 12.3 The ratio of the area of a region to the area of a square enclosing the region can be approximated by a Monte Carlo method. So Monte Carlo methods can be used to approximate the area of a region in 2D (and volumes of regions in 3D). Use a Monte Carlo method to estimate the area of the region defined by the inequalities $x^2 + y^2 \leq 1$ and $y \geq x^2 - 1$.

Problem 12.4 Since a definite integral can be interpreted as a net area, we can use Monte Carlo methods to approximate a definite integral. Use both numerical integration method and a Monte Carlo method to approximate the definite integral

$$\int_{-1}^{1} e^{\sin x} dx$$

Appendices

A

An Introduction of MATLAB for Scientific Computing

We give a short tutorial about MATLAB for scientific computing. We cover MATLAB basics toward MATLAB programming.

A.1 What is MATLAB?

MATLAB is a computer language with many ready-to-use powerful and reliable algorithms for doing numerical computations. You can use it in an interactive manner like an advanced calculator, or you can build up your own set of functions and programs. Excellent graphics facilities are also available for visualizing your results.

You may purchase and download a personal copy of MATLAB at

https://www.mathworks.com/academia/student_version/

The student version is available for Windows, Mac and Linux computers. Many universities provide *free* installation or access of MATLAB. Please check the IT office of your university.

You can take online MATLAB tutorials at

http://www.mathworks.com/academia/student_center/tutorials/

These interactive tutorials are designed for students and are *free*. Please register to start learning MATLAB on your own pace. There are also many books with instructions on using and programming in MATLAB.

A.1.1 Starting MATLAB

On Windows systems, MATLAB is started by double-clicking the MATLAB icon on the desktop or by selecting MATLAB from the start menu. On OS X systems, MATLAB is installed in the main `Applications` folder, and can be opened by double-clicking the MATLAB icon there. On Linux or UNIX systems, MATLAB may be started by typing `MATLAB &` in the terminal.

On all systems, starting up MATLAB will take you to the Command window where the Command line is indicated with the prompt ». You can use MATLAB like a calculator by entering and executing MATLAB commands in this window. Later, you will learn how to write and save a MATLAB program and run your program like a command.

A.1.2 MATLAB as an Advanced Calculator

You can use MATLAB as an advanced calculator in the Command window by typing commands at the » prompt. Try out each of the follows and see what happens.

```
>> 1+4/2*5
>> 1+4/(2*5)
>> 1+(4/2)*5
>> 9^2
>> sqrt(81)
>> exp(1)
>> pi
>> sin(pi/2)
>> cos(pi/2)
>> tan(pi/4)
>> cot(pi/4)
```

Note that in the above commands, pi is reserved by MATLAB for π, exp(1) produces the approximate value of Euler's number e, and MATLAB's trigonometric functions are permanently set to **radians mode**.

Here is a useful tip. You can bring up a previous command by hitting the up-arrow key ↑ repeatedly until the command you want appears at the » prompt. You can then use keys ←, → and 'Delete' to edit the command. If you don't want to cycle through every command you've used, you can start typing the same letters and then use the up-arrow to cycle through only your previous commands that started with the same letters.

A.1.3 Order of Operations

MATLAB follows the standard mathematical order of operations in performing calculations. In order of precedence, these are:
- operations in parentheses, () (outward if nested parentheses)
- exponentiation, ^
- multiplication and division, * and / (left to right if together)
- addition and subtraction, + and - (left to right if together)

For example, try the following operations

```
>> 4 - 2 * 3^2
ans =
    -14
>> (4 - 2) * 3^2
ans =
```

```
    18
>> 4 - (2 * 3)^2
ans =
    -32
>> (4 - 2 * 3)^2
ans =
    4
```

A.1.4   MATLAB Built-in Functions and Getting Help

Many mathematical functions are already built into MATLAB and can be directly used as on a calculator. The following is a list of some common functions.

```
sin(x), cos(x), tan(x), cot(x), sec(x), csc(x)
asin(x), acos(x), atan(x), atan2(y,x)
exp(x), log(x), log10(x), log2(x)
sqrt(x), nthroot(x,n)
factorial(n), abs(x)
max(x), min(x)
```

In the above list, the trigonometric functions preceded by an a are the inverse trigono-metric functions, i.e. `asin(x)` is the same as $\arcsin(x)$ (or $\sin^{-1}(x)$). Also, `log(x)` is the natural log function, while `log10` is the customary base 10 (common) logarith-mic function, and `log2` is the base 2 logarithmic function. Finally, `factorial(n)` corresponds with $n!$ and `abs(x)` computes $|x|$.

When using MATLAB to write programs, the following functions are also useful:

```
sign(x), ceil(x), floor(x), round(x), mod(x,y)
```

You will see some useful functions associated with vectors and matrices later.

You can learn how to use a function if you know its MATLAB name. Just type `help` plus the function name at the prompt ». For example

```
>> help atan2
```

A help document will appear in the Command window telling you how to use `atan2`. For more detailed and interactive help on a given function, type `doc` plus the function name at the prompt, e.g.

```
>> doc atan2
```

Sometimes, you may use `help` or `doc` to find out what related functions are available. For example, if you know `plot` is a function to draw *x–y* plots, and you want to know what other plotting functions are available, you can issue `help plot` or `doc plot` and find the names of other plotting functions at the end of the help information.

A.1.5   Keeping a Record for the Command Window

Issuing the `diary` command,

```
>> diary mysession.txt
```

will cause all subsequent text that appear on the screen in the Command window, including commands and their outputs, to be **expanded** into the text file `mysession.txt` under the current working directory. The recording may be stopped by the command

```
>> diary off
```

and resumed by

```
>> diary on
```

It is recommended that if you wish to create a diary, you should use the file manager buttons at the top of the MATLAB desktop to change directories to the location where you want the diary locates *before* you issue the `diary` command.

A.1.6   Making M-Scripts

You can store your MATLAB commands (codes) into a file called an m-script. You can then run your file at any time you want. Basically, your file acts as a list of MATLAB commands. As you run the file, the MATLAB commands in your m-script are executed in the order from top to bottom. An m-script is very useful because it allows you to pack many commands with logical order into one file to fulfill a complicated task.

To create a new m-script file, click the icon 'New Script' in the 'Home' tab. To edit an old m-script file, click the icon 'Open' on the tool bar. These operations will open the **MATLAB Editor** for you to input or edit commands. Note that a m-script file must be saved with a ".m" extension. You can run your file (i.e. the commands contained in the file) by typing the file name without .m extension at the prompt » in the Command window. Below is an example of an m-script, which may be saved with the file name "TestMScript.m".

```
% Evaluate a numerical expression
1/2020-sin(2020)+cos(2020)
```

A line in an m-file starting with % is just a comment that is not executed by MATLAB. Comments should be used to help others and yourself understand your codes. A comment line starting with %% opens a new **cell** so that you can organize your codes into different cells according to their functionality. A MATLAB m-script is just a plain text file with a .m extension, so if you prefer you can use a different text editor (say vim, Geany or Sublime) to write and edit.

An m-script can be published to a webpage as a .html file by clicking the 'Publish' tab and then the 'Publish' icon. The webpage includes the commands in the m-script followed by the displayed results of these commands. A line starting with %% that opens a new cell shows up as a link at the beginning of the published webpage.

You can also choose to publish an m-script into other file formats, such as **pdf**, latex and xml (live scripts, see Section A.7).

A.2   Variables, Vectors and Matrices

A.2.1   Variables

Not all calculations can or should be performed on a single line; instead it may be helpful to store the results of calculations in variables for later use.

You can use a custom name for a variable. For example

```
>> a = 1+4/2*5
a =
    11
>> A = a*5
A =
    55
>> a
a =
    11
```

As shown above, MATLAB is **case-sensitive**, in that the variable a is different from the variable A. Also as seen above, you can use a stored result in subsequent calculations. If no name is given to a result, MATLAB uses the default variable ans to store the result, for example

```
>> 1+4/2*5
ans =
11
>> ans*5
ans =
55
```

These are examples of assignment statements: values are assigned to variables. Each variable must be assigned a value before it can be used on the right of another assignment statement. Please use meaningful names for your variables, such as `weight`, `position`, `speed`, etc., and avoid using names that correspond with MATLAB's built-in functions and variables, such as `pi` and `cos`. Finally, MATLAB variable names must not include spaces; if you wish to separate portions of a variable name you may use an underscore (_), as in `x_velocity`.

You can use the MATLAB command whos to show you the active variables and their sizes currently used in MATLAB.

```
>> whos
  Name       Size            Bytes  Class     Attributes

  A          1x1                 8  double
  a          1x1                 8  double
  ans        1x1                 8  double
```

You can clear all the variables using the MATLAB command `clear all`. It is a good practice to issue `clear all` before you start a new MATLAB session or program to avoid the accidental use of a previous variable active in your current session or program.

A.2.2   Suppressing Output

You may not want to see the results of intermediate calculations, especially if those calculations occur repeatedly within loops (more on loops later). If this is the case, you can suppress the output of an assignment statement or an expression by adding a semi-colon ; at the end of the statement or expression. For example,

```
>> x = 11; y = 5*x, z = x^2
y =
55
z =
121
```

With ; the value of x is hidden, while without ; the values of y and z are displayed. Note also that you can place several statements on one line separated by commas or semi-colons.

A.2.3   Vectors and Matrices

The name "MATLAB" stands for "MATrix LABoratory", since it was initially designed to simplify matrix arithmetic. Here is an example to input a row vector r.

```
>> r = [1 3, sqrt(49)]
r =
   1 3 7
```

The entries of the row vector may be separated by either a space or a comma, and are enclosed in square brackets.

Here is an example to input a column vector c.

```
>> c = [1; 3
       sqrt(49); 5]
c =
   1
   3
   7
   5
```

The entries of a column vector must be separated by semi-colons or by pressing `Enter` between rows of the vector.

Try the following MATLAB commands.

```
>> r = [1; 3; sqrt(49)];
>> length(r)
ans =
    3
>> norm(r)
ans =
   7.6811
```

Note that the MATLAB command `length(r)` gives the number of entries in the vector r. To compute the mathematical length/magnitude of the vector, you can use the command `norm(r)`, which computes $\|r\|$ as the standard square root of the sum of the squares of each vector entry. The `length` and `norm` functions work for both row and column vectors.

Here is an example of how to input a 3×2 matrix A.

```
>> A = [1 5; 7 9
        -3 -7]
A =
     1     5
     7     9
    -3    -7
```

As with row vectors, entries in each row are separated by spaces or commas. As with column vectors, different rows are separated by semi-colons or by pressing Enter.

Try the commands

```
>> A = [5 7 9; 1 -3 -7]
A =
     5     7     9
     1    -3    -7
>> size(A)
ans =
     2     3
```

The command `size(A)` gives a row vector, the first entry shows the number of rows and the second the number of columns in the matrix A.

A.2.4 Special Matrices

Some matrices are used frequently in scientific computing, so MATLAB provides functions to construct such matrices directly. To create a $n \times n$ identity matrix:

```
>> I = eye(3);
I =
     1     0     0
     0     1     0
     0     0     1
```

To create a $m \times n$ matrix of zeros:

```
>> A = zeros(2,3);
A =
     0    0    0
     0    0    0
```

To create a $m \times n$ matrix of ones:

```
>> A = ones(3,2);
A =
     1    1
     1    1
     1    1
```

To create a $m \times n$ matrix full of uniformly-distributed random numbers between 0 and 1:

```
>> A = rand(1,5);
A =
   0.8147    0.9058    0.1270    0.9134    0.6324
```

Finally, to create a $m \times n$ matrix full of normally-distributed random numbers with mean 0 and standard deviation 1:

```
>> A = randn(3,1);
A =
   -1.3077
   -0.4336
    0.3426
```

A.2.5 The Colon Notation and `linspace`

MATLAB provides a quick approach for creating a row vector with evenly-spaced entries. Try the commands.

```
>> a = 2:5
a =
     2    3    4    5
>> b = 1:2:8
b =
     1    3    5    7
>>
```

Generally, s:c:e produces a row vector of entries starting with the value s, incremented by the value c until it gets to the value equal or closest to e and in the range between s and e. If the increment c=1, you can omit it and just write s:e as in the first example above, but you have to include c if c=-1. For example,

```
>> 5:-1:1
ans =
  5  4  3  2  1
>>
```

Here are two more examples:

```
>> -1.4: -0.3: -2.1
ans =
     -1.4000    -1.7000   -2.0000
>> 1.4: 0.3: 2.1
ans =
1.4000   1.7000   2.0000
```

Note that the increment need not be an integer, nor must it be positive. Also pay attention to the last entry in the examples.

Another quick way to generate a row vector with evenly-spaced entries is to use the command linspace. For example

```
>> linspace(1.1,1.5,5)
ans =
    1.1000   1.2000   1.3000   1.4000   1.5000
>>
```

In general, linspace(s,e,n) generated a row vector with n evenly-spaced entries that start with s and end with e (the increment is computed automatically from these inputs).

A.2.6   Accessing Entries in a Vector or Matrix

The i-th entry of either a row or column vector v can be accessed through the index i by v(i). For example

```
>> a = 2:5
a =
    2  3  4  5
>> a(3)
ans = 4
```

The entry at row i and column j of a matrix A can be accessed as A(i,j). For example

```
>> A = [5 7 9; 1 -3 -7; 6 8 10]
A =
     5   7   9
     1  -3  -7
     6   8   10
>> A(2,3)
ans =
    -7
```

Not only can you retrieve a specific value of a matrix or vector, but this same notation may be used to change vector or matrix values. For example, following the previous commands, we may do

```
>> A(2,3) = 5
A =
        5     7     9
        1    -3     5
        6     8    10
```

You can access a range of entries of a vector or matrix by using index vectors generated by colon notation. For example

```
>> A = [5 7 9; 1 -3 -7; 6 8 10]
A =
        5     7     9
        1    -3    -7
        6     8    10
>> A(2:3,1:2:3)
ans =
        1    -7
        6    10
```

In this example, the accessed rows are specified by the vector $2:3$, which are row 2 and 3, and the accessed columns are specified by the vector $1:2:3$, which are columns 1 and 3.

To access a whole row or column, refer to the following example

```
>> A = [5 7 9; 1 -3 -7; 6 8 10]
A =
        5     7     9
        1    -3    -7
        6     8    10
>> A(2,:)
ans =
        1    -3    -7
>> A(:,3)
ans =
        9
       -7
       10
```

Note that $A(2,:)$ retrieves the second row of the matrix A, and $A(:,3)$ returns the third column of the matrix A.

We note that the use of colon notation to access subsets of a vector or matrix is called **array slicing**, and is a unique benefit of MATLAB over most other programming languages.

A.3   Matrix Arithmetic

As long as the dimensions of a matrix arithmetic operation are allowable, MATLAB can perform matrix operations using natural mathematical notation.

A.3.1   Scalar Multiplication

Scalar multiplication refers to the product of a scalar and a matrix or vector. For example

```
>> A = [5 7 9; 1 -3 -7; 6 8 10]
A =
        5     7     9
        1    -3    -7
        6     8    10
>> -A
ans =
       -5    - 7    -9
       -1      3     7
       -6     -8   -10
>> 0*A
ans =
        0     0     0
        0     0     0
        0     0     0
>> 2*A
ans =
       10    14    18
        2    -6   -14
       12    16    20
```

A.3.2   Matrix Addition

Matrices or vectors may be added or subtracted as long as each object has the same size and shape. For example

```
>> A = [5 7 9; 1 -3 -7; 6  8  10], B = [1 3 2; 2  3  6;-4 -2  11]
A =
        5     7     9
        1    -3    -7
        6     8    10
B =
        1     3     2
        2     3     6
```

```
          -4    -2    11
>> A+A
ans =
       10    14    18
        2    -6   -14
       12    16    20
>> A+B
ans =
        6    10    11
        3     0    -1
        2     6    21
>> A-B
ans =
        4     4     7
       -1    -6   -13
       10    10    -1
```

A.3.3   Matrix Multiplication

Matrices and vectors may be multiplied as long as the inner dimensions of the operands agree. For example

```
>> A = [5 7 9;1 -3 -7],B = [1 3;2 3;-4 -2],x = [1;2;3],y = [4 5]
A =
        5     7     9
        1    -3    -7
B =
        1     3
        2     3
       -4    -2
x =
        1
        2
        3
y =
        4     5
>> A*B
ans =
      -17    18
       23     8
>> A*x
ans =
       46
      -26
>> y*A
ans =
```

```
25    13    1
```

MATLAB requires that all matrix arithmetic operations adhere to the correct mathematical rules, meaning that when multiplying two matrices, the number of columns of the first must equal the number of rows in the second. When you violate this rule, MATLAB will issue an error message:

```
>> B*x
Error using  *
Inner matrix dimensions must agree.
```

The error message text is usually red, and MATLAB will typically beep on errors.

A.3.4 Transpose

The transpose of a matrix A, denoted as A^T in linear algebra, is obtained in MATLAB by A'. For example

```
>> A = [5 7 9; 1 -3 -7]
A =
        5    7    9
        1   -3   -7
>> A'
ans =
        5    1
        7   -3
        9   -7
>> (A')'
ans =
        5    7    9
        1   -3   -7
```

You can combine the transpose with matrix multiplication in interesting ways. For example, you can compute the inner product of the column vectors a and b using the command a'*b (or equivalently b'*a). Similarly, you can compute the outer products of these column vectors as a*b' and b*a'. For example

```
>> a=[1;2;3]
a =
        1
        2
        3
>> b=[4;5;6]
b =
        4
        5
        6
```

```
>> a'*b
ans =
    32
>> b'*a
ans =
    32
>> a*b'
ans =
     4     5     6
     8    10    12
    12    15    18
>> b*a'
ans =

     4     8    12
     5    10    15
     6    12    18
>>
```

A.3.5   Entry-wise Convenience Operations

In addition to supporting standard matrix arithmetic operations, MATLAB also supports a variety of convenient operators that are not standard mathematical operations. However, while these do not correspond to correct mathematics, they may be very useful for some programming purposes.

For example, to add a number to every entry of a vector or matrix:

```
>> A = [5 7 9; 1 -3 -7]
A =
     5     7     9
     1    -3    -7
>> A+100
ans =
   105   107   109
   101    97    93
```

If you want to apply entry-wise operations such multiplication, division or power on matrices, you can supply a . preceding the arithmetic operator. For example

```
>> A = [5 7 9; 1 -3 -7]
A =
     5     7     9
     1    -3    -7
>> B = [1 7 3;-1  3  7]
B =
     1     7     3
```

```
         -1    3    7
>> A.*B
ans =
          5   49   27
         -1   -9  -49
>> A./B
ans =
          5    1    3
         -1   -1   -1
>> A.^2
ans =
         25   49   81
          1    9   49
```

Note that A.*B produces a matrix (with the same size as A and B) with each entry being the product of the corresponding entries in A and B.

Here is another example of a "dot" operation.

```
>> r = [1 0.5 0.25 2]
r =
        1.0000    0.5000    0.2500    2.0000
>> r1 = 1./r
r1 =
        1.0000    2.0000    4.0000    0.5000
```

This example shows how to generate a vector with each entry being the reciprocal of the corresponding entry of another vector.

A.4   Outputting/Plotting Results

A.4.1   disp

The MATLAB function disp(x) displays the array x, without printing the array name. If x is a string, the text is displayed. For example

```
>> r = [1 0.5 0.25 2];
>> disp(r)
    1.0000    0.5000    0.2500    2.0000

>> disp('Hello!')
Hello!
>> s = 'Hello again!';
>> disp(s)
Hello again!
```

A.4.2 `fprintf`

The MATLAB function `fprintf` writes data in a controlled format on the screen or into a file. If you have ever used the `printf` command in C or C++, the syntax in MATLAB is identical. For example

```
>> balance = 50;
>> fprintf('The balance is %d $. \n',balance)
The balance is 50 $.
>>
```

In this example, the value of the variable `balance` is shown in the format controlled by `%d` (meaning decimal number) at the location of the format control `%d`. The special format `\n` means linefeed (another subsequent output will be on a new line). The other texts (except `%d` and `\n`) are shown as they are within the quotation marks. Some other often-used **formatting specifiers** include:

- `%d` – scientific notation for decimal numbers, normal notation for integers
- `%e` – decimal number with seven significant digits, using scientific notation
- `%f` – decimal number with six digits after the decimal, using fixed-point notation
- `%g` – most appropriate notation for decimal numbers (six significant digits), normal notation for integers
- `%i` – same as `%d`
- `\t` – horizontal tab

For more information, type `help fprintf` or `doc fprintf`, or look this up on the web.

A.4.3 `plot`

The MATLAB function `plot` produces $x - y$ plots. For example

```
>> a = [1 2 3];
>> b = [1 4 9];
>> plot(a,b)
```

The plot in Fig. A.1 is generated. What the MATLAB function `plot` does in the example is to draw a line that connects the three points (1,1), (2,4) and (3,9) on an $x - y$ coordinate system. The x-coordinates of the points are given by the vector a, and the y-coordinates of the points are given by the vector b.

You can change in what shape the points are drawn, how they are connected, and what color is used (by default, the points are drawn as dots and are connected by solid lines in blue). For example

```
>> a = [1 2 3];
>> b = [1 4 9];
>> plot(a,b,'k--o')
>> xlabel('a')
>> ylabel('b')
>> title('b = a^2')
```

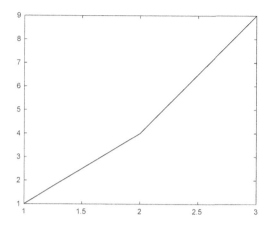

FIGURE A.1
An $x - y$ plot of three points (1,1), (2,4) and (3,9) by MATLAB.

The added string

    'ko--'

in the arguments of `plot` changes the:
- color to black with the `'k'`,
- symbol to circle with the `'o'` and
- line style to a dashed line with the

    '--'

When plotting data, you may want to add labels for the x and y axes, and add a title to the plot, as shown in this example. See the effect in Fig. A.2.

The final example below demonstrates one way to draw multiple curves in a single plot. Pay attention to `legend`, which is used to name and distinguish each curve from one another.

```
>> t = linspace(-pi,pi,200);
>> s = sin(t);
>> c = cos(t);
>> plot(t,s,'b-',t,c,'r--')
>> xlabel('t')
>> legend('sin(t)','cos(t)')
```

The resulting plot is shown in Fig. A.3.

By default, MATLAB generates one plot in a figure. If the function `plot` is used for multiple times, only the last plot is shown in the figure because a current plot replaces the previous plot. If you want to show all the plots in the figure, you can use `hold on` after `plot` as in the following example

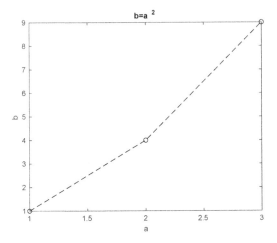

FIGURE A.2

An $x - y$ plot of three points $(1,1)$, $(2,4)$ and $(3,9)$ with changed specified color, symbol and line style, and added labels and title.

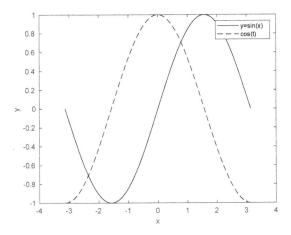

FIGURE A.3

Multiple curves on an $x - y$ plot.

```
>> t = linspace(-pi,pi,200);
>> s = sin(t);
>> c = cos(t);
>> plot(t,s,'b-')
>> hold on
>> plot(t,c,'r--')
```

```
>> xlabel('t')
>> ylabel('y')
>> hold off
```

You can generate multiple figures using the function figure as in the following example

```
>> t = linspace(-pi,pi,200);
>> s = sin(t);
>> c = cos(t);
>> figure(1)
>> plot(t,s,'b-')
>> xlabel('t')
>> ylabel('y')
>> figure(2)
>> plot(t,c,'r--')
>> xlabel('t')
>> ylabel('y')
```

You can also generate multiple plots in one figure using the MATLAB subplot command (try doc subplot for details).

A.5   Loops and Decisions

A.5.1   for Loops

The primary way to do a repeated task in any programming language is through a loop. The idea with a loop is that it will do a task repeatedly at every **iteration** of the loop.

The main loop structure in MATLAB is the for loop. Below is a simple example:

```
>> for i=1:2:5
fprintf('the value of i at the current step is: i = %d\n',i)
end
the value of i at the current step is: i = 1
the value of i at the current step is: i = 3
the value of i at the current step is: i = 5
>>
```

Note that the first line of the loop begins with the keyword for. The contents (i.e. the command fprintf in this example) of the loop follow this first line, and cease just before the keyword end. The syntax for i=1:2:5 indicates that the loop variable i takes in turn each value in the vector 1:2:5, and the contents of the loop are repeatedly executed for each value of i.

Below is another example:

```
>> v=[50 40 30 20];
>> for k=v
        disp(k)
   end

   50

   40

   30

   20
>>
```

Now let's use a loop to compute the sum $100 + 99 + 98 + \cdots + 2 + 1$. The trick here is to define a variable, say s, initialize it to 0, and then add in turn each entry in the array $100 : -1 : 1$ to s. The code is

```
>> s = 0;
>> for i=100:-1:1
        s = s+i;
end
>> disp(s)
        5050
>>
```

Note that s is initialized to be zero (s = 0), and then it is updated with the command s = s+i by adding the current value of i (taking one by one from the array 100:-1:1) to the current value of s. You should understand this trick and know how to use it.

Suppose you need to generate an $m \times n$ matrix H with the entry $h_{ij} = 1/(i+j-1)$. You can do so using nested for loops as follows:

```
>> m=3;
>> n=4;
>> H=zeros(m,n);
>> for i=1:m
       for j=1:n
           H(i,j)=1/(i+j-1);
       end
   end
>> H
H =

    1.0000    0.5000    0.3333    0.2500
    0.5000    0.3333    0.2500    0.2000
    0.3333    0.2500    0.2000    0.1667
>>
```

Before the nested for loops, H=zeros(m,n) sets up the size of the array H and allocates the memory for the array by initializing it.

A.5.2 Logicals and Decisions

As with other programming languages, MATLAB supports logic statements:

```
>> m = 1; n = 2;
>> m == n
   ans =
        0
>> m ~= n
   ans =
        1
>> m < n
   ans =
        1
>> m > n
   ans =
        0
>> m <= n
   ans =
        1
>> m >= n
   ans =
        0
>> ((m==m) && (n==n))
   ans =
        1
>> ((m~=n) || (m>n))
   ans =
        1
```

In the above logical statements, "true" equates to 1 and "false" equates to 0. The **comparison operations** in logical statements are
- == "is equal to"
- ~= "is not equal to"
- < "is less than"
- > "is greater than"
- <= "is less than or equal to"
- >= "is greater than or equal to"

Moreover, logic statements can be combined with && (and) or || (or).

Sometimes whether some commands will be executed depends on certain conditions. The if statement can help in this situation. For example, suppose a bank applies 5% annual interest rate on an account if the account balance is above 5000$.

Otherwise, the interest rate is only 2%. To compute the new account balances for a set of four accounts after one year, the bank can use a MATLAB code as:

```
>>  bal = [6000 5000 20000 200];
>>  for i = 1:length(balance)
        if bal(i)>5000
           bal(i)=bal(i)*(1+0.05);
        else
           bal(i)=bal(i)*(1+0.02);
        end
    end
>> disp(bal)
        6300          5100        21000         204
>>
```

The general form of the if statement is

```
if logical test 1
   Commands to be executed if test 1 is true
elseif logical test 2
   Commands to be executed if test 2 is true but test 1 is false
.
.
.
else
  Commands to be executed if all the previous tests are false
end
```

For complicated logical statements, it is standard to enclose the logic test in parentheses.

You can exit (jump out of) a for loop using break after a logical statement. For example

```
>> for i=1:100
       if i^2+i>1234
          break
       end
    end
>> disp(i)
   35
```

In the above example, when i moves to 35, the logical statement is true, and the loop is terminated by break at $i = 35$. The break statement exits a for loop completely. To skip the rest of the instructions in a loop and begin the next iteration, use a continue statement as in the following example

```
>> for n = 1:50
      if mod(n,7)
```

```
        continue
      end
      disp(['Divisible by 7: ' num2str(n)])
   end
Divisible by 7: 7
Divisible by 7: 14
Divisible by 7: 21
Divisible by 7: 28
Divisible by 7: 35
Divisible by 7: 42
Divisible by 7: 49
```

In this example, if the remainder after the division of the current n by 7 is not zero, the loop begins the next iteration (the next n) by skipping the disp instruction. So we only see the display of the values of n that are divisible by 7 (with zero remainder).

A.5.3 while Loops

Sometimes you want to repeatedly execute a section of codes until a condition is satisfied, but you do not know in advance how many times you will need to repeat the codes. Since a for loop needs to know how many times it will be evaluated beforehand, MATLAB supplies another looping mechanism, a while loop. For example,

```
>> i=0;
n=11;
while n<100
      n=3*n;
      i=i+1;
      fprintf('at step i = %d: n = %d\n',i,n)
end
at step i = 1: n = 33
at step i = 2: n = 99
at step i = 3: n = 297
```

Note that the commands of the while loop are executed 3 times, as shown by the counter i, and when the loop finishes, $n = 297$ and the condition n<100 is no longer satisfied.

A.6 Functions

You have seen some built-in MATLAB functions such as sin, cos and so on. If you want to use sin to compute $\sin(\pi)$ and assign the result to a variable, you can do the following

```
>> y=sin(pi)
y =
   1.2246e-16
```

In the above example, the function has the name `sin`, it receives the value `pi` as an input in the parentheses following its name, and it returns the result $\sin(\pi)$, which is assigned to the variable y.

> **Remark** Note that in this example the value of $\sin \pi$ given by MATLAB is very close to zero, but not exactly zero. This is due to approximation and roundoff **errors** (see Chapter 3).

MATLAB enables you to create your own functions which can be used in a similar way as the above example.

A.6.1 M-Functions

For example, the following m-function is created to find the two roots of the quadratic equation $ax^2 + bx + c = 0$ using the quadratic formula, under the assumptions that $a \neq 0$ and that the equation has real-valued roots:

```
function [x1 x2] = quadroots(a,b,c)
sdelta = sqrt(b^2-4*a*c);
x1 = (-b+sdelta)/(2*a);
x2 = (-b-sdelta)/(2*a);
end
```

As you can see, unlike a MATLAB script, the m-function starts with the keyword `function` and ends with the keyword `end`. Here, the created function is named quadroots – you should use meaningful names.

The above function receives the coefficients a, b and c, and it returns the two roots x1 and x2, which are listed between the square brackets at the left of the assignment (i.e. the left of the equal sign =). The main body of the function m-file applies formulas to compute the two roots using the received coefficients.

This m-function must be saved with the file name `quadroots.m`, i.e. the file name must be the same as the function name, such that other m-scripts or m-functions know in which file to find the codes when they call (use) the created function by the name quadroots. Furthermore, as with MATLAB variable names, MATLAB function names *must not* contain spaces.

Sometimes, you may terminate your m-function under a specified condition with the command `return` before all the lines of the m-function are executed, as in the following example

```
function [x1 x2] = quadroots(a,b,c)
if a==0
```

```
    disp('a can NOT be zero!')
    return
end
sdelta = sqrt(b^2-4*a*c);
x1 = (-b+sdelta)/(2*a);
x2 = (-b-sdelta)/(2*a);
end
```

After the function `quadroots` is created and saved, you can use it in an m-script or another m-function, as long as the m-files are located in the same directory on the computer. Below is an example of a test m-script `testquadroots.m`, which passes the values of the coefficients to the function `quadroots` and receives the roots from it.

```
aa=1;
bb=0;
[root1 root2] = quadroots(aa,bb,-4);
fprintf('the roots are: %d and %d\n',root1,root2)
```

The caller `testquadroots.m` passes values to and receives values from the callee `quadroots` as shown in Fig. A.4 The variables (aa and bb) passed to `quadroots`

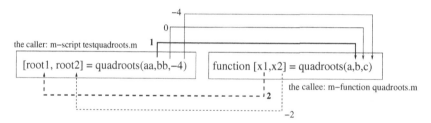

FIGURE A.4
Pass by value: the caller and the callee have two independent variables with the same value, and changes made to one variable do not affect the other variable.

can have different names from those in the input list (a and b) of the function m-file `quadroots.m`, but they must have been assigned values and must be in the correct order (i.e. aa=1 corresponds to a and so on). Likewise, the values returned from `quadroots` are assigned to the variables `root1` and `root2` in the order determined by the m-function, and the names of the variables that are assigned to can be different from the names in the output list of the m-function.

The variables declared inside a function m-file exist only within that function. In the above example, although `quadroots` created a local variable `sdelta`, since it was not included in the output list for the function, the caller `testquadroots.m` cannot access that value since it only existed within the scope of the callee `quadroots`. Likewise, although `testquadroots.m` declared the variables aa and bb, the `quadroots` function itself cannot reference variables with those names, and must instead reference the variables (a, b and c) that were passed with values.

A function m-file has the following general form:

```
function [output1, output2, ...] = name(input1, input2, ...)
% description of the function, including input, output and
% method
...
output1 = ...;
...
output2 = ...;
...
end
```

A.6.2 Anonymous Functions

Many functions do not require much space, or you may want them to change as a program proceeds. In such cases, creating an entire function m-file to hold the small function, or creating many function m-files to hold each version of a function, may be unnecessary. In such situations, MATLAB allows you to declare an anonymous function. These may be defined in a script m-file, a function m-file, or even in an interactive MATLAB session. For example, to create a simple function that implements $g(x) = \cos(\pi x)$ you can use the MATLAB code

```
g = @(x) cos(pi*x);
```

This function will persist in the namespace of the function, script or session where it is defined, similar to a MATLAB variable.

Here are a few rules about anonymous functions:

- The entire function contents must consist of a single line of execution.
- The function name is listed at the left of the equal sign to the left of the ampersand.
- All arguments to the function must be listed in the parentheses following the ampersand.
- The function cannot create temporary variables.
- The function can see and use any other variables defined previously within the file.
- While it is possible for anonymous functions to return multiple outputs, it is somewhat challenging and is not recommended.

For example, here is a more complicated example using anonymous functions:

```
a = 2;
b = -3;
c = 1;
d = 8;
f = @(x,y) a*(x-b)^2 + c*(y-d)^2;
clear a b c d
x = linspace(-5,-1,100);
y = linspace(4,12,200);
z = zeros(200,100);
for i=1:100
```

```
   for j=1:200
       z(j,i) = f(x(i),y(j));
   end
end
surf(y,x,z)
```

Because a, b, c and d were available at the time f was defined, the anonymous function handle includes those values even though they were deleted prior to using f.

A.6.3 Passing Functions to Functions

Sometimes we need to pass functions to other functions. For example, if the m-function s = integrate(f,a,b) is written to approximate the definite integral $\int_a^b f(x)dx$, we need to pass the integrand $f(x)$ to this m-function. We have two options for this:

(1) Define the integrand as an anonymous function and then pass its name, for example

```
>> myfunction  = @(x) exp(x)-x+cos(x);
>> s = integrate(myfunction,-1,1);
```

(2) Create an m-function func.m for the integrand and then pass the corresponding function handle, for example

```
function [y] = func(x)
y = exp(x)-x+cos(x);
end
```

```
>> s = integrate(@func,-1,1);
```

Note that the function handle is obtained by the symbol @ in front of the function name.

Below is another example. The function m-file myplot receives the function handle of a function to plot the function on a specified interval $[a, b]$ using the specified number n of points.

```
function myplot(myfunction,a,b,n)
% plot the graph of a function on [a,b] with n points
% myfunction is name of the function m-file that
% defines the function to be plotted
% myfunction must accept a vector input and return a vector
% output

x=linspace(a,b,n);
y=myfunction(x);
% or y=feval(myfunction,x);

plot(x,y,'b-')
```

```
xlabel('x')
ylabel('y')
title('the graph of y=f(x)')
end
```

The m-script that tests `myplot` is given as the following.

```
a = 1-0.005;
b = 1+0.005;
n = 200;

myfun = @(x) (x-1).^6;

figure(1)
myplot(myfun,a,b,n)
hold on
myplot(@samefun,a,b,n)
hold off
```

where the function m-file that defines the function to be plotted is

```
function y = samefun(x)
y = x.^6-6*x.^5+15*x.^4-20*x.^3+15*x.^2-6*x+1;
end
```

> **Remark** Note that
>
> $$(x-1)^6 = x^6 - 6x^5 + 15x^4 - 20x^3 + 15x^2 - 6x + 1$$
>
> However, if you try the above example, you will see the two curves are strangely different. The reason for the difference is catastrophic cancellation (see Chapter 3).

A.7 Creating Live Scripts in the Live Editor

MATLAB has the feature of live scripts in 2016R and newer versions.

Live scripts are program files that you write to include your code, output from the code, formatted text, images and equations together in a single interactive environment called the Live Editor. They are a powerful way to present and share your work.

A live script can be created in the Live Editor of MATLAB. Go to the 'Home' tab and click 'New Live Script' to launch the Live Editor. A live script in the Live Editor is saved with a .mlx file extension. You can open an existing .mlx live script. You can also save a .m m-script as a .mlx live script by choosing .mlx extension using the menu 'Save As', which leaves original .m m-script intact but generates a corresponding .mlx live script.

To run the code, go to the 'Live Editor' tab and click 'Run'. By default, MATLAB displays output to the right of the code. To move the output in line with the code, you can go to the 'View' tab and click the 'Output Inline' button.

You can add formatted text, hyperlinks, images and equations to your live scripts at suitable places to create a presentable document to share with others. You can go to the 'Insert' tab to do such tasks.

Finally, click the icon Save in the live editor, you can export your live script as a pdf, html or Latex file.

A.8   Concluding Remarks

MATLAB is a powerful computing environment that has been built up over many years as a tool for scientific computing. As a result, no tutorial can give an exhaustive account of all MATLAB functionality, though this tutorial should be enough to help you get started. As with any technology tool, the best way to learn is through practice, sharing tips with others, and reading the documentation.

Below are some general suggestions on programming.

- **Write pseudocode** Write down the pseudocode for an algorithm to identify inputs, outputs, data structures, modules and flow of the algorithm.
- **Break the problem** Break the problem into small self-contained tasks and write a function to complete each small task. If a built-in function is available for a task, use it.
- **Be careful and check** Pay attention to all details (such as syntax, variable types, order of operations and updates of values, etc.) when writing a program. Check a program by reading its codes line by line.
- **Use meaningful names** A variable name should reflect the meaning and type of the variable and be easy to distinguish.
- **Include comments** Include a preface to each program to briefly describe the program (such as its purpose, inputs and outputs and major method). Separate a code into functional blocks and provide concise comments for each block.
- **Show intermediate results** Sometimes it is helpful to show intermediate results to debug a program (find mistakes).

Have fun!

A.9   Programming Problems

Problem A.1 Input the following matrices and vectors in MATLAB.

$$A = \begin{bmatrix} 0 & -6 & 8 & 1 \\ -2 & 5 & 5 & -3 \\ 7 & 8 & 0 & 3 \end{bmatrix}, \quad B = \begin{bmatrix} 0 & 1 \\ 5 & -3 \\ 7 & 3 \\ -2 & 12 \end{bmatrix},$$

$$a = \begin{bmatrix} 3 & 2 & -13 \end{bmatrix}, \quad b = \begin{bmatrix} 3 \\ 2 \end{bmatrix}, \quad c = \begin{bmatrix} -6 \\ 11 \\ 9 \end{bmatrix}.$$

(1) Test the commands `length` and `size` on A, b and c. What do the commands do?
(2) Write a command to retrieve the entry of A at row 2 and column 4, and write a command to retrieve the third entry in c.
(3) Use the colon notation to assign the 2×3 lower right part of the matrix A to the matrix C.
(4) Write commands to swap the second and third rows of B and show the new B.
(5) Test the commands `3*A`, `A/2,-A`, `3+A`, `a+1`. What do `A/2` and `3+A` do?
(6) Test the commands `A*B`, `B*A`, `a*A`, `A*a`, `b*B`, `B*b`, `a*c` and `c*a` and try to understand the results.
(7) Test the commands `diag(a)` and `diag(b)`. What kind of matrix does each command generate?

Problem A.2 Test the command `eye(4)`, `zeros(2,5)`, `ones(3,4)` and `rand(3,2)`. What kind of matrix does each command generate?

Problem A.3 (1) Use the colon notation to generate a row vector with entries that are even numbers between 10 and 0 (including 10 and 0) in the decreasing order, then generate the same vector using the command `linspace`.
(2) Use the command `linspace` and the transpose operation to generate a column vector with 9 equally spaced entries between 0 and 1 (including 0 and 1), and then use the colon notation to generate the same vector.

Problem A.4 Input the matrix

$$A = \begin{bmatrix} 0 & -6 & 8 & 1 \\ -2 & 5 & 5 & -3 \\ 7 & 8 & 0 & 3 \end{bmatrix},$$

(1) Use the "dot" operation to generate a matrix with each entry the cube of the corresponding entry of A.
(2) Use the "dot" operation to generate a matrix with each entry the reciprocal of the corresponding entry of A.

Problem A.5 Issue the command doc fprintf in MATLAB to learn more about fprintf and then complete the following practice.
(1) Display the MATLAB value of π such that your output looks as the following:
The MATLAB value of pi = 3.14159265358979
(Hint: Use %16.14f)
(2) Display the MATLAB value of e (given by exp(1)) such that your output looks as:
The MATLAB value of e = 2.71828182845905
(3) Display the value of the universal gravitational constant $G = 6.674 \times 10^{-11} [m^3 kg^{-1} s^{-2}]$ such that your output looks as:
The universal gravitational constant G = 6.67e-11 m^3/kg/s^2
(Hint: Use %5.2e or %g)

Problem A.6 Let x = 0:.1:1;. Generate a table of values by A = [x; exp(x); log(x+1)]. Then save the table to a file named 'mytable.txt' using fprintf. Open your file and check if it stores the table as you want. (Hint: doc fprintf)

Problem A.7 Plot the graphs of the functions $y = e^x$ and $y = 1 + x + x^2/2 + x^3/6$ in the same figure using 21 data points for each graph with 21 equally spaced values of x in the interval $[-2, 2]$. In your figure, you need to distinguish the two graphs with different colors, different line styles and different symbols for the data points. You also need to annotate your plot with labels, title and legend.

Problem A.8 Use a for loop to add all the odd numbers between 1 and 99: $1, 3, 5, \ldots, 97, 99$. Use fprintf to output your final result. Remember to use semicolons at the end of lines in your loop to suppress garbage output.

Problem A.9 Use nested for loops to assign entries of a 5×5 matrix A such that $A(i, j) = ij$.

Problem A.10 The Fibonacci sequence $\{F_n\}_{n=0}^{\infty} = \{1, 1, 2, 3, 5, 8, \ldots\}$ satisfies the recurrence relation $F_{n+2} = F_n + F_{n+1}$, $n = 0, 1, 2, \ldots$. Given $F_0 = F_1 = 1$, use a for loop but without an array to implement the recurrence relation to compute and display the first 20 entries in the sequence.

Problem A.11 If an bank account balance is less than 5000, the annual interest rate is 1%, if it is in the range $[5000, 20000)$, the interest rate is 2%, otherwise the annual interest rate is 5%. Generate 5 account balances as bal=[0 1000 5000 10000 200000 600000], then use if statement to calculate the new balances of the 5 accounts after a year.

Problem A.12 The variable *epsilon* is initially equal to 1. Use a while loop to keep dividing *epsilon* by 2 until *epsilon* $< 10^{-6}$, and determine how many divisions are used until *epsilon* $< 10^{-6}$ using MATLAB. Verify your result by algebraic derivation.

Problem A.13 Write an m-function that accepts one numeric parameter. This parameter will be the measure of an angle in radians. The function should convert the radians into degrees and then return that value. Write an m-script to test your function on angles in $\pi/6$, $\pi/4$, $\pi/3$, $\pi/2$ and π radians.

Problem A.14 Write a MATLAB m-function which converts temperature between Celsius (C) and Fahrenheit (F). The function determines the direction of the conversion (i.e. from F to C or C to F) by an input. Add comments in your m-function to describe how this input indicates the conversion direction. Write an m-script to test your m-function to make two plots, one for F versus C (i.e. by converting a vector of temperature in C to F) and the other for C versus F. Annotate each of your plots. Use the command `grid` on after each plotting command to turn on the grid in each plot.

Problem A.15 Write a MATLAB m-function function
`[fg gf] = compositefun(f,g,x)`
that receives the function handles `f` and `g` to compute the values of the composite functions $f(g(x))$ and $g(f(x))$. Define one mathematical function as an m-function and the other mathematical function as an anonymous function in an m-script, and test `compositefun` by plotting the graphs of the two composite functions of your mathematical functions. Do not forget to annotate your plots.

Problem A.16 The factorial of n is defined as $n! = 1 \times 2 \times 3 \times \cdots \times n$. Write a MATLAB m-function `myfactorial` to compute factorials in a recursive manner by calling the m-function `myfactorial` itself. Test your m-function `myfactorial` for $n = 0, 1, 2, 3, 4, 5, 6, 7, 8, 9, 10$. Make sure your recursive call does not go on forever.

B

An Introduction of Python for Scientific Computing

We give a short tutorial about Python for scientific computing. We introduce Python basics toward Python programming using elucidating examples.

B.1 What is Python?

Python is a programming language that can be installed and run on all the major operating systems (Windows, Linux and OS X). It has high-level data types and many ready-to-use modules (a module stores functions and classes in Python). You can use it in an interactive manner like an advanced calculator, or you can build up your own programs and modules.

You can take an online Python tutorial at

https://www.learnpython.org/

B.1.1 Starting Python

Python has different versions. This tutorial is for Python 3. Python can be launched from a **terminal window** in a Linux, OS X or Windows system. Following the prompt of the terminal window of your operating system, issue the command `python3` to launch the Python interpreter. After you launch Python, you see the Python prompt `>>>` after the version information

```
Python 3.6.9 (default, Jan 26 2021, 15:33:00)
[GCC 8.4.0] on linux
Type "help", "copyright", "credits" or "license" for more info.
>>>
```

You can use Python like a calculator by entering and executing statements immediately after the prompt `>>>` in the window. Later, you will learn how to write and run a Python program which includes many lines of a Python code.

You can terminate Python by entering the function `quit()` (a function has a name, carries out a specific task, and is used by calling its name) or typing an end-of-file key combination `Control-Z`.

B.1.2   Python as an Advanced Calculator

You can use Python as an advanced calculator in the terminal window by typing a Python statement (a statement can be a numerical expression, a definition, an assignment of a variable, a function call, etc.) at the >>> prompt (without indentation before the statements). Try out each of the follows.

```
>>> 1+4/2*5
11.0
>>> 1+4/(2*5)
1.4
>>> 5%2
1
>>> 6%3
0
>>> 5//2
2
>>> 5//-2
-3
```

where % is the modulo operator which returns the remainder of a usual division, and // is the integer division that returns the value of the so-called floor function (i.e. maximum integer less than or equal to the result of the usual division).

Python follows the standard mathematical order of operations in performing calculations. In order of precedence, these are:
- operations in parentheses, () (outward if nested parentheses)
- exponentiation, **
- multiplication and division, * and / (left to right if together)
- addition and subtraction, + and - (left to right if together)

Try the following examples (besides the previous examples)

```
>>> 4-2*3**2
-14
>>> (4-2)*3**2
18
>>> 4-(2*3)**2
-32
>>> (4-2*3)**2
4
```

Python calls a number with a decimal point a **float**. Try the following

```
>>> 0.1+0.2
0.30000000000000004
```

We end up with a result slightly different from the exact result 0.3. This difference is due to the inexact representations of floats by a computer, as explained in Chapter 3.

Besides numbers, Python can also manipulate strings (a string is a series of characters). A string can be enclosed in single quotes or double quotes. For example

```
>>> "Hello world!"
'Hello world!'
>>> 'Hello world!'
'Hello world!'
```

The flexibility to use both kinds of quotes allows us to use quotes and apostrophes within a string as follows

```
>>> 'He said "Hello!"'
'He said "Hello!"'
>>> "It's he who said 'hello'."
"It's he who said 'hello'."
```

The character combination \t in a string adds a tab when the string is printed by the Python function `print`. For example

```
>>> print('Hello\tworld!')
Hello world!
```

The character combination \n in a string adds a newline when the string is printed by the Python function `print`. For example

```
>>> print('Hello\nworld!')
Hello
world!
```

Python has many methods to operate on strings. The Python function `str()` uses an integer as a string by wrapping the integer (or its variable) in this function. For example

```
>>> year = 2021
>>> print('Hello year'+str(year))
Hello year 2021
```

where the + sign combines strings. Without using the function `str()`, you get a type error

```
>>> year = 2021
>>> print('Hello year'+year)
Traceback (most recent call last):
File "<stdin>", line 1, in <module>
TypeError: must be str, not int
```

B.1.3 Python Programs

You can store your Python statements (i.e. code) into a file called a Python program. You can then run your program at any time you want. A Python program allows you to pack many lines of code with a logical order to fulfill a complicated task.

You can create a Python program in any text editor. Some text editors (such as Geany and Sublime) use syntax highlighting to color your Python programs and allow you to run your Python programs directly from the editor instead of through a terminal window. Below is a Python program which uses the Python function `print` to display a string

```
# Greeting message
print('Hello world!')
```

where the first line is just a comment. A comment in Python starts with the hash mark #. Comments in a code are written to explain the code to its readers and are ignored by Python when the code is run. You can create this program in your favorite text editor and save it in a file with *.py* as the file extension (for example, *hello_world.py*). If your text editor allows you to run Python programs directly, you can run your *hello_world.py* and get the following output

```
Hello world!
```

To run your Python program *hello_world.py* from a terminal window, you issue in the directory of your program the following command following the prompt of your terminal window (in the example below, the prompt is $)

```
sheng@1711338:~/PythonLearning$ python3 hello_world.py
```

Again you get the output

```
Hello world!
```

B.2 Variables, Lists and Dictionaries

In this section, we introduce variables and some important data types.

B.2.1 Variables

Generally speaking, a variable is room with a name to store a value, which can be a numerical value (an integer or float), a string, or other types of data. You assign a value to a variable using the = sign. For example

```
>>> x = 1
>>> y = 1/2
>>> message = "Hello world!"
```

The variables x, y and `message` are assigned with their corresponding values. You can display the value of a variable using the Python function `print`. For example

```
>>> x = 1
>>> print(x)
1
>>> y = 1/2
>>> print(y)
0.5
>>> message = "Hello world!"
>>> print(message)
Hello world!
```

The value of a variable can be updated (i.e. replaced) with a new value. For example

```
>>> x = 1
>>> x = x+2
>>> print(x)
3
```

where x=x+2 evaluates the right-side x+2 using the old value 1 of the variable x and then assigns the calculated result 3 back to the variable x. The variable x now stores the value 3 and its previous value 1 is replaced/lost. This trick is widely used to do summation and **iteration** in scientific computing (see examples later).

A variable name can contain only letters, numbers and underscores but no spaces (underscores are used as spaced to separate words, for example *greeting_message*). A variable name should be short but descriptive.

B.2.2 Lists

A list is a set of items in the order as seen from left to right in the list. The items can be strings, numbers or even lists (nested lists). The items do not have to be related and need not to have the same type. A list is created by including in square brackets its items separated by commas. The created list can be assigned to a name (as a value is assigned to a variable). For example

```
>>> var = 3.0
>>> mixed=['item 0', 1, "2", var]
>>> print(mixed)
['item 0', 1, '2', 3.0]
```

In the above example, the name of the list is `mixed`; from left to right the items are respectively a string, an integer, a string and a float (stored in the variable `var`); and the list is printed out, including the square brackets, by the command `print`.

A list is an ordered collection of items. So you can access it any item through the index of the item. Note that the **index** for the leftmost item is 0 (not 1). To access an item, you write the name of the list followed by the item index enclosed in square brackets. For example

```
>>> mixed=['item 0', 1, "2", 3.0]
>>> beginning_item = mixed[0]
>>> print(beginning_item)
item 0
>>> print(mixed[3])
3.0
>>> ending_item = mixed[-1]
>>> print(ending_item)
3.0
```

As can be seen, the last item in the list can be accessed by the index 3 as expected, and it can also be accessed by the index -1 in Python. We can also access a subset (slice) of the items in a list using the colon notation : (**slicing** a list) as demonstrated by the following

```
>>> mixed=['item 0', 1, "2", 3.0]
>>> print(mixed[1:3])
[1, '2']
```

Note that $1:3$ represents the interval $[1,3)$, excluding 3. The subset forms a new list. Below are more examples about slicing a list.

```
>>> mixed=['item 0', 1, "2", 3.0]
>>> print(mixed[:3])
['item 0', 1, 2]
>>> print(mixed[2:])
[2, 3.0]
```

If the first index in a slice is omitted, Python automatically starts the slice from the beginning of the list. If the second index is omitted, then Python ends at the last item (including the last item).

Knowing how to access an item in a list, we can use the item. For example

```
>>> mixed=['item 0', 1, "2", 3.0]
>>> print('The beginning item is '+mixed[0])
>>> The beginning item is item 0
>>> x = mixed[-1]+1
>>> print(x)
4.0
```

The function range() in Python is a handy way to generate a series of numbers. We can create a numerical list by using the functions range() and list(). For example

```
>>> list(range(-2,4))
[-2, -1, 0, 1, 2, 3]
```

where `range(-2,4)` generates the arithmetic sequence of integers starting with -2 (included) and ending before 4 (excluded) with 1 as the common difference. The integers are supplied to the function `list()` to make a numerical list. If the common difference is not the default 1, we can specify the common difference as in the following example

```
>>> list(range(4,-5,-3))
[4, 1, -2]
```

where the sequence of integers starts at 4 (included) and end before 1 (excluded) with -3 as the common difference.

We can change, add and remove items in a list. For example,

```
>>> mixed=['item 0', 1, "2", 3.0]
>>> mixed[2] = 2
>>> print(mixed)
['item 0', 1, 2, 3.0]
```

In this example, we change an item from the string "2" to the integer 2 and print the new list. We can insert a new item using the method (a method is a function associated with an object such as the string here) `insert()` as in the following example.

```
>>> mixed=['item 0', 1, "2", 3.0]
>>> mixed.insert(1,0.5)
>>> print(mixed)
['item 0', 0.5, 1, 2, 3.0]
```

In this example, the method `insert()` is issued to the list `mixed` by `mixed.insert`; and a new item 0.5 is added before the old `mixed[1]` by `mixed.insert(1,0.5)`, where the inputs to the method 1 and 0.5 are enclosed in parentheses. Other methods for lists include `append()` (to add an item at the end), `del` (to delete an item), `pop()` (to pop out the last item) and `remove()` (to remove an item by its value instead its index).

Sometimes we want to create a list of items that cannot be changed. We say a value that cannot be change as **immutable**. An immutable list is called a **tuple**. A tuple looks like a list except that it uses parentheses instead of square brackets to enclose items. Accessing an item in a tuple is done by using the index of the item just as accessing an item in a list. For example

```
>>> space_dimensions=(1,2,3)
>>> print(space_dimensions[-1])
3
>>> print(space_dimensions[0])
1
```

B.2.3 Dictionaries

A dictionary in Python is a collection of **key-value pairs** which can be used to model many real-world situations. A dictionary is wrapped in braces { }, which enclose a series of key-value pairs separated by commas. A key-value pair consist of a key (a string for description and indexing) and its value and connect them by a colon. Below is an example of a dictionary.

```
>>> authors = ['A','B']
>>> book = {'name':'Sci. Comput.','author': authors,'page':300}
>>> print(book)
{'name': 'Sci. Comput.', 'author': ['A', 'B'], 'page': 300}
```

In the example, the keys are 'name', 'author' and 'pages' and their associated values are respectively 'Sci. Comput.' (a string), ['AB','CD'] (a list) and 300 (a number).

- To access the value associated with a key, place the key inside square brackets [] following the name of the dictionary. For example

```
>>> book = {'author': ['A','B'],'page':300}
>>> print(book['author'])
['A', 'B']
>>> print(book['page'])
300
```

- We can modify a value in a dictionary as in the following example.

```
>>> book = {'name':'Sci. Comput.','author': 'SX','page':250}
>>> book['page'] = 300
>>> print(book)
{'name': 'Sci. Comput.', 'author': 'SX', 'page': 300}
```

- We can add new key-value pairs in an existing dictionary as a dictionary is a **dynamic structure**. For example

```
>>> book = {'name':'Sci. Comput.','author': 'SX'}
>>> book['page']=300
>>> print(book)
{'name': 'Sci. Comput.', 'author': 'SX', 'page': 300}
```

- We can remove a key-value pair. For example

```
>>> book = {'name':'Sci. Comput.','author': 'SX','page':300}
>>> del book['page']
>>> print(book)
{'name': 'Sci. Comput.', 'author': 'SX'}
```

We can use (nest) a dictionary as an item in a list, a list as a value in a dictionary and a dictionary as a value in a dictionary.

B.3 Looping and Making Decisions

This section describes some important mechanisms to conduct repeated tasks and make decisions based on conditions. From now on, instead of using Python as an advanced calculator interactively (by entering statements following the prompt <<<), we present demos by displaying Python codes (without providing files names) and then the outputs of the codes.

B.3.1 `for` Loops

When we want to do the same action on every item in a list or every key-value pair in a dictionary, we use `for` loops in Python. Below is an example to print each item in a list by a `for` loop.

```
mixed=['item 0', 1, "2", 3.0]
for each_item in mixed:
    print(each_item)
```

The second line of the above code defines a `for` loop. Note the use of colon : in the loop. It tells Python to pull an item from the list `mixed` one by one with the left-to-right order and store it in the loop variable `each_item`. Remember to include the colon : at the end of this line. The third line tells Python to print (do the same action on) the value just stored in the loop variable `each_item`. The repeated actions form the body of the loop. In this case, the body is just the third line. We may read the loop as: for each item in the list, print the item. The output of the code is

```
item 0
1
2
3.0
```

Note that the body of a `for` loop must be indented as **indentation** (four spaces per indentation level are recommended) is Python's way of grouping statements. At the interactive prompt, we have to use spaces (or a tab) for each indented line and enter a blank line to end the indentation. For example

```
>>> authors = ['AB','CD']
>>> for author in authors:
...     print(author)
...
AB
CD
```

In practice, Python codes are written in a text editor, and many text editors have an auto-indent facility.

The general structure of a `for` loop looks like

```
for item in list_of_items:
    action 1
    action 2
    more actions
```

where the same actions are conducted on each item to form the indented body of the loop. All lines in the body need to be indented.

Here is a more comprehensive example to build a list of cube numbers using a `for` loop.

```
cubes = []
for value in range(1,11):
    cube = values**3
    cubes = cubes.append(cube)
print(cubes)
```

We first create an empty list and then use the method `append()` to append each cube number to the list. Each cube number is computed as the cube of a number in s sequence generated by the function `range`. Note that the `print` function is not indented and is therefore outside the loop (not in its body). The output from the program is

```
[1, 8, 27, 64, 125, 216, 343, 512, 729, 1000]
```

Below is an example to do **summation** using a `for` loop by adding each number from 1 to 100 to the same variable named s and initialized as 0. This trick of summation is very useful in scientific computing.

```
s = 0
for k in range(1,101):
    s = s + k
print('The sum of 1, 2, ... , 100 is '+str(s))
```

The output from this program is

```
The sum of 1, 2, ... , 100 is 5050
```

We can form nested `for` loops. For example

```
A = [[0, 0, 0],[0, 0, 0]]
for i in range(0,2):
    for j in range(0,3):
        A[i][j] = 1/(i+j+1)
print(A)
print(A[0])
print(A[0][0])
```

In this program, we first define A as a list of 2 lists (a nested list) with 3 zero items each to model a 2×3 zero matrix, and we then use nested `for` loops to assign each entry in this modeled matrix. The inner loop with the loop variable j assigns each entry (column by column) in a row, and the outer loop with the loop variable i sweeps down the rows. The output from this program is

```
[[1.0, 0.5, 0.3333333333333333], [0.5, 0.3333333333333333, 0.25]]
[1.0, 0.5, 0.3333333333333333]
1.0
```

Python doesn't have a built-in type for matrices. NumPy is a Python package (containing many ready-to-use methods) for scientific computing which has support for array objects. We will describe how to use this package in Python later.

We can loop through a dictionary by key-value pairs with the method `items()`, by keys with the method `keys()`, or by values with the method `values()`, as demonstrated by the following example.

```
methods = {
'root': 'Newton',
'interpolation': 'Lagrange',
'integration': 'Simpson'}

for prob, method in methods.items():
    print('key = '+prob+', value = '+method)

for prob in methods.keys():
    print('\tkey = '+prob)

for method in methods.values():
    print('\t\tvalue = '+method)
```

Note that two loop variables `prob` and `method` are used when looping through key-value pairs, and the first variable `prob` takes the key and the second `method` takes the value of each key-value pair. The output from the code is

```
key = root, value = Newton
key = interpolation, value = Lagrange
key = integration, value = Simpson
    key = root
    key = interpolation
    key = integration
        value = Newton
        value = Lagrange
        value = Simpson
```

B.3.2 `if` Statements

In programming, we often need to examine conditions to decide which actions to take. The `if` statements in Python allow us to examine conditions (logical statements) for decision making. A logical statement has a Boolean value that is either `True` or `False`. Below are some examples of logical statements involving **comparison operations**.

```
>>> 1==1.0
```

```
True
>>> 'ab'=="ab"
True
>>> 1==-1
False
>>> 'ab'=='AB'
False
>>> 1!=-1
True
>>> 1>0
True
>>> 1>=0
True
>>> -1<0
True
>>> -1<=0
True
```

The comparison operations are
- == "is equal to"
- != "is not equal to"
- < "is less than"
- > "is greater than"
- <= "is less than or equal to"
- >= "is greater than or equal to"

Moreover, logic statements can be combined using and, or, as shown in the following examples.

```
>>> 'ab'=="ab" and 'ab'=='AB'
False
>>> 'ab'=="ab" or 'ab'=='AB'
True
```

Now we show how to use if statements. The simplest if statement has one test and one action. For example

```
age = 20
if age>=18:
    print("You are old enough to vote.")
```

Note the use of colon : in the if statement.

Here is a little more complex example. Suppose a bank applies 5% annual interest rate on an account if the account balance is above 5000$. Otherwise, the interest rate is only 2%. To compute the new account balances for a set of four accounts after one year, the bank can use the following Python code:

```
oldbal = [6000,5000,20000,200]
```

```
if oldbal:
    print('Apply interests to accounts.')
else
    print('No accounts')

newbal = []
for b in oldbal:
    if b>5000:
        newbal.append(b*1.05)
    else:
        newbal.append(b*1.02)
print(newbal)
```

Note that Python returns True if a list contains at least one item and False if a list is empty. The output from this code is

```
Apply interests to accounts.
[6300.0, 5100.0, 21000.0, 204.0]
```

The general form of the if statement is

```
if logical_test_1:
    actions if test 1 is true
elif logical_test_2:
    actions if test 2 is true but test 1 is false
else:
    actions if previous tests are all false
```

For example

```
age = 15;
if age<=3:
    print('Free admission')
elif age<=18:
    print('$10 admission fee')
elif age<=65:
    print('$15 admission fee')
else:
    print('$5 admission fee')
```

B.3.3 while Loops

A for loop executes a block of code (the body of the loop) once for each item in a finite collection. A while loop runs as long as a certain condition is true. Below is an example to use a while loop to find the maximum integer whose cube is less than or equal to 2021.

```
imax = 1
while imax**3 <= 2021:
    imax += 1
imax -= 1
print('The max integer whose cube <= 2021: '+str(imax))
```

The output from the code is

```
The max integer whose cube <= 2021: 12
```

We start with a small nmax = 1. We use a while loop to keep increasing imax by 1 as long as the cube of imax is less than or equal to 2021. The condition to be tested by the while loop is imax**3 <= 2021, and the body of the loop (actions to be repeated) is the indented line imax += 1, which is shorthand for imax = imax + 1. Note that when the condition in the while loop becomes False (which happens when imax is increased to 13 and imax**3 is larger than 2021), the loop stops. We therefore need to subtract 1 from imax (by imax -= 1, a shorthand for imax = imax - 1) immediately after the loop to recover the previous value of imax, which is the largest integer with cube less than or equal to 2021.

When using a while loop, we need to make sure the condition to keep running the loop becomes False at some point. Otherwise, the loop runs forever and becomes an infinite loop. If we are stuck in an infinite loop, we can press CTRL-C to terminate it.

B.3.4 break **and** continue **in Loops**

The break statement in a for or while loop terminates the loop without running any remaining code in the loop body. In the following example, we test if an integer bigger than 2 is a prime number.

```
n = 2021
last = 0
for m in range(2,n):
    if n%m == 0:
        print(n, ' = ', m, '*',n//m)
        break
    last = m
if last==n-1:
    print(n, ' is prime number.')
```

We use a for loop to divide a given integer n by the loop variable m from 2 to n-1 to test if n has a factor between 2 and n-1. If yes, we use the break statement to terminate the loop, and the tracking variable last has its initial value 0. If no, the tracking variable last is updated with the last value n-1 of the loop variable m, and we know the integer n is a prime. Note the use of % and \\ operators. The output of the above code is

```
2021  =  43 * 47
```

If we change the value of *n* to 2027, the output becomes

```
2027  is prime number.
```

In the following code, we use a `while` loop to receive an integer and test if the integer is a prime number, and we terminate the test (the `while` loop) by the `break` statement when 0 is received. The method to test if an integer is a prime number is from the above example.

```
prompt = "\nEnter a positive integer to test if it is prime"
prompt += "\n(Enter 0 when you are done): "

while True:
    n = input(prompt)
    n = int(n)
    if n == 0:
        break
    else:
        last = 0
    for m in range(2,n):
        if n%m == 0:
            print(n, ' = ', m, '*',n//m)
            break
        last = m
    if last==n-1:
        print(n, ' is prime number.')
```

In the code, we use the Python function `input()` to receive an input from an user. The `input` function can take a string as its own input/argument which is displayed to the user to serve as the prompt (a prompt is an instruction for users). After the function is executed, the prompt is displayed, and the program waits for the user to enter an input and continues after the user presses ENTER. The input from the user can be assigned to a variable (n here). It is interpreted as a string, so we use another Python function `int()` to convert a string representation of a number to its numerical value.

Rather than terminating a loop entirely as the `break` statement does, the `continue` statement in a loop skip the remaining code in the loop body and returns to the beginning of the loop to continue the loop with the next value of the loop variable. Here is an example.

```
i = 0
while i < 10:
    i += 1
    if i%2 == 0:
        continue
    print(i, ' is odd.')
```

The code display the odd numbers between 1 and 10 using a `while` loop. If a number is an even number, we use the `continue` statement to return to the beginning of the loop and skip the reaming code (the `print` command here) in the loop body. We may change the `while` loop to a `for` loop in this case as the following.

```python
for i in range(1,10):
    if i%2 == 0:
        continue
    print(i, ' is odd.')
```

The `continue` statement is used to return to the beginning of the `for` loop to run with the next value of the loop variable.

B.4 Functions

We have used a few Python built-in functions in previous sections, including `print`, `range()`, `str()`, `int()` and `input()` as well as the built-in methods `insert()` and `append()`. A function is a named block of code that carries out a specific task. A method is a function associated with an object (e.g. a list or a dictionary) and operates on the object.

The built-in function `str()` is used as in the following example.

```python
year_in_string = str(2021)
```

The function `str()` has the name `str`. It receives the numerical value 2021 as an input/argument (a piece of information passed to a function) in the parentheses following its name and it returns the string `'2021'` as the result, which is assigned to the variable `year_in_string`.

We can build our own functions. For example, we can create the following function to find the two roots of the quadratic equation $ax^2 + bx + c = 0$ using the quadratic formula, under the assumptions that $a \neq 0$ and that the equation has two real-valued roots.

```python
def quadroots(a,b,c):
    """Finding the roots of a quadratic equation ax^2+bx+c=0.

    Input: coefficients a, b, c
    Output: two real roots x1, x2
    Algorithm: quadratic formula"""

    discriminant = b**2-4*a*c
    x1 = (-b+discriminant**0.5)/(2*a)
    x2 = (-b-discriminant**0.5)/(2*a)
    return x1,x2
```

```
aa=1
bb=0
roots = quadroots(aa,bb,-4)
print('the roots are:', roots)
```

The function starts with the key word def followed by the function name quadroots and the **parameters** a,b,c in parentheses. A parameter receives information (from a function call) and is used by the function to do the job (e.g. compute returned values). Note the colon : at the end. The indented lines that follow the def line make up the body of the function. The text included between triple quotes (""") is called a **docstring** (documentation string), which describes what the function does and is also used by Python to generate documentation for the function.

The function quadroots receives the coefficients a,b and c, and it returns the two roots x1 and x2. The main body of the function applies the quadratic formulas to compute the two roots using the received coefficients and return the two roots in a tuple.

After the function quadroots is created, we use it in a function call which explicitly passes an **argument** to each corresponding parameter in the function and executes the code in the function. The second last line in the above code calls the function quadroots by passing the values aa=1,bb=0,-4 to the function parameters a,b,c respectively (in the same order). The function then uses a=1,b=0,c=-4 to compute and return the two root x1=2,x2=-2 as a tuple (2,-2). The tuple (2,-2) is assigned to roots. The output of the code is

```
the roots are: (2.0, -2.0)
```

Functions should have descriptive names composed of lowercase letters and underscores.

B.4.1 Passing Arguments

Because a function can have multiple parameters to receive inputs, a function call may need multiple arguments to be passed to the function. We can use positional arguments, which are in the same order as the parameters, as in the above example.

We can also use keyword arguments, which are **key-value pairs** to be passed to a function. A key-value pair associates the parameter name of the function with the argument value of the function call. Key-values pairs in a function call clarify the role of each value and free us from worrying about the correct order of the arguments. We can rewrite the function call in the previous example as

```
def quadroots(a,b,c):
    """Finding the roots of a quadratic equation ax^2+bx+c=0.

    Input: coefficients a, b, c
    Output: two real roots x1, x2
    Algorithm: quadratic formula"""
```

```
    discriminant = b**2-4*a*c
    x1 = (-b+discriminant**0.5)/(2*a)
    x2 = (-b-discriminant**0.5)/(2*a)
    return x1,x2

aa=1
bb=0
root1,root2 = quadroots(c=-4,a=aa,b=bb)
print('the roots are:', root1, root2)
```

The output of the code is

```
the roots are: 2.0 -2.0
```

We can define a **default value** for each parameter in a function. If an argument for a parameter is provided in the function call, Python uses the argument value; and if not, it uses the default value. Note that parameters that do not have default values must be listed before those that have so that positional arguments can still be used for the parameters without default values. For example, we can assign a default value for the parameter c in the function quadroot as

```
def quadroots(a,b,c=-4):
    """Finding the roots of a quadratic equation ax^2+bx+c=0.

    Input: coefficients a, b, c
    Output: two real roots x1, x2
    Algorithm: quadratic formula"""

    discriminant = b**2-4*a*c
    x1 = (-b+discriminant**0.5)/(2*a)
    x2 = (-b-discriminant**0.5)/(2*a)
    return x1,x2

aa=1
bb=0
root1,root2 = quadroots(aa,bb)
print('the roots are:', root1, root2)
roots = quadroots(aa,bb,-1)
print('the roots are:',roots)
```

The output is

```
the roots are: 2.0 -2.0
the roots are: (2.0, -2.0)
```

Note that when specifying a default value, no spaces should be used on either side of = sign.

B.4.2 Passing Lists

It is useful to pass a list or a more complex object (e.g. a dictionary) to a function. We need to realize the difference between passing an integer, a float, a string or a tuple to a function (in previous examples) and passing a list or a dictionary.

A variable assigned with an integer, a float, a string or a tuple is called **immutable** and is passed to a function by copying its value to the corresponding parameter in the function (the so-called **pass-by-value** model). A change made to the parameter inside the function does not change the original variable outside the function, as demonstrated in the following example.

```
def passbyvalue(a):
    a = 1
    print('inside the function, parameter a =',a)

a = 0
print('before the function call, variable a =',a)
passbyvalue(a)
print('after the function call, variable a =',a)
```

The output of the code is

```
before the function call, variable a = 0
inside the function, parameter a = 1
after the function call, variable a = 0
```

A list or a dictionary is called a **mutable** object. When a list is passed to a function, the function can gain the direct access to the list so that the changes made to the list inside the function become permanent and are visible outside the function (as in the so-called **pass-by-reference** model). For example

```
def pass_a_list(list_received):
    list_received.append(1)
    list_received[0] = -1

list_sent = [0]
print('before the function call, list = ',list_sent)
pass_a_list(list_sent)
print('after the function call, list = ',list_sent)
```

The output of the code is

```
before the function call, list =  [0]
after the function call, list =  [-1, 1]
```

Remark Python passes an object using neither the pass-by-value model or the pass-by-reference model. Python uses the pass-by-object-reference model to pass a mutable object.

Remark Sometimes we do not know in advance how many arguments a function needs to accept. We can use an asterisk * before (without a space) a parameter name in the input list of the function to create an empty tuple in the function. The function packs whatever values it receives from a function call into this tuple. Similarly, we can use double asterisks ** before (without a space) a parameter name in the input list of a function to create an empty dictionary in the function. The function pack whatever name-value pairs it receives from a function call into this dictionary.

B.5 Classes

Classes are a key concept in the so-called **object-oriented programming (OOP)**. A class can be regarded as a template for creating real-world objects. A class tells us what data an object should have, what are the initial/default values of the data, and what methods are associated with the object to take actions on the objects using their data. An object is an instance of a class, and creating an object from a class is called **instantiation**. Below is an example of a class `Polynomial` (by convention, a class name in Python is written in CamelCaps by capitalizing the first letter of each word in the name without underscores) and an object/instant of the class `mypolynomial`.

```python
class Polynomial():
    """A class of polynomials"""

    def __init__(self,coefficent):
        """Initialize coefficient attribute of a polynomial."""
        self.coeff = coefficient

    def degree(self):
        """Find the degree of a polynomial"""
        return len(self.coeff)-1

mypolynomial = Polynomial([1,2,3])
print(mypolynomial.coeff)
print(mypolynomial.degree())
```

The first two lines above define a class called `Polynomial` with a docstring to describe the class. The parentheses in the class definition are empty because we create this class from scratch (later we include a parent class in parentheses when creating its child class for inheritance). A function that is part of a class is a method. In the above example, there are two methods, `__init__()` and `degree()`.

The `__init__()` method (with two leading underscores and two trailing underscores) is a special method that Python runs automatically to construct an object when the object is created from the class. So it is also called the `__init__()` **constructor**. The `self` parameter is required and must come first before the other parameters in each method. With this `self` parameter, every call of the method to act on an object automatically passes the reference of the object itself to the method. The other parameter in the method `__init__()` is `coefficient`, which is a list containing the coefficients of a polynomial in this example. Whenever we make an object from the `Polynomial` class, we provide arguments for the parameter `coefficient`, but we do not need to pass an argument for `self`. The variable `self.coeff` prefixed with `self` is available to every method in the class and is accessible by any object created from the class, and it takes the value from the parameter `coefficient` through the assignment `self.coeff = coefficient`. Variables prefixed with `self` are called **attributes**.

The other method `degree()` uses the built-in function `len()` to find the number of entries in a list. It uses the number of coefficients to return the degree of a polynomial

$$c_0 + c_1 x + c_2 x^2 + \cdots + c_n x^n$$

The line `mypolynomial = Polynomial([1,2,3])` in the above code makes an object `mypolynomial` (a polynomial $1 + 2x + 3x^2$) from the class `Polynomial` by passing the coefficients a list `[1,2,3]`. When Python reads this line, it calls the method `__init__()` in the class `Polynomial` with the list as an argument and creates the object named `mypolynomial` that represents this particular polynomial $1 + 2x + 3x^2$.

Variables defined in the `__init__()` constructor (prefixed with `self`) that are accessed by an object using the **dot notation** with the object name are called object attributes. See `mypolynomial.coeff` in the above example. Python also use the dot notation with an object name to apply a method of a class to the object created from the class. See `mypolynomial.degree()` in the above example.

B.5.1 Attributes

Unlike an object attribute which belongs to a particular object, a class attribute is a variable that belongs to a class but not a particular object. All objects of this class share this same variable (the class attribute). Class attributes are usually defined outside the `__init__()` constructor. For example

```
class Polynomial():
    """A class of polynomials"""
```

```
    count = 0

    def __init__(self,coefficient):
        """Initialize the coefficient attributes."""
        self.coeff = coefficient
        Polynomial.count += 1

    def __del__(self):
        """Delete a polynomial object"""
        Polynomial.count -= 1

print('number of polynomials before creation:',Polynomial.count)
poly1 = Polynomial([1])
poly2 = Polynomial([2,2])
poly3 = Polynomial([3,3,3])
print('number of created polynomials:',Polynomial.count)
print(Polynomial.count)
del poly3
print('number of polynomials after a deletion:',Polynomial.count)
print(Polynomial.count)
```

The output of the code is

```
the number of polynomials before creation: 0
the number of created polynomials: 3
the number of polynomials after a deletion: 2
```

In this example, we create three objects, polynomials poly1, poly2 and poly3. The variable count is a class attribute that belongs to the class Polynomial to track the number of polynomials (objects). It is accessed by the dot notation with the class name Polynomial. Whenever a new polynomial (object) is created, it is increased by 1. Whenever an existing polynomial (object) is deleted, it is decreased by 1. We define the __del__() method in the class for the deletion of objects. See the line del poly3.

Every object or class attribute in a class needs an initial value. Object attributes are initialized by the __init__() constructor. We can set a default value for an object attribute in the __init__() constructor, and we then do not have to include a parameter for that attribute in the constructor. See the following example.

```
class Polynomial():
    """A class of polynomials"""

    count = 0

    def __init__(self):
        """Initialize the coefficient attributes."""
        self.coeff = [1]
```

```
                Polynomial.count += 1

mypolynomial = Polynomial()
print('default coefficients:',mypolynomial.coeff)
mypolynomial.coeff = [2,2]
print('coefficients after reset:',mypolynomial.coeff)

print('polynomial count before reset:',Polynomial.count)
Polynomial.count = 0
print('polynomial count after reset:',Polynomial.count)
```

The output of the code is

```
default coefficients: [1]
coefficients after reset: [2, 2]
polynomial count before reset: 1
polynomial count after reset: 0
```

In this example, the class attribute count is initialized to be 0, and the object attribute self.coeff has a default value [1]. So a constant polynomial $p(x) = 1$ is created by default by the instantiation mypolynomial = Polynomial(), which does not pass any argument for coeff. We can modify an attribute directly, as in the lines mypolynomial.coeff = [2,2] and Polynomial.count = 0.

B.5.2 Methods

After we create an object from a class, we can use dot notation to call any method in the class to act on the object. To call a method, just give the name of the object and the method separated by a dot. We can add a method called evaluate in the class Polynomial to evaluate a polynomial, as shown below

```
class Polynomial():
    """A class of polynomials"""

    count = 0

    def __init__(self,coefficient):
        """Initialize the coefficient attributes."""
        self.coeff = coefficient
        Polynomial.count += 1

    def __del__(self):
        """Delete a polynomial object"""
        Polynomial.count -= 1

    def degree(self):
        """Find the degree of a polynomial"""
```

```
            return len(self.coeff)-1

    def evaluate(self,x):
        """Evaluate a polynomial."""

        n = self.degree()
        p = []
        for xi in x:
            s = 0
            for k in range(0,n+1):
                s += self.coeff[k]*xi**k
            p.append(s)
        return p

mypolynomial = Polynomial([1,6,15,20,15,6,1])
print(mypolynomial.evaluate([-2,-1,0,1,2]))
```

The output of the code is

```
[1, 0, 1, 64, 729]
```

which are values of the polynomial

$$p(x) = 1 + 6x + 15x^2 + 20x^3 + 15x^4 + 6x^5 + x^6 = (1+x)^6$$

at $x = -2, -1, 0, 1$ and 2, respectively.

B.5.3 Inheritance

If we want to write a class that is just a specialized version of another class, we do not need to write the class from scratch. We call the specialized class we want to write a **child class** and the other general class a **parent class**. The child class can inherit all the attributes and methods form the parent class but defines its own special attributes and methods (that we need but the parent class does not have) or even overrides methods of the parent class. For example, we may write a child class called Quadratic_Polynomial which inherits from the parent class Polynomial to create quadratic polynomials and has its own method roots to find the zeros of a quadratic polynomial. Below is the code

```
class Polynomial():
    """A class of polynomials"""

    def __init__(self,coefficient):
        """Initialize the coefficient attributes."""
        self.coeff = coefficient

    def degree(self):
```

```
        """Find the degree of a polynomial"""
        return len(self.coeff)-1

class Quadratic(Polynomial)
    """A class of quadratic polynomial"""

    def __init__(self,coefficient,power_increase):
        """Initialize the coefficient asstributes ."""
        super().__init__(coefficient)

    def roots(self):
        c = self.coeff[0]
        b = self.coeff[1]
        a = self.coeff[2]
        if self.power_increase == -1:
            a = self.coeff[0]
            c = self.coeff[2]
        discriminant = b**2-4*a*c
        r1 = (-b+discriminant**0.5)/(2*a)
        r2 = (-b-discriminant**0.5)/(2*a)
        return r1,r2

    def degree(self):
        return 2

quadpoly1 = Quadratic([2, -3, 1],1)
print('roots:', quadpoly1.roots())
quadpoly2 = Quadratic([2, -3, 1],-1)
print('roots:',quadpoly2.roots())

poly = Polynomial([2, -3, 1, 0])
print('degree:',poly.degree())
quadpoly = Quadratic([2, -3, 1,0],1)
print('degree:',quadpoly2.degree())
```

The output of the code

```
roots: (2.0, 1.0)
roots: (1.0, 0.5)
degree: 3
degree: 2
```

As in the above example, the parent class Polynomial must appear before the child class Quadratic in the same file. We must include the name of the parent class in the parentheses óf the definition of the child class (to indicate the parent-child relation for inheritance). The __init__() constructor in the child class has the line `super().__init__(coefficient)` that uses the `super()` function to give an child object all the attributes defined in the parent class.

Besides the attribute `self.coeff`, the child class defines its own attribute `self.power_increase` and its own method `roots()`. The method `roots()` applies the quadratic formula to find the two roots of a quadratic polynomial equation. Because we are used to write a quadratic polynomial as $ax^2 + bx + c$ in which the powers of x are decreasing from left to right, we use the attribute `self.power_increase` to specify if the supplied coefficients in a list are for increasing (default) or decreasing (`self.power_increase` is set to -1) powers of x. In the method `roots()`, we set the values of a, b and c in the quadratic formula according to the value of `self.power_increase` (with the `if` statement). When the first three coefficients are 2, -3 and 1, if `self.power_increase` is not equal to -1, we solve $2 - 3x + 1x^2 = 0$ and get the two roots `(2.0, 1.0)`; and if `self.power_increase` is set to -1, we solve $2x^2 - 3x + 1 = 0$ and get the two roots `(1.0, 0.5)`.

The parent class has the method `degree()`, but the child class also has the method `degree()` (with the same name). When the method `degree()` is applied to an object created from the parent class, the parent's method `degree()` is used. When the method `degree()` is applied to an object created from the child class, the child's method `degree()` **overrides** the parent's method and is used instead. When we use the list `[2,-3,1,0]` to create an object from the parent class, we create a polynomial of degree 3, and the parent's method `degree()` correctly gives the degree 3. When we use the same list `[2,-3,1,0]` to create an object from the child class, we use only the first three entries in the list to create a quadratic polynomial, and the child's method `degree()` correctly give the degree 2.

B.5.4 Objects as Attributes

When modeling complex things in the real world, we need to think at a high logical level rather than a syntax-focused level, and we may want to break a large class into smaller classes that work together. For example, we may want to know the properties of the leading term (the term of the maximum power of x) of a polynomial because the leading term characterizes the end behavior of the polynomial, so we can define a small class called `Monomial` to describe the leading term and introduce a `Monomial` object as an attribute (say `leading_term`) in the `Polynomial` class.

```python
class Monomial():
    """A class of monomial"""

    def __init__(self,a=1,n=0):
        self.a = a
        self.n = n

    def property(self):
        if self.a>0:
            print('positive coefficient:',self.a)
        if self.a<0:
```

```
            print('negative coefficient:',self.a)
        if self.a==0:
            print('zero coefficient:',self.a)
        if self.n%2 == 0:
            print('even power:',self.n)
        else:
            print('odd power:',self.n)

class Polynomial():
    """A class of polynomials"""

    def __init__(self,coefficient):
        """Initialize the coefficient attributes."""
        self.coeff = coefficient
        n = len(coefficient)-1
        a = coefficient[n]
        self.leading_term = Monomial(a,n)

class Quadratic(Polynomial):
    """A class of quadratic polynomial"""

    def __init__(self,coefficient,power_increase):
        """Initialize the coefficient asstributes ."""
        super().__init__(coefficient)
        self.power_increase = power_increase
        self.leading_term = Monomial(coefficient[2],2)
        if power_increase == -1:
            self.leading_term = Monomial(coefficient[0],2)

mono = Monomial()
print('default monomial:')
mono.property()
poly=Polynomial([1, 2, 3, 4, 5])
print('leading term of a polynomial:')
poly.leading_term.property()
quadpoly = Quadratic([1, 2, 3],-1)
print('leading term of a quadratic:')
quadpoly.leading_term.property()
```

The output of the code is

```
default monomial:
positive coefficient: 1
even power: 0
leading term of a polynomial:
positive coefficient: 5
```

```
even power: 4
leading term of a quadratic:
positive coefficient: 1
even power: 2
```

In this example, the attribute `leading_term` of a polynomial object is a mono-mial object, as defined by `self.leading_term = Monomial(a,n)`. The construc-tor of the `Monomial` class sets the default values `a=1,n=0` in the monomial ax^n. When we create a monomial without passing any values to a and n as in the line `mono=Monomial()`, we get the default monomial 1. When we create a polynomial by the line `poly=Polynomial([1, 2, 3, 4, 5])`, we can describe its leading term by the line `poly.leading_term.property()`, which tells Python to look at the ob-ject (a polynomial) `poly`, find its `leading_term` attribute (a monomial object), and call the method `property()` that is associated with the `leading_term` monomial object. The object `quadpoly` created from the child class `Quadratic` needs to reset its `leading_term` attribute (using an `if` statement) in the constructor of the child class `Quadratic` because we reverse the order of the coefficients by the attribute `power_increase`. So the quadratic polynomial created by the line `quadpoly = Quadratic([1, 2, 3],-1)` has the leading term $1x^2$ instead of $3x^2$, as described by `quadpoly.leading_term.property()`.

B.6   Modules

We know how to create our own functions and classes. We can store our functions and classes in separate files called modules. If we think of a function or class as a tool, a module is like a toolbox full of tools. We can repeatedly pull out tools from the same toolbox to work on many different jobs, and we can share our toolboxes with others or use others' toolboxes. Similarly, we can reuse functions and classes from the same modules in many different main programs, and we can share our modules with others or use others' modules. In addition, the separation of modules from main programs makes the main programs easy to follow and allows us to focus on high-level logic of the main programs. In this section, we learn how to create and use modules.

A module is a file with the extension *.py* that is packed with functions and/or classes that are somehow related. For example, we can make a module saved with the file name *solver.py* that contains the function `quadroots()`.

```
def quadroots(a,b,c=-4):
    """Finding the roots of a quadratic equation ax^2+bx+c=0.

    Input: coefficients a, b, c
    Output: two real roots x1, x2
    Algorithm: quadratic formula"""
```

```
discriminant = b**2-4*a*c
x1 = (-b+discriminant**0.5)/(2*a)
x2 = (-b-discriminant**0.5)/(2*a)
return x1,x2
```

We can make a separate main program saved with the file name *solve_quad.py* in the same directory to import the module `solver` and use the function `quadroots()` in the module.

```
import solver

aa=1
bb=0
roots = solver.quadroots(aa,bb)
print('the roots:', roots)
root1,root2 = solver.quadroots(c=-4,a=aa,b=bb)
print('the roots:', root1, root2)
```

The output of the code is

```
the roots: (2.0, -2.0)
the roots: 2.0 -2.0
```

When Python reads the main program *solve_quad.py*, the beginning line `import solver` in the program tells Python to open the file *solver.py* and make its code available to the program *solve_quad.py*. To call a function in an imported module, we enter the name of the module followed by the name of the function, separated by a dot, as in the line `root1,root2 = solver.quadroots(c=-4,a=aa,b=bb)`. We can also provide an **alias** for a module name, such as an alias `sl` for `solver` by `import solver as sl`, and use the alias to call functions in the module, such as in `root1,root2 = sl.quadroots(c=-4,a=aa,b=bb)`.

Note that the line `import solver` (or `import solver as sl`) imports the entire module `solver` to make everything in the module available. We can also import a specific function from a module and we can also give the function an alias. For example

```
from solver import quadroots as qr

aa=1
bb=0
roots = qr(aa,bb)
print('the roots:', roots)
root1,root2 = qr(c=-4,a=aa,b=bb)
print('the roots:', root1, root2)
```

With this syntax, we do not need to use the dot notation when calling the function `quadroots()`. We call it using its alias `qr`. If we import the function without an alias, for example `from solver import quadroots`, we then call it by name, for example root1,root2 = quadroots(c=-4,a=aa,b=bb). In either case, we get the same output as above.

Similarly, we can store classes in a module and import the entire module or specific classes of the module to our main program. For example, we can create a module with the file name *polynomial.py* that include the following classes

```python
class Monomial():
    """A class of monomial"""

    def __init__(self,a=1,n=0):
        self.a = a
        self.n = n

    def property(self):
    if self.a>0:
        print('positive coefficient:',self.a)
    if self.a<0:
        print('negative coefficient:',self.a)
    if self.a==0:
        print('zero coefficient:',self.a)
    if self.n%2 == 0:
        print('even power:',self.n)
    else:
        print('odd power:',self.n)


class Polynomial():
    """A class of polynomials"""

    count = 0

    def __init__(self,coefficient):
        """Initialize the coefficient attributes."""
        Polynomial.count += 1
        self.coeff = coefficient
        n = len(coefficient)-1
        a = coefficient[n]
        self.leading_term = Monomial(a,n)

    def __del__(self):
        """Delete a polynomial object"""
        Polynomial.count -= 1

    def degree(self):
        """Find the degree of a polynomial"""
        return len(self.coeff)-1
```

```python
    def evaluate(self,x):
        """Evaluate a polynomial."""

        n = self.degree()
        p = []
        for xi in x:
            s = 0
            for k in range(0,n+1):
                s += self.coeff[k]*xi**k
                p.append(s)
        return p

class Quadratic(Polynomial):
    """A class of quadratic polynomial"""

    def __init__(self,coefficient,power_increase):
        """Initialize the coefficient asstributes ."""
        super().__init__(coefficient)
        self.leading_term = Monomial(coefficient[2],2)
        # power_increas: attribute specific to the child class
        self.power_increase = power_increase
        if power_increase == -1:
            self.leading_term = Monomial(coefficient[0],2)

    def degree(self):
        return 2

    def roots(self):
        c = self.coeff[0]
        b = self.coeff[1]
        a = self.coeff[2]
        if self.power_increase == -1:
            a = self.coeff[0]
            c = self.coeff[2]
        discriminant = b**2-4*a*c
        r1 = (-b+discriminant**0.5)/(2*a)
        r2 = (-b-discriminant**0.5)/(2*a)
        return r1,r2
```

We can make a new file *my_polynomial.py* in which we import the entire the module.

```python
import polynomial as p

quadpoly = p.Quadratic([1, 2, 3],-1)
print('leading term of a quadratic:')
quadpoly.leading_term.property()
```

We access the classes in the module through *module_name.class_name* or *module_alias.class_name*. We can also import a single or multiple classes (separated by commas) specifically from a module. For example

```
from polynomial import Monomial,Polynomial

mono = Monomial()
print('default monomial:')
mono.property()
poly=Polynomial([1, 2, 3, 4, 5])
print('leading term of a polynomial:')
poly.leading_term.property()
```

We do not need the dot notation to make objects from the specifically imported classes.

Sometimes we may need to import one module into another module. For example, we may find one class in one module depends on a class in another module, so we need to import the required class in the second module into the first module.

Python comes with the **standard library** of modules. We can use any function or class in the standard library by adding a simple import statement at the top of our programs. For example, the math module gives access to many floating point math functions.

```
import math
a = math.cos(math.pi/3)
print(a)
b = math.exp(1)
print(b)
print(math.log(b))
```

The output of the code is

```
0.5
2.71828182846
1.0
```

B.7 numpy, scipy, matplotlib

There are some very useful Python packages (a package is a collection of Python modules) for scientific computing.

B.7.1 numpy

numpy is the numerical Python package that offers numerical foundation that more sophisticated numerical methods rely on. It provides high-dimension array objects, matrix operations, random numbers and many other numerical capabilities. Below are some examples of numpy usage.

(1) The first example shows some basics about arrays

```
import numpy as np

# Inputing an array
A = np.array([[1,2,3],[4,5,6]])
print('matrix A:',A)
print('dimension of A:', A.ndim)
print('shape of A:',A.shape)
print('number of elements in A:',A.size)
print('data type of A:',A.dtype)

# Slicing an array
# row 1 column 1
A[0,0]
# 2rd row, all columns
A[1,:]
# 1nd-2rd rows, all columns
A[0:2,:]
# all rows, 2nd-3rd columns
A[:,1:3]

# Arrays are mutable
# 2 by 2 zero matrix
Z = np.zeros((2,2))
print('zero matrix Z:',Z)
C = Z
C[0,0] = 1
print('copy of Z:',C)
print('array Z:',Z)
```

The output of the code is

```
('matrix A:', array([[1, 2, 3],
[4, 5, 6]]))
('dimension of A:', 2)
('shape of A:', (2, 3))
('number of elements in A:', 6)
('data type of A:', dtype('int64'))
('zero matrix Z:', array([[ 0.,   0.],
[ 0.,   0.]]))
('copy of Z:', array([[ 1.,   0.],
[ 0.,   0.]]))
('array Z:', array([[ 1.,   0.],
[ 0.,   0.]]))
```

(2) The second example shows generation of some special arrays.

```python
import numpy as np

u = np.arange(0,1,0.2)
print('equally spaced entries with increment 0.2:', u)
v = np.linspace(0,2,6)
print('6 equally spaced entries between 0 and 2:',v)

# 3 by 2 random array
R = np.random.random((1,2))
print('random matrix R:',R)
# 2 by 2 matrix with entries 1
W = np.ones((2,2))
print('matrix W:',W)
# 2 by 2 identity matrix
I = np.eye(2)
print('identity matrix I:',I)
# 2 by 2 zero matrix
Z = np.zeros((2,2))
print('zero matrix Z:',Z)
```

The output of the code is

```
equally spaced entries with increment 0.2: [0.  0.2 0.4 0.6 0.8]
6 equally spaced entries between 0 and 2: [0.  0.4 0.8 1.2 1.6 2. ]
random matrix R: [[0.66048715 0.21493249]]
matrix W: [[1. 1.]
[1. 1.]]
identity matrix I: [[1. 0.]
[0. 1.]]
zero matrix Z: [[0. 0.]
[0. 0.]]
```

(3) The third example shows some operations on arrays.

```python
import numpy as np

# Reshape an 2 by 3 array to 3 by 2 array
A = np.array([[1,2,3],[4,5,6]])
print('2 by 3 A:',A)
B = A.reshape((3,2))
print('2 by 3 reshaped to 3 by 2:', B)

# Matrix arithmetic
# transpose
print('transpose of A:',A.T)
```

```
# scalar multiple and adding a constant
print('2*A-1 = ',2*A-1)
# matrix product
print('A*B = ',np.dot(A,B))
print('B*A = ',np.dot(B,A))

# Vector arithmetic
# numpy automatically convert lists to vectors
u = [1,2,3]
v = [1,1,1]
print('inner product:',np.inner(u,v))
print('outer product:',np.outer(u,v))
```

The output of the code is

```
2 by 3 A: [[1 2 3]
 [4 5 6]]
2 by 3 reshaped to 3 by 2: [[1 2]
 [3 4]
 [5 6]]
transpose of A: [[1 4]
 [2 5]
 [3 6]]
2*A-1 =  [[ 1  3  5]
 [ 7  9 11]]
A*B =  [[22 28]
 [49 64]]
B*A =  [[ 9 12 15]
 [19 26 33]
 [29 40 51]]
inner product: 6
outer product: [[1 1 1]
 [2 2 2]
 [3 3 3]]
```

(4) In the last example, we demonstrate how numpy saves and reads a text file.

```
import numpy as np

W = np.random.random((3,2))
print('3 by 2 random matrix:',W)
# save to a file
np.savetxt("aout.txt",W)
# read from a file
R = np.loadtxt("aout.txt")
print('loaded array:',R)
```

The output of the code is

```
3 by 2 random matrix: [[0.27350679 0.66719967]
[0.3769092  0.66180467]
[0.14327177 0.88523306]]
loaded array: [[0.27350679 0.66719967]
[0.3769092  0.66180467]
[0.14327177 0.88523306]]
```

The package numpy provides many other numerical capabilities. Its linear algebra module numpy.linalg has functions such as solve(A,b) to solve a full rank linear system $Ax = b$, lstsq(A,b) to solve a least squares problem $\min \|Ax - b\|_2$, qr to compute QR factorization, and many more. Its FFT module numpy.fft has functions such as fft and ifft to do DFT and inverse DFT in 1 dimension and others in high dimensions. For more information about numpy, please visit the web site https://numpy.org

B.7.2   scipy

scipy is a Python library of algorithms and tools that are built upon numpy to work with arrays in numpy. It includes modules scipy.linalg and scipy.optimize, etc. that does numerical linear algebra, optimization and many more. For more information about scipy, please visit the web site https://scipy.org

B.7.3   matplotlib

matplotlib is a plotting library for Python that works with numpy. It has the plotting syntax very similar to MATLAB. For more information about matplotlib, please visit the web site https://matplotlib.org/

Below is an example to use matplotlib to plot the graphs of functions. The graph of a function is sketched by plotting the points created by the function. The *x*- and *y*-coordinates of all the points are stored in two vectors, respectively.

```
import numpy as np
import matplotlib.pyplot as plt

x = np.linspace(0,2,11)
y1 = np.exp(x)-1
y2 = np.power(x,2)
plt.plot(x,y1,'bk-',label='$y=e^x-1$')
plt.plot(x,y2,'ko--',label='$y=x^2$')
plt.xlim((-1,3))
plt.ylim((-1,7))
plt.xlabel('x')
plt.ylabel('y')
plt.title('graph of and $y=e^x-1$ and $y = x^2$')
plt.legend()
plt.savefig('line_plot_1.pdf')
```

The plot generated by the above code is shown in Fig. B.1. `xlim` and `ylim` set the

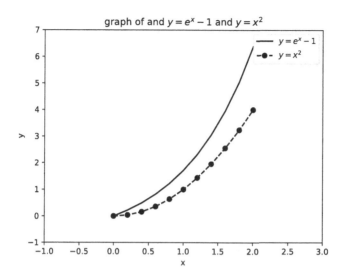

FIGURE B.1

An $x - y$ plot generated by `matplotlib`.

ranges for the x- and y-axes. The string `'ro-'` in the arguments of `plot` specifies the

- color to red with the `'r'`,
- symbol to circle with the `'o'` and
- line style to a dashed line with the `'-'`

When plotting data, it is always appreciated if we add labels to the x- and y-axes, a title to the plot, and the legend if there are multiple curves as shown in this example.

B.8   Jupyter Notebook

The Jupyter notebook is a web application that allows you to create and share documents that contain live Python code, the output from code, narrative text, mathematical equations and images. It shares many similarities with the Live Editor in MATLAB. For more information about how to install and run the Jupyter notebook, please visit the web site `https://jupyter.org/`

B.9   Concluding Remarks

Python is a very efficient and powerful language. It can do more in fewer lines of code than many other languages; and it can be used to develop applications for business purpose and academic research. A Python code is easy to read, debug and extend.

Below are some general suggestions on programming for scientific computing.

- **Write pseudocode** Write down the pseudocode for an algorithm to identify inputs, outputs, data structures, modules and flow of the algorithm.
- **Break the problem** Break the problem into small self-contained tasks. Write a function to complete each small task.
- **Be careful and check** Pay attention to all details (such as syntax, variable types, order of operations and updates of values, etc.) when writing a program. Check a program by reading its codes line by line.
- **Use meaningful names** A variable name should reflect the meaning and type of the variable and be easy to distinguish.
- **Include comments** Include a preface to each program to briefly describe the program (such as its purpose, inputs and outputs and major method). Separate a code into functional blocks and provide concise comments for each block.
- **Show intermediate results** Sometimes it is helpful to show intermediate results to debug a program (find mistakes).

B.10   Programming Problems

Problem B.1

$$A = \begin{bmatrix} 0 & -6 & 8 & 1 \\ -2 & 5 & 5 & -3 \\ 7 & 8 & 0 & 3 \end{bmatrix}, \quad b = \begin{bmatrix} 3 & 2 & -13 \end{bmatrix}$$

(1) Generate the above matrix A and vectors b.

(2) Write commands to find the sizes of A and b?

(3) Write a command to retrieve the entry of A at row 2 and column 3 and a command to retrieve the third entry in b.

(4) Write commands to swap the second and third rows of A.

Problem B.2 Write a numpy program to get the powers of array values element-wise.

Problem B.3 Generate a row vector with entries that are even numbers between 10 and 0 (including 10 and 0) in the decreasing order

Problem B.4 Plot the graphs of the functions $y = e^x$ and $y = 1 + x + x^2/2 + x^3/6$ in the same figure using 21 data points for each graph with 21 equally spaced values of x in the interval $[-2, 2]$. In your figure, you need to distinguish the two graphs with different colors, different line styles and different symbols for the data points. You also need to annotate your plot with labels, title and legend.

Problem B.5 Use a for loop to add all the odd numbers between 1 and 99: $1, 3, 5, \ldots, 97, 99$.

Problem B.6 Use nested for loops to assign entries of a 5×5 matrix A such that $A(i, j) = ij$.

Problem B.7 The Fibonacci sequence $\{F_n\}_{n=0}^{\infty} = \{1, 1, 2, 3, 5, 8, \ldots\}$ satisfies the recurrence relation $F_{n+2} = F_n + F_{n+1}$, $n = 0, 1, 2, \ldots$. Given $F_0 = F_1 = 1$, use a for loop but without an array to implement the recurrence relation to compute and display the first 20 entries in the sequence.

Problem B.8 If an bank account balance is less than 5000, the annual interest rate is 1%, if it is in the range $[5000, 20000)$, the interest rate is 2%, otherwise the annual interest rate is 5%. Generate 5 account balances and then use if statement to calculate the new balances of the 5 accounts after a year.

Problem B.9 Write a function that takes as input three variables and returns the largest of the three. Do this without using the Python max() function.

Problem B.10 The variable *epsilon* is initially equal to 1. Use a while loop to keep dividing *epsilon* by 2 until *epsilon* $< 10^{-6}$, and determine how many divisions are used until *epsilon* $< 10^{-6}$ using Python. Verify your result by algebraic derivation.

Problem B.11 Write a program to count the total number of digits in a number using a while loop.

Problem B.12 Write a function that accepts one numeric parameter. This parameter will be the measure of an angle in radians. The function should convert the radians into degrees and then return that value. Test your function on angles in $\pi/6$, $\pi/4$, $\pi/3$, $pi/2$ and π radians.

Problem B.13 Write a function which converts temperature between Celsius (C) and Fahrenheit (F). The function determines the direction of the conversion (i.e. from F to C or C to F) by an input. Add comments in your function to describe how this input indicates the conversion direction. Test your function by making two plots, one for F versus C (i.e. by converting a vector of temperature in C to F) and the other for C versus F. Annotate each of your plots.

Problem B.14 The factorial of n is defined as $n! = 1 \times 2 \times 3 \times \cdots \times n$. Write a function to compute factorials in a recursive manner by calling the function itself. Test your function to compute $n!$ for $n = 0, 1, 2, 3, 4, 5, 6, 7, 8, 9, 10$. Make sure your recursive call does not go on forever.

Problem B.15 (1) Create a `Vehicle` class with `max_speed` and `mileage` instance attributes. (2) Create a child class `Bus` that will inherit all of the variables and methods of the `Vehicle` class.

Problem B.16 Write a Python class named `Rectangle` constructed by a length and width and a method which will compute the area of a rectangle.

Problem B.17 Write a Python class named `Circle` constructed by a radius and two methods which will compute the area and the perimeter of a circle.

Index

Abel-Ruffini theorem, 87
absolute error, 3
absolutely stable, 246
accuracy, 5
activation function, 288
adaptive integration, 217
algorithm stability, 33
algorithms, 2
aliasing phenomenon, 169
alternating series theorem, 19
Amdahl's law, 300
analytical solution, 2
artificial neural network, 287
autonomous ODEs, 241

backpropagation (BP), 289
backward finite difference, 226
backward substitution, 45
basic Gaussian quadrature rule, 188
basic midpoint rule, 188, 191
basic quadrature rules, 188
basic Simpson's rule, 188, 205
basic trapezoidal rule, 188, 200
basis functions, 9
big O notations, 11
bisection method, 90
bisection method theorem, 93
boundary conditions, 163, 238
boundary value problem (BVP), 237
Buffon's needle problem, 295

catastrophic cancellation, 32
centered finite difference, 226
change of intervals, 189
Chebyshev nodes, 136, 156
classical RK method, 251
composite Gaussian quadrature rule, 213

composite midpoint rule, 192
composite quadrature rules, 188
composite Simpson's rule, 205
composite trapezoidal rule, 201
condition numbers, 76
conditioning, 70
consistency, 244
cubic spline, 161

degree of precision (DOP), 192
diagonally dominant matrices, 272
diagonally dominant, 67
direct methods, 90
Dirichlet boundary conditions, 238
discrete Fourier transform (DFT), 167,
    231
discretized equations, 240

early stopping, 289
efficiency, 5
error, 3
error estimate, 91
Euler's method, 239
explicit methods, 247
extrapolation, 154

finite difference methods, 252
fixed point iteration, 263
fixed points, 88, 105
fixed-point iteration, 105
fixed-point iteration theorem , 106
floating-point numbers, 26
forward finite difference, 226, 240
function spaces, 12
fundamental theorem of algebra, 10
fundamental theorem of linear algebra,
    284

Gauss nodes, 210
Gauss weights, 210
Gauss-Lobatto quadrature, 216
Gauss-Radau quadrature, 216
Gauss-Seidel (G-S) method, 265
Gaussian elimination, 46
Gaussian quadrature rules, 209
Gaussian quadrature theorem, 211, 213
global error, 243
gradient descent, 121

Hessian matrix, 118
high-performance parallel computers
(HPC), 298
Horner's method, 10

ill-conditioned problems, 69
implicit methods, 247
initial conditions, 237
initial value problem (IVP), 237
integration with singularities, 216
intermediate value theorem (IVT), 88
interpolants, 132
interpolating conditions, 132
interpolation, 132
interpolation error, 149, 230
iteration, 3, 91
iteration matrix, 263
iterative methods, 91

Jacobi iteration, 264
Jacobi method, 264
Jacobian matrix, 115

Kronecker delta, 144

Lagrange interpolation, 143
Lax equivalence theorem, 256
least squares (LS) solution, 278
least squares fit, 278
line search, 120
linear convergence, 93
linear regression, 280
Lipschitz condition, 238
local truncation error (LTE), 242
logistic function, 287

logistic sigmoid function, 288
LU decomposition, 55

machine epsilon, 30
machine learning, 287
machine numbers, 25
Maclaurin series, 13
mantissa, 27
mathematical approximation errors, 5
matrix norms, 75
mean value theorem for integrals, 186
Message-Passing Interface (MPI), 299
method of undetermined coefficients, 227
midpoint rule, 191
midpoint rule theorem, 195
modeling errors, 5
monomial interpolation, 141
monomials, 9
Monte Carlo methods, 296
Monte Carlo simulation, 295
Monte Carlo methods, 297
multiplicity, 87

natural cubic spline, 163
nested multiplication, 10
Neumann boundary conditions, 238, 258
Newton's interpolation, 145
Newton's iteration, 94
Newton's method, 94
Newton's method theorem, 96
Newton-Cotes rules, 209
Newton-Raphson method, 94
normal equation, 282
norms, 73
numerical methods, 2
numerical solution, 2

order of accuracy, 242
order of convergence, 92
over-fitting, 289

parallel computing, 298
partial pivoting, 59
partition, 184
Peano kernel theorem, 206
polynomial interpolation, 132

polynomial space, 133
polynomials, 9

QR decomposition, 286
quadratic convergence, 93
quadrature error, 189

random number generator, 295
relative error, 3
residual, 266
residual correction method, 266
residual vector, 77
residuals, 95
Richardson extrapolation, 232
Rolle's theorem, 90
roots, 86
rounding unit, 29
roundoff error, 28
roundoff errors, 5, 25
Runge function, 155
Runge's phenomenon, 155
Runge-Kutta (RK) methods, 249

scientific notation, 27
secant method, 99
secant method theorem, 101
sequential computing, 300
sigma notation, 9
significant digits, 27
Simpson's rule, 205
Simpson's rule theorem, 208

simultaneous relaxation, 264, 265
space step size, 253
sparse matrix, 67
spectral accuracy, 173
spectral radius, 271
speedup, 300
spline interpolation, 159
stability, 5
stationary iteration, 263
stopping criterion, 91
superlinear convergence, 93

Taylor polynomials, 12
Taylor series methods, 248
Taylor series/expansion, 13
Taylor's series expansion, 227
Taylor's theorem, 11, 14, 110
the mean value theorem, 17
time step size, 239
time-marching methods, 239
trapezoidal rule, 200
trapezoidal rule theorem, 203
trigonometric interpolant, 169, 231
truncation error, 227

uniform partition, 184
upper bounds, 4

Vandermonde matrix, 142
vector norms, 73

well-conditioned problems, 70

Printed in the United States
by Baker & Taylor Publisher Services